AF348835

Analytic Learning Methods for Pattern Recognition

Kar-Ann Toh · Huiping Zhuang · Simon Liu ·
Zhiping Lin

Analytic Learning Methods for Pattern Recognition

 Springer

Kar-Ann Toh
School of Electrical and Electronic
Engineering
Yonsei University
Seoul, Korea (Republic of)

Huiping Zhuang
Shien-Ming Wu School of Intelligent
Engineering
South China University of Technology
Guangzhou, China

Simon Liu
School of Electrical and Electronic
Engineering
Nanyang Technological University
Singapore, Singapore

Zhiping Lin
School of Electrical and Electronic
Engineering
Nanyang Technological University
Singapore, Singapore

ISBN 978-981-96-2150-7   ISBN 978-981-96-2151-4 (eBook)
https://doi.org/10.1007/978-981-96-2151-4

This Springer imprint is published by the registered company Springer Nature Singapore Pte Ltd.
The registered company address is: 152 Beach Road, #21-01/04 Gateway East, Singapore 189721,
Singapore

If disposing of this product, please recycle the paper.

Preface

According to Henri Bergson, "An absolute can only be given in an intuition, while all the rest has to do with analysis." The significance of analysis becomes evident when the absolute remains unknown. Therefore, as Herbert Simon observed, "the more relevant patterns at your disposal, the better your decisions will be." This sets the background for the topic.

Presently, courses in the field of machine learning and artificial intelligence place significant emphasis on deep learning. Such emphasis assumes that students already possess a solid foundation in the classical aspects of machine learning. However, essential and potent topics like analytic learning and ensemble learning seemingly receive less attention in most curricula. In particular, analytic learning offers distinct advantages, including determinism, analytical rigor, and interpretability. Unfortunately, these aspects are often overlooked. As deep learning systems grow increasingly complex, understanding their behavior becomes more challenging. The trend toward intricate models inevitably leads to solutions that defy human intuition and control. Given these considerations, we propose the creation of a new book specifically focused on analytic learning methods to provide readers with a comprehensive perspective in this field.

Zooming in on the analytic aspects of machine learning, the information appears to be relatively fragmented for the application community. For example, when it comes to linear regression, practitioners are often led to a less than satisfactory approach according to the fragmented number of references they have at hand. The over-determined system appears to be the default starting point for solving such a baseline system. When facing systems of large dimensions but with a small number of samples, such a default solution is deemed to encounter the singularity problem. The emergence of "small sample size problem" thence addressing the problem through regularization or data augmentation. Many seem to be unaware of the bigger picture of the equal-determined, over-determined, and under-determined systems to cater for the various scenarios of data with different combinations of dimension size and sample size. If such knowledge of the complete systems of linear equations was known at the beginning, the journey of solving the singularity problem would be straightforward. The first intent of this book is to fill this gap for engineering practitioners.

Moving a step forward, polynomial regression is a simple yet effective method for pattern recognition that is comparable with many nonlinear models and shallow networks. However, due to its explosive number of expansion terms for high-dimensional features, it is relatively less applied except for lower order models. By exploiting the much lower resolution of the classification target compared with that of the regression target, polynomial expansion features can be reduced while preserving their nonlinear mapping capabilities for pattern classification. This book includes a collection of polynomial expansion-based algorithms to outstretch the linear regression model for pattern classification.

Next, minimization of the classification error counting loss is a fundamental problem in pattern classification. However, as the classification error counting loss constitutes a nonlinear step function, the optimization inevitably leads to a numerical solution based on an iterative search whereby global optimality cannot be guaranteed. It has been found that by matching between the link and the loss functions, the minimization of misclassification error (MCE) problem can be solved analytically. When a linear link is utilized with a quadratic loss, the MCE problem has a global and unique solution. This book introduces such an analytic solution, showcasing the use of polynomial models as the link function and the use of a biased quadratic as the loss function to analytically solving the MCE problem.

Compressive learning is another subject of great importance in machine learning. From the perspective of resource utilization, a sparse solution is more cost-effective. From the perspective of learning, a sparse solution offers better prediction generalization. Compressive learning can be formulated by including a parametric penalty term in the optimization objective. When the penalty term uses an L_p-norm of the parameters, it is known as bridge regression. It has been found that bridge regression can be formulated and solved analytically as well. This book offers solutions to bridge regression in primal form for over-determined systems and in dual form for under-determined systems.

Deep learning is an important topic in artificial intelligence. However, interpreting the resultant learning is becoming more difficult due to its ever-increasing complexity. In this book, we offer a stacking approach to analytically solve the Multilayer Perceptron and Convolutional Neural Network learning problems, paving the way toward interpretability of network learning.

Ensemble learning, with a structure akin to shadow networks but employing an analytic fusion learning strategy, leverages the collective strength of multiple models to enhance performance. At its core, ensemble learning constructs a composite model from a group of base learners or models often referred to as "weak learners" when their individual performance is modest. However, when combined, these "weak learners" form a "strong learner" capable of achieving superior accuracy. The fundamental premise is that by aggregating predictions from multiple models, the ensemble compensates for individual weaknesses, reduces the risk of overfitting, and manages the bias-variance trade-off more effectively than single models. Ensembles have found applications in finance, where interpretability is crucial. For example, ensemble models like Random Forest or XGBoost offer predictive power similar to deep learning models while maintaining a high level of interpretability, which satisfies financial regulators.

Chapters 1–9 each contain 10 exercises, making the book suitable for use as a textbook. Additionally, every chapter includes bibliographic notes that offer insights into the development trends. However, the authors do not claim to provide exhaustive coverage of these trends.

The corresponding author has been working in the field of pattern recognition and machine learning for over three decades. He worked in pattern-based process control for his Ph.D. thesis and then spent 5 years with the Institute for Infocomm Research, Singapore, before becoming a professor at Yonsei University, South Korea in 2005. A significant proportion of this book is a record of his core research and teaching over the years.

All the co-authors are either collaborators or co-workers of the main author. Two of them are academics, serving, respectively, as professors at Nanyang Technological University, Singapore, and at South China University of Technology, China. They have been working in the field of machine learning and pattern classification for over 35 years and over 10 years, respectively. The remaining co-author comes from the industry with a partial commitment to teaching.

Seoul, Korea (Republic of)   Kar-Ann Toh
Guangzhou, China   Huiping Zhuang
Singapore   Simon Liu
Singapore   Zhiping Lin

Acknowledgements

Through writing this book, I have had the opportunity to connect with talented and engaging individuals on a deeper level. This journey began long before my formal research career, and I would like to express my deep appreciation to Mr. Yu-Long Tu, my primary school form teacher, whose guidance helped me choose a secondary school well suited to my strengths. I am also deeply thankful to Master Hock-Meng Siew and Master Seng-Yong Tan, who have been mentors to me in both life and art. Regarding this book, first and foremost, I extend my heartfelt gratitude to Prof. Zhiping Lin, my long-term research collaborator and co-author, for his unwavering support. I am also fortunate to have worked with Dr. Huiping Zhuang, now a professor, and Dr. Simon Liu, in integrating the book with more complex and practical systems. Special thanks go to Prof. Giuseppe Molteni and Dr. Zhengguo Li for their research contributions related to bridge regression. Additionally, I appreciate the various forms of assistance and friendships offered by Prof. Anil Kumar Jain, Prof. Jaihie Kim, Dr. Wei-Yun Yau, Prof. Bah-Hwee Gwee, Dr. Geok-Choo Tan, Prof. Donghyun Kim, Prof. Jong-Hyun Ahn, Prof. Andrew Beng Jin Teoh, Prof. Sangyoun Lee, Prof. Zhongke Wu, Prof. Lei Sun, Prof. Beom-Seok Oh, Prof. Xudong Jiang, Prof. Chien-Chern Cheah, Dr. How-Lung Eng, Dr. Deruo Cheng, Dr. Cheng-Yaw Low, Dr. Anh-Duc Nguyen, Dr. Kangrok Oh, Dr. Han Cheol Moon, Dr. Jooyoung Kim, Mr. Jaekwon Lee, Mr. Lu Zhang and Ms. Hari Kang. Last but not least, I express my gratitude to my past and present students, whose invaluable feedback has field-tested a significant portion of the content now presented in this book. This book is dedicated to my mother, my family and the cherished memory of my beloved grandparents and father.

Kar-Ann Toh

I would like to express my heartfelt gratitude to my supervisor, Prof. Zhiping Lin, for his unwavering support throughout my Ph.D. journey, which has continued into my faculty time. I would like to thank Prof. Kar-Ann Toh, my guide to analytic learning, where my research truly starts. It is an honor to participate in this book as a co-author.

And of course, my family always offers the support I need. I thank everything (good or harsh) that makes me who I am.

Huiping Zhuang

I am immensely grateful to Prof. Kar-Ann Toh, Prof. Zhiping Lin, and Prof. Huiping Zhuang for the opportunity to contribute to this book and for their patient guidance. My inspiration for discussing ensemble learning stems from years of industrial practice, for which I am thankful to Dr. Mark Engel, Pubudu Premawardena, and Carrie Chai, who allowed me to explore these techniques in real industrial applications. Much of the material and presentation style were adapted from lessons I gave to my Data Science team at Scotiabank, and I thank them for their patience and insightful feedback. Lastly, I appreciate Tian Tian for reviewing my initial manuscript and providing thoughtful suggestions.

Simon Liu

I would like to especially thank Prof. Kar-Ann Toh, my research collaborator for a decade and also a good friend, for many inspiring discussions and fruitful collaborations, and for his leading efforts in writing this book. I appreciate the joint efforts with Huiping Zhuang, my former Ph.D. student and now a professor himself, and with Dr. Simon Liu, for writing this book and also for other productive collaborations in research or teaching. I thank Profs. Lei Sun, Beom Seok Oh, Giuseppe Molteni, and Dr. Zhengguo Li, for collaborations which led to some joint papers related to this book.

Zhiping Lin

Contents

Symbols

x	A *scalar* data variable
w	A *scalar* parameter variable
$\exp(x)$ or $e^{(x)}$	The natural exponential function of x
$\min(a, b, c)$	The smallest among a, b and c
$\max(a, b, c)$	The largest among a, b and c
$\boldsymbol{x}$	A data *column vector* of d dimension, i.e., $\boldsymbol{x} = \begin{bmatrix} x_1 \\ \vdots \\ x_d \end{bmatrix}$
$\boldsymbol{x}^T$	Transposition of *vector* $\boldsymbol{x}$, i.e., $\boldsymbol{x}^T = [x_1, ..., x_d]$
$\boldsymbol{w}$	A parameter *column vector* of d or D dimension, $\boldsymbol{w} = [w_1, ..., w_d]^T$
$\min_w$	*Minimization* with respect to parameter $\boldsymbol{w}$
$\langle \boldsymbol{x}, \boldsymbol{y} \rangle$	The inner product between $\boldsymbol{x}$ and $\boldsymbol{y}$, i.e., $\boldsymbol{x}^T \boldsymbol{y}$
$\boldsymbol{x} \circ \boldsymbol{y}$	The Hadamard product between $\boldsymbol{x}$ and $\boldsymbol{y}$, i.e., $\boldsymbol{x} \circ \boldsymbol{y} = [x_1 y_1, ..., x_d y_d]^T$, also known as the Schur product or *entrywise* product
$f(\boldsymbol{x})$ or $g(\boldsymbol{x})$	A *scalar* function of $\boldsymbol{x}$
$\frac{\partial f(\boldsymbol{x})}{\partial \boldsymbol{x}}$	The partial derivative of f with respect to $\boldsymbol{x}$
$\boldsymbol{f}(\boldsymbol{x})$ or $\boldsymbol{g}(\boldsymbol{x})$	A *vector* function of $\boldsymbol{x}$
X	A data matrix of $m \times (d + 1)$ dimension, where m denotes the *number* of data samples and d denotes the *dimension* of each sample
W	A parameter matrix of $(d + 1) \times h$ or $(D + 1) \times h$ dimension, where h denotes the *number* of parameter columns and d or D denotes the *dimension* of each column
$\boldsymbol{p}$ or $\boldsymbol{p}(\boldsymbol{x})$	A projected data *column vector* of $(D + 1)$ dimension, $\boldsymbol{p} \triangleq \boldsymbol{p}(\boldsymbol{x}) = [p_0, ..., p_D]^T = [p_0(\boldsymbol{x}), ..., p_D(\boldsymbol{x})]^T$
P	A projected data *matrix* of $m \times (D + 1)$ dimension
$\|\boldsymbol{x}\|$	The ℓ_1-norm (also called the 1-norm, the L_1-norm, or the magnitude) of $\boldsymbol{x}$ given by $\sum_{i=0}^{d} |x_i|$

$\|\boldsymbol{x}\|_2$ or $\|\boldsymbol{x}\|$	The ℓ_2-norm (also called the 2-norm, the L_2-norm or the Euclidean distance) of $\boldsymbol{x}$ given by $\sqrt{\boldsymbol{x}^T\boldsymbol{x}}$
$\|\boldsymbol{x}\|_p$	The ℓ_p-norm (also called the p-norm or the L_p-norm) of $\boldsymbol{x}$ given by $\left(\sum_{i=0}^{d}|x_i|^p\right)^{1/p}$
$u(x)$	A unit *step* function given by $u(x) = \begin{cases} 1 \text{ if } x \geqslant 0 \\ 0 \text{ if } \text{otherwise} \end{cases}$
$\text{sgn}(x)$ or $\text{sign}(x)$	The *sign* or *signum* function given by $sgn(x) = \begin{cases} 1 & \text{if } x > 0 \\ 0 & \text{if } x = 0 \\ -1 & \text{if } x > 0 \end{cases}$
$\text{sigm}(x)$ or $\sigma(x)$	The *sigmoid* function given by $\text{sigm}(x) = \frac{1}{1+e^{-x}}$

Chapter 1
Introduction

Having access to relevant information and patterns helps in making good decisions. Particularly when different patterns are better understood, situations can be analyzed more effectively, leading to more accurate predictions of outcomes and better choices of action. This chapter introduces the subjects of *pattern recognition* and *analytic learning*, providing the necessary terminological background before detailing the pattern recognition pipeline. Additionally, it presents the relationships among the modular chapters.

1.1 What Is Pattern Recognition?

Human beings inherently possess the ability to recognize shapes, patterns, and trends. However, replicating this task in a machine is not straightforward. Multiple processing steps are involved in achieving this goal. The initial step is machine perception or image acquisition, where sensors such as cameras or radio receivers come into play. During this stage, the sensed signals are converted into numerical values. The subsequent processing step leverages these numerical values for shape or pattern inference. This step constitutes the core of machine learning and artificial intelligence, where meaningful interpretations can be derived from the information processing. Finally, the results are utilized for subsequent actions. This entire process is referred to as pattern recognition, which is commonly found in machine learning applications.

Example 1.1: Digit Recognition

In this example, the goal is to recognize handwritten digits captured as images, as illustrated in Fig. 1.1. We present two sets of data samples, each with a different resolution.

© The Author(s), under exclusive license to Springer Nature Singapore Pte Ltd. 2025
K.-A. Toh et al., *Analytic Learning Methods for Pattern Recognition*,
https://doi.org/10.1007/978-981-96-2151-4_1

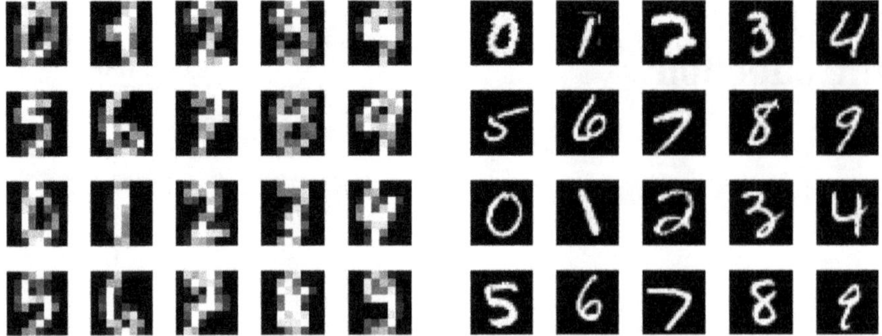

Fig. 1.1 Samples of optdigit (left panel) and MNIST (right panel) data sets. Samples shown in the upper two rows are taken from the training set and samples shown in the bottom two rows are taken from the test set

1. Optdigits Database:[1] The sample images on the left panel are drawn from the Optdigits database. These images are 8×8 pixels grayscale maps, where each pixel's intensity ranges from 0 to 16. The total number of images in this database is 5,620.
2. MNIST Database LeCun et al. (1998):[2] The sample images on the right panel are sourced from the MNIST database. These images are 28×28 pixels grayscale maps, with each pixel represented by a floating-point number ranging from 0 (black) to 1 (white). The total number of images in this database is 70,000.

Both of these data sets contain handwritten digits numbered from 0 to 9.

These databases are divided into two sets: one for training and the other for test evaluation. The primary task is to predict the images in the test set by learning from the training set. If the recognition accuracy reaches a sufficient level, the machine can be applied to automatically sort envelopes with handwritten postal codes. Furthermore, when extended to recognize alphabets, the same machine can be employed for handwritten document reading. This example illustrates an application with relatively constrained information variation on a two-dimensional surface.

□

---

[1] This data set was created by E. Alpaydin and C. Kaynak, and being introduced in Kaynak (1995). The data set is licensed under a Creative Commons Attribution 4.0 International (CC BY 4.0) license (https://creativecommons.org/licenses/by/4.0/legalcode). The data set is available at https://archive. ics.uci.edu/dataset/80/optical+recognition+of+handwritten+digits.

[2] Yann LeCun (Courant Institute, NYU) and Corinna Cortes (Google Labs, New York) hold the copyright of MNIST data set, which is a derivative work from the original NIST data sets. MNIST data set is made available under the terms of the Creative Commons Attribution-Share Alike 3.0 license (https://keras.io/api/datasets/mnist/). The data set is available at https://yann.lecun.com/ exdb/mnist/ and https://archive.ics.uci.edu/dataset/683/mnist+database+of+handwritten+digits.

Fig. 1.2 Samples of ten identities drawn from the database of CMU face images with quarter resolution (30 × 32 pixels per image)

Example 1.2: Face Recognition

In this example, the task is to recognize images of faces captured under variations in pose, expression, and the presence of glasses. Compared with handwritten digits, the degree of variation is larger because it involves imaging three-dimensional objects (faces). Figure 1.2 shows several face image samples drawn from the database of CMU face images.[3] This database contains 640 face images from 20 subjects with each subject contributing 32 different face images. Each face image of full resolution has 120 × 128 pixels, with 256 gray levels per pixel. Images of half and quarter resolutions are also available. Face recognition plays a crucial role in various applications, ranging from highly accurate custom and banking systems to more user-friendly mobile device control applications.

$$\square$$

Example 1.3: Financial Application—Credit Risk Modeling

In the domain of finance, credit risk modeling is a critical tool for assessing the probability that a borrower may default on their loan commitments. This evaluation is pivotal in managing the level of risk involved in extending credit, influencing a wide range of financial decisions from personal loans and credit card approvals to broader institutional lending. Credit risk modeling aligns with the broader theme of recognizing patterns in data, similar to digit and face recognition examples previously discussed. However, instead of identifying digits or facial features, here the focus is on discerning patterns in financial activities that indicate potential default risks.

Traditionally, credit scoring has been predominantly executed using logistic regression. This method is favored for its simplicity and the clarity with which it conveys information, making the interpretation and explanation of results straightforward. In financial modeling, explainability is crucial as stakeholders, including regulators and loan officers, require clear justifications for decisions that affect financial risk management.

---

[3] This data set was created by Tom Mitchell. The data set is licensed under a Creative Commons Attribution 4.0 International (CC BY 4.0) license (https://creativecommons.org/licenses/by/4.0/legalcode). The data set is available at https://archive.ics.uci.edu/dataset/124/cmu+face+images.

However, with the advancement in data storage capabilities and computational power, financial institutions now harness large volumes and diverse types of historical customer data. This shift has led to the adoption of more sophisticated machine learning techniques. Among these, ensemble learning methods like Random Forests and XGBoost have gained prominence. These models have consistently outperformed traditional logistic regression in terms of precision and accuracy, providing more reliable assessments of credit risk.

A significant advantage of ensemble methods like Random Forest and XGBoost is their inherent explainability. Despite their complexity, these models facilitate a degree of transparency that allows users to understand the factors driving the predictions. This feature is particularly beneficial in financial applications where the rationale behind a credit scoring decision must be clear and justifiable.

In Chap. 8, we will delve deeper into these ensemble learning methods, exploring how they enhance predictive performance while maintaining the necessary transparency for financial applications. This exploration will include detailed discussions on the implementation and advantages of using these advanced techniques in credit risk modeling.

$\square$

1.2 What Is Analytic Learning?

According to Gogus (2012), *"The word analytic comes from the Greek word 'analytikos', which means having the ability to analyze and divide the whole into its components or elements"*.

In our context, *analytic learning* refers to learning with a solution that has an exact, analytical, or closed form. The solution comes in a well-defined form which allows analysis and interpretation. In contrast to numerical method which searches based on solution guess and tests whether the problem is solved well enough for search termination, analytic learning arrives at solution without needing such a search process.

1.3 Terminological Background

It is difficult to find a clear boundary among the terms *pattern recognition*, *machine learning*, and *data mining* due to their highly overlapping scopes of coverage. The origins of these terms can be traced to diverse disciplines covering engineering, computer vision, statistics, and artificial intelligence. The term *pattern recognition* is inclined toward engineering and computer vision applications whereas the term *machine learning* is more toward computer science, statistical, and mathematical developments. On another hand, *data mining* involves the process of discovering patterns in large data sets under the umbrella of computer science. Although these terms bear subtle difference under focused disciplines, to a certain extend they share common functional components such as preprocessing, feature extraction (also known as dimension reduction), learning, classification, and performance evaluation. While greater details along the line of analytic learning shall be presented in respective chapters, an introduction to these functional components is provided in Sect. 1.5 right after a brief account of several common terminologies related to recognition.

Under the *pattern recognition* paradigm, a distinction among several common terminologies such as classification, identification, and verification cannot be neglected. Particularly, while the term "classification" refers to either a *binary* decision or *n-ary* decision, the term "verification" only refers to binary decision. On another hand, the terms "identification" and "recognition" carry the meaning of n-ary decisions. A variable is *binary* or *dichotomous* when it has only two categories, values or levels. A binary decision is a choice between two alternatives. A dichotomy refers to a partition of a set into two subsets which are jointly and mutually exclusive. For instance, the choice between a "yes" and a "no" action is a dichotomous or binary decision. For a variable or a decision with more than two categories, the terms *multi-category*, *multi-class*, and *polychotomous* can be used interchangeably. Table 1.1 summarizes the underlying decision connotation among these terminologies.

Under the *machine learning* paradigm, the learning goal or target can be discrete or continuous. When the learning is performed with a known target label, the learning is *supervised*. For discrete target labeling or grouping, such a learning task is called a *classification* or a categorization task. For learning of continuous target values, the learning process is known as *regression*. On the other hand, when the target label is not available, the learning can be *unsupervised*. Here, learning toward a discrete grouping is called *clustering* and learning toward a smaller feature space is called *dimensionality reduction*. Table 1.2 summarizes the relationship among these terminologies.

In addition to the underlying organizational information acquisition based on pattern recognition and machine learning, the *data mining* performs information *association* and *concept description*. Table 1.3 summarizes the hierarchical construction of concepts in data mining.

Table 1.1 Common Terminologies

Decision connotation	
Binary	n-ary
Dichotomous	Polychotomous
Two-category	Multi-category
Two-class	Multi-class
Classification	Classification
Verification	Identification
–	Recognition

Table 1.2 Machine learning problems

	Supervised learning	Unsupervised learning
Discrete	Classification	Clustering
Continuous	Regression	Dimension reduction

Note For dimension reduction, the resolution of "continuous" variable has been limited to continuity of integer values

Table 1.3 Data mining functionalities

Concept description	*Generalize, summarize,* and contrast data characteristics
Association	Correlation and causality analysis such as *association rule* and *sequential pattern* discoveries
Classification and prediction	Construct models or functions that describe and distinguish among classes or concepts for future prediction

1.4   Scope and Coverage of This Book

This book focuses on learning-based methods for pattern recognition, with particular emphasis on analytic or deterministic learning approaches. The pipeline of learning and recognition process will be introduced, highlighting common building blocks such as learning models, learning cost functions, analytic learning, and performance evaluation. Different formulations within each building block can be adapted across the first three fundamental blocks to form an analytic learning algorithm for either regression or classification purposes.

The book dedicates separate chapters to penalized learning and network learning. The former specializes in parameter compression, while the latter delves into deeper models with structural differences. Additionally, a relatively large chapter is devoted to ensemble learning, which capitalizes on an analytic fusion strategy. The final chapter provides application examples to aid in understanding the subject. In addition to regression applications, classification problems involving both binary and multi-category outcomes are also addressed.

1.5   Pattern Recognition Pipeline

In general, the pipeline of learning and recognition can be categorized into the following three building blocks, namely, (a) preprocessing and feature extraction, (b) classifier/predictor learning, and (c) classifier/predictor testing. These three building blocks are shown in Fig. 1.3 with their main components within. The main components of building block (a) are preprocessing and feature extraction and the main components of building block (b) are model selection, model learning, and prediction

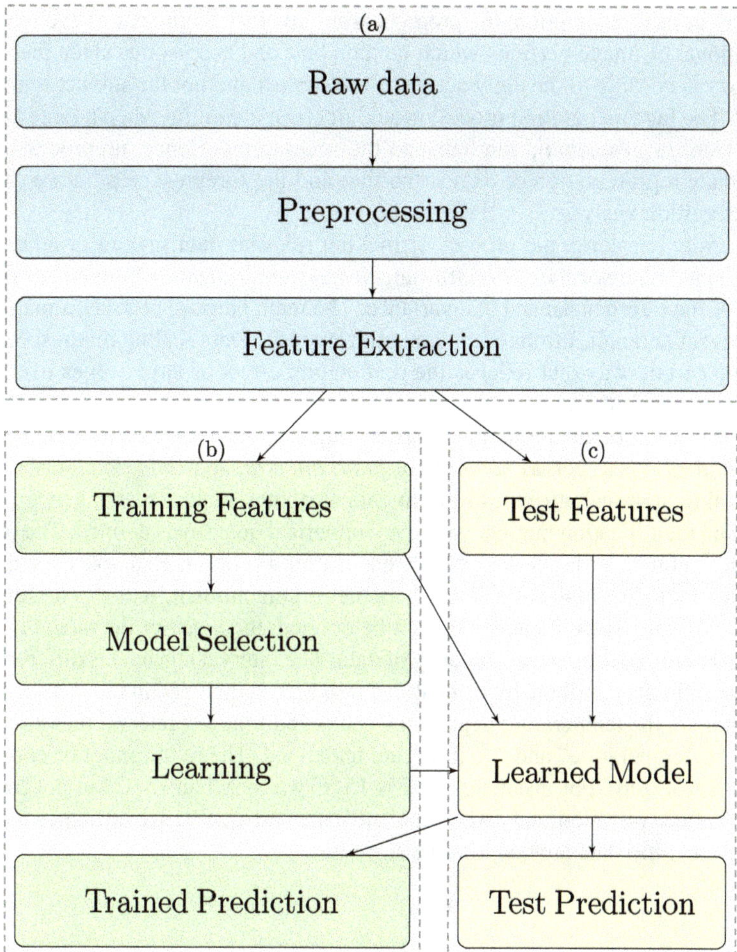

Fig. 1.3 The building blocks of pattern recognition process

of trained data. The building block (c) consists of testing the unseen data based on the trained model. The outcome of this testing stage is often used for classifier/predictor performance evaluation.

1.5.1 Preprocessing

Data preprocessing is an important step in pattern recognition because it is the very beginning step of the recognition process right after data acquisition. Essentially, the preprocessing stage prepares the data for meaningful representation. For instance, in face recognition, the preprocessing involves alignment of eye locations and removal of image portions which contain hair and accessories since these contents vary according to the daily action of the subject and not the subject himself or herself. The face recognition process would not be meaningful when a large amount of the face images contain the hair and the accessories. Hence, preprocessing for appropriate representation of data is the first and the foremost step before running any recognition analysis.

Generally speaking, the process of making relevant data scaling or adjustment can be called data *normalization*. In statistics, a *standardization* amounts to *scaling* of data to have zero mean and unit variance. The main purpose of data normalization is to prevent anomalies from dominating the analysis. Data scaling might spread the data in a certain way that reduces the dominating effect of large values over small values in numerical analysis.

Apart from normalization, the preprocessing step can include conversion among different data types such as *nominal, ordinal, interval,* and *ratio data* attributes. In data mining, nominal attributes refer to data attributes such as name, gender, color, place, and identification number where no numerical quantity is defined. The ordinal attributes refer to ordered data which has a ranking order. Examples for ordinal attributes include satisfaction level, happiness, contentment, and discomfort. For this type of data, since the *mean* cannot be defined, the *mode* or the *median* can be used to determine the central tendency of data. The interval attribute is differentiated from the ratio data attribute by whether the true zero is included in the measurement. For example, the temperature degree in Celsius and time are interval measurements where "no temperature" and "no time" are undefined. The ratio cannot be computed without a true zero. For instance, 30 °C is 15 °C warmer than 15 °C but not twice as warm. Collectively, nominal and ordinal attributes are qualitative attributes whereas interval and ratio data are quantitative attributes.

1.5.2 Feature Extraction

Feature extraction is the process of extracting informative and relevant components from the bulk of data to facilitate machine learning and interpretation. Essentially, it involves a reduction of the amount of resources required for describing a large set of data. This process often largely reduces the dimension of data and is thus related to dimension reduction. While *feature extraction* often involves the process of data transformation, *feature selection* need not. Feature selection is also known as attribute selection or variable subset selection. It involves selection of useful subsets of features for model construction or machine learning.

In general, the approaches to feature extraction can be broadly categorized into the approach of generic *dimension reduction* and the approach of extracting *heuristic* or *semantic features*. Examples for the dimension reduction approach include *principal component analysis* (PCA), *independent component analysis* (ICA), *isomap, multilinear subspace learning,* and *autoencoder*. On the other hand, examples for the approach of heuristic or semantic features extraction include *edge detection, corner detection, blob detection, ridge detection*, and *scale-invariant feature transform* which are typically found in image processing literature.

1.5.3 Learning

Essentially, machine learning in our context can be summarized as learning a function f that maps input variables to output or response variables, i.e., $Y = f(X)$ where X and Y are, respectively, generic symbols for input and output variables which can take either scalar or vector quantities. The form of f, if it exists, is unknown in practice and the job is to evaluate those algorithms at hand that best describe the problem.

Depending on the assumptions made regarding the form of the function f, learning can be by *parametric* or *nonparametric* means. A parametric learning model assumes a fixed and finite set of parameters whereas a nonparametric learning model does not make strong assumptions about the form of the mapping function. Examples of parametric models include logistic regression, linear discriminant analysis, perceptron, naive Bayes, and basic neural networks. Examples of nonparametric learning algorithms include k-nearest neighbors, decision trees, and Parzen windows. In our context, those parametric learning models which can produce an analytic solution shall be our subject of interest.

The approaches to machine learning can generally be classified into *supervised learning, unsupervised Learning, semi-supervised learning*, and *reinforcement learning*. In supervised learning, a "teacher" or a target output or response label is available where the goal is to learn a general rule that maps the inputs to the outputs. In unsupervised learning, no labels are given and the algorithm is to find the structure of the given input features. In semi-supervised learning, both labeled data and unlabeled data are considered for learning. In reinforcement learning, the algorithm learns by interacting with the dynamic environment without a teacher having an explicit learning goal. The main focus of this book is on supervised learning algorithms which has an analytic solution form.

To facilitate learning, a learning cost function $J(\cdot)$ which embeds the mapping function $f(X)$ is frequently required. In parametric learning, the learning cost function can be written as $J(f(X, \Theta))$ where Θ is a set of parameters of fixed size. Naturally, it is desired that the learning cost function matches or aligns with the learning objective or goal.

While learning is a mathematical optimization that is performed on a predictive model of fixed structure, classification is a categorization decision process upon the output of the learned model. For binary classification, when the model output is expressed as a *distance metric* or *confidence level*, the process of categorization boils down to the problem of finding an appropriate decision threshold for data grouping. Alternatively, the model output can be *ordinal* or *nominal* where the process of categorization boils down to a *voting* problem. Since the multi-category problem can be treated as a set of multiple binary problems, the binary classification can be extended to multi-category classification by adopting a *one-versus-all* or *winner-takes-all* decision technique on the multiple binary decisions.

1.5.4   Performance Evaluation

The essence of learning is to model the behavior of data through an optimization process. Such behavior of data can be statistically described by a *distribution function* called *probability density function* (pdf) in continuous sense and probability mass function (pmf) in discrete sense. In statistical learning, we assume that data samples are drawn from such hypothesized density or mass function which may not even exist. If the drawn data samples are representative, where a small quantity accurately reflects the larger entity, then learning is meaningful since the learned model can predict well the unseen data. According to the *Central Limit Theorem*, which states that, *under various conditions, the aggregate effect of the sum of a large number of small, independent random disturbances will lead to a Gaussian or normal distribution*, the density function can be estimated accurately and hence no representation issue arises in such a case. However, in real-world applications, the size of samples is often limited due to measurement and cost constraints. The goodness of learning is thus hinged upon the *predictivity* or *generalization* of the learned model based on the limited size of learning data samples.

Although full information regarding the underlying data distribution is not known, the goodness of learning can nevertheless be assessed through an appropriate performance evaluation. This is frequently achieved by isolating a portion out of the entire data set to test the generalization ability of the learning algorithm. To capitalize on all the available data, an N-fold cross-validation test can be performed. This is done by partitioning the entire data set into N portions and then use the $(N - 1)$ portions for learning and the remaining portion for testing. The process is repeated to test each and every portion of the N portions. The final result is often recorded in averaged form for the N number of tests.

1.6   Organization

For beginners, reading the book chapter by chapter should flow well since the beginning chapters pave the essential preliminaries for the later chapters where the topics run from simple ones to difficult ones. In this chapter, a general introduction to the subject is provided, while Chap. 2 details the data preparation process. Chapters 3–5 cover the fundamental modules of analytic learning, which can be combined to form the learning pipeline. For example, a learning model from Chap. 3 can be paired with a cost function from Chap. 4, with the corresponding solution found in Chap. 5. Chapters 6 and 7 delve into two important topics that build upon and extend beyond

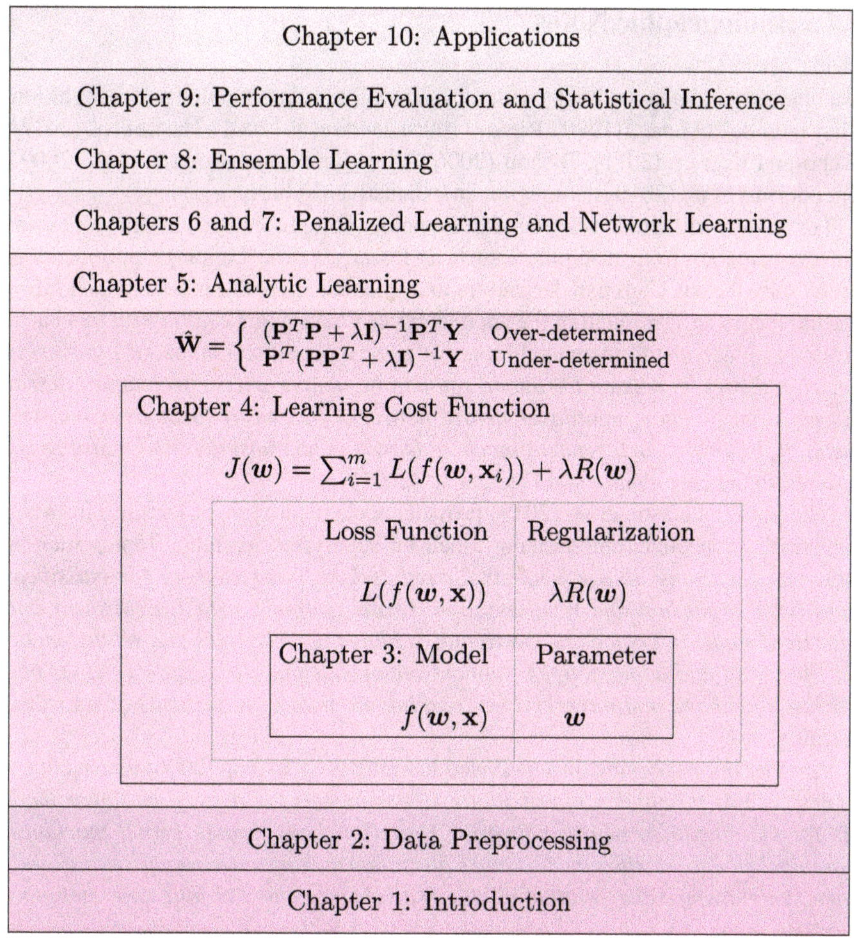

Fig. 1.4 Organization of topics and link among chapters

these fundamental modules. Chapter 8 introduces advanced methods of ensemble learning, crucial for field applications requiring high reliability. Chapter 9 outlines common practices for performance evaluation and statistical inference, and Chap. 10 presents various application examples. For use as a quick reference, a map for linking the mix of different chapters is provided in Fig. 1.4 as reading guide.

"What information consumes is rather obvious: it consumes the attention of its recipients. Hence, a wealth of information creates a poverty of attention and a need to allocate that attention efficiently among the overabundance of information sources that might consume it."—Herbert Simon

1.7 Bibliographic Notes

There are several popular books with focused topics in classification and recognition. They include Fukunaga (1990), Ripley (1996), Duda et al. (2001), Hastie et al. (2017), Webb and Copsey (2011), Bishop (2006), Theodoridis and Koutroumbas (2009), Theodoridis et al. (2010), though not an exhaustive list here.

For readers with little background in the field, a comprehensive summary of essential mathematics from disciplines such as linear algebra, Lagrange optimization, probability theory, Gaussian derivatives and integrals, hypothesis testing, and information theory can be found in the appendices of Duda et al. (2001). With this background in place, the fundamental topics of pattern classification, namely, *Bayesian decision theory, maximum likelihood estimation, nonparametric techniques, linear discriminant function, multilayer neural networks, stochastic methods, nonmetric methods, algorithm-independent machine learning*, and *unsupervised learning* are covered in the core chapters of Duda et al. (2001).

The book by Hastie et al. (2017) provides a classical view of learning from the statistical perspective, emphasizing much on supervised learning. Topics such as *supervised learning, linear methods for regression, linear methods for classification, basis expansion and regularization, kernel methods, model assessment and selection, model inference and averaging, additive models, trees and related methods, boosting and additive trees, support vector machines, prototype methods and nearest neighbors*, and *unsupervised learning* are treated as elements of statistical learning.

Another book focusing on supervised learning is by Bishop (2006) which covers topics such as *probability distributions, linear models for regression, linear models for classification, neural networks, kernel methods, sparse kernel machines, graphical models, mixture models and expectation maximization, approximate inference, sampling methods, continuous latent variables, sequential data*, and *combining models*.

The book by Webb and Copsey (2011) starts with an introduction to statistical pattern recognition, and then followed by topics such as *parametric* and *nonparametric density estimation, linear discriminant analysis, kernel methods, projection methods, tree-based methods, performance, feature selection and extraction*, and finally *clustering*. An earlier book which covers similar topics is that by Fukunaga (1990). For readers who need more background on clustering, the book by Theodoridis and Koutroumbas (2009) comes with six chapters devoted to clustering.

1.8 Exercises

1.1 What is the relationship among AI (artificial intelligence), ML (machine learning), and DL (deep learning)?

1.2 A computer program is said to learn from experience E with respect to some task T and some performance measure P, if its performance on T, as measured by P, improves with experience E. Suppose we feed a learning algorithm a lot of historical weather data, and have it learn to predict weather. In this setting what is T?

(a) The historical weather data
(b) The probability of it correctly predicting a future data's weather
(c) The weather prediction task
(d) None of these

1.3 What is data preprocessing and what are the typical processing tasks?

1.4 In standardization, the data features will be re-scaled with

(a) Mean 1 and Variance 1
(b) Mean 0 and Variance 1
(c) Minimum 0 and maximum 1
(d) Minimum -1 and maximum $+1$

1.5 Which of the following is/are example(s) of feature extraction?

(a) Measure the length, the width, the texture, and the number of fins of fishes for classification using computer vision
(b) Imputation of missing data
(c) Principal component analysis
(d) Construct histogram of data values for each sample

1.6 Suppose you are working on stock market prediction. Typically tens of millions of shares of a company's stock are traded each day. You would like to predict the number of shares that will be traded tomorrow for the company.

(i) Would you treat this as a classification or a regression problem?
(a) Regression
(b) Classification
(c) Clustering
(d) None of these
(ii) If the data you have collected involved millions of attributes, what would you do?

1.7 Suppose you are working on weather prediction and use a learning algorithm to predict tomorrow's temperature (in degrees Centigrade/Fahrenheit).

(i) Would you treat this as a classification or a regression problem?
(a) Regression
(b) Classification
(c) Clustering
(d) None of these
(ii) What kind of data should you gather?

1.8 What is/are the differences between supervised and unsupervised learning?

1.9   In a report, a research team claimed that their method is the best based solely on the low training error evaluation. The claim in the report is trustworthy. True or false?

1.10   A research team analyzed information from several earlier studies involving more than 188 million of adults. They observed the link between sitting and risk of mortality over 20 to 30 years. In the studies, about 28% of deaths could be attributed to sitting, 21% to television viewing, the researchers said. The researchers concluded the analysis with the following statement: "Sit less than 5 years, add 2 years to your life". This report is trustworthy as it involved a huge database. True or false?

Chapter 2
Data Preprocessing

Data types vary widely and depend on the specific domain. It is thus impossible to provide an exhaustive coverage of data types across all existing domains. However, some general guidelines upon typical data types and their preprocessing are nevertheless possible. In this chapter, a generic coverage of data types, related distance metrics, and their preprocessing are presented.

2.1 Data Types

2.1.1 Nominal Data

Nominal data attributes, also known as categorical or qualitative data attributes, commonly occur in natural settings as labels or names. While some of these labels possess an intrinsic order where a quantitative measurement can be established, others may not have any computable measurement at all. Examples of nominal attributes which do not possess an intrinsic order or measurable property include gender (male or female), race, blood type, place, identification number, name of fishes, name of fruits, and etc. These data can be represented in aggregate form, either in percentage or in terms of frequency of occurrence. However, it is not possible to define a *mean* value for these labels. On the other hand, examples of nominal attributes with an intrinsic order include color where the underlying electromagnetic wavelength can be measured, age where the measurement is time based, and educational level where an intrinsic order can be established.

There are two general ways to convert nominal attributes into numerical values for computation. The first way is to assign arbitrary numbers to represent these attributes. For example, one can assign a "1" value for male and a "2" value for female for the case of gender attribute. However, the label which is assigned a higher value may have

K.-A. Toh et al., *Analytic Learning Methods for Pattern Recognition*,
https://doi.org/10.1007/978-981-96-2151-4_2

a greater influence than the one with a lower value when extremely large values and extremely small values are involved along the computational process. Moreover, the intrinsic relationship among the nominal attributes may be lost when the numerical assignment is arbitrary. For example, when representing the directional attributes "north", "east", "south", and "west" using "1", "2", "3", and "4", respectively, the inherent spatial relationship among the attributes is lost.

The second way of representing the nominal attributes is by means of binary coding. Depending on the number of representing bits used, the binary coding offers a rich source to represent nominal attributes in terms of binary numbers. Common examples of binary coding schemes include *binary-coded decimal, n-ary gray codes*, and *one-hot encoding*. Sophisticated coding schemes take into account the probability distribution of each attribute during conversion.

After conversion, the binary feature vectors can be compared based on various distance measures such as Hamming distance and Spearman distance (see Sect. 2.1.2 for examples and see Sect. 2.2 for definition of distance metrics).

2.1.2　Ordinal Data

For data of ordinal scale, the exact numerical value has no significance over the ranking order that it carries. Such data is often seen in social and medical sciences where certain survey or health conditions are coded in importance level or strength level. For example, in the patient queue management system, patients who are either in pain or in frail health are often given a higher priority than those without severe symptoms. In a typical survey, several levels of response preferences are often assigned to represent the order of preference in the observation.

Similar to the nominal data, the percentage or the frequency of occurrence can be quantified. Unlike the nominal case, the *mean, median*, and *mode* values can be calculated when the ranks are assigned a numerical value. However, care should be taken for using the *mean* value since there is no knowledge regarding the distance among the ranks.

The rating and ranking orders can be transformed into quantitative variables through normalization. Suppose the ranks are given by $r = 1, \ldots, R$. Then, the ranks can be normalized into standardized distance values (d) which fall within $[0, 1]$ using

$$d = \frac{r - 1}{R - 1}. \tag{2.1}$$

When comparing the similarity or dissimilarity between two ranks, several distances such as Spearman distance, Hamming distance, and Chebyshev distance can be used. The Spearman distance is proportional to the square of Euclidean distance (see Sect. 2.2) between two rank vectors. For example, given that the preferences of person $\mathbf{x}$ and person $\mathbf{y}$ on fruits given by {1:apple, 2:orange, 3: kiwi} are, respectively, [2, 3, 1] and [3, 2, 1]. Then, the Spearman distance of preference between $\mathbf{x} = [x_1, x_2, x_3]^T$ and $\mathbf{y} = [y_1, y_2, y_3]^T$ is proportional to

$$
\begin{aligned}
d(\mathbf{x}, \mathbf{y}) &= (x_1 - y_1)^2 + (x_2 - y_2)^2 + (x_3 - y_3)^2 \\
&= (2 - 3)^2 + (3 - 2)^2 + (1 - 1)^2 = 2.
\end{aligned}
\tag{2.2}
$$

The Hamming distance between two strings of ranks of equal length is the number of positions at which the corresponding ranks are different. For example, the Hamming distance between [2, 3, 1] and [3, 2, 1] is 2 since the first two rank values are different. The Chebyshev distance is also called *maximum value distance* which measures the maximum absolute difference between two ordinal vectors. Other distances for ordinal variables include Kendall distance, Cayley distance, and Ulam distance.

2.1.3 Interval and Ratio Data

Both interval attributes and ratio attributes of data are represented by numerical values. However, they are different in terms of whether the natural (absolute) zero point is included. The ratio scale has an absolute zero point. For example, age, height, and weight never fall below zero. For the interval scale, the zero point is arbitrary where negative values are permitted. For instance, the temperature degree in Celsius is considered an interval attribute, since the "zero-degree" does not correspond to "no temperature" where we cannot say that $60\,^\circ$C is twice as hot as $30\,^\circ$C. However, for the Kelvin scale, the zero point represents the complete absence of molecular motion where we can say that 60 Kelvins is twice as hot as 30 Kelvins. Hence, the Kelvin scale is a ratio attribute. Although a distance measure between two points can be made for both ratio and interval attributes, scaling of these two attributes produces different interpretations. Hence, for these two reasons (absolute zero value and the interpretation of scales), special attention needs to be paid for the effect of scaling during data normalization.

2.2 Distance Metrics

A *metric* or a *distance function* refers to a function that defines a distance between each pair of elements of a set. Mathematically, the distance mapping function can be written as

$$
d(\mathbf{x}, \mathbf{y}) : \mathcal{X} \times \mathcal{X} \mapsto [0, \infty),
\tag{2.3}
$$

where $[0, \infty)$ is the set of nonnegative real numbers. The following conditions are satisfied for all $\mathbf{x}, \mathbf{y}, \mathbf{z} \in \mathcal{X}$.

$$\text{Non-negativity axiom:}\quad d(\mathbf{x}, \mathbf{y}) \geq 0, \tag{2.4}$$

$$\text{Identity of indiscernibles:}\quad d(\mathbf{x}, \mathbf{y}) = 0 \quad \Leftrightarrow \quad \mathbf{x} = \mathbf{y}, \tag{2.5}$$

$$\text{Symmetry:}\quad d(\mathbf{x}, \mathbf{y}) = d(\mathbf{y}, \mathbf{x}), \tag{2.6}$$

$$\text{Triangle inequality:}\quad d(\mathbf{x}, \mathbf{z}) \leq d(\mathbf{x}, \mathbf{y}) + d(\mathbf{y}, \mathbf{z}). \tag{2.7}$$

A set with a metric is called a metric space. Figure 2.1 illustrates the three distances between each pair of points $\mathbf{x}, \mathbf{y}, \mathbf{z}$.

The Hamming distance is a distance metric since it fulfills all the four metric conditions. Common examples of distance metrics include the Euclidean metric and the Manhattan or taxicab metric. The Euclidean metric is also known as the 2-norm or L_2-norm distance. It is the geometrical distance between two points in space. The taxicab metric is also known as the 1-norm or L_1-norm distance. A generalization of both the Euclidean distance and the taxicab distance is called the Minkowski distance or p-norm metric. These metrics are defined as follows:

$$\text{1-norm distance:}\quad \sum_{i=1}^{d} |x_i - y_i|$$

$$\text{2-norm distance:}\quad \left(\sum_{i=1}^{d} |x_i - y_i|^2 \right)^{1/2}$$

$$\text{p-norm distance:}\quad \left(\sum_{i=1}^{d} |x_i - y_i|^p \right)^{1/p}$$

$$\text{infinity-norm distance:}\quad \lim_{p \to \infty} \left(\sum_{i=1}^{d} |x_i - y_i|^p \right)^{1/p}$$

$$= \max \left(|x_1 - y_1|, |x_2 - y_2|, \ldots, |x_d - y_d| \right).$$

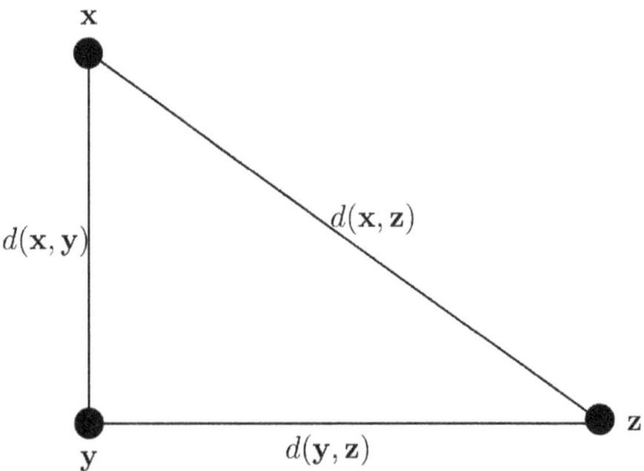

Fig. 2.1 Three points form a triangle: $d(\mathbf{x}, \mathbf{z}) \leq d(\mathbf{x}, \mathbf{y}) + d(\mathbf{y}, \mathbf{z})$

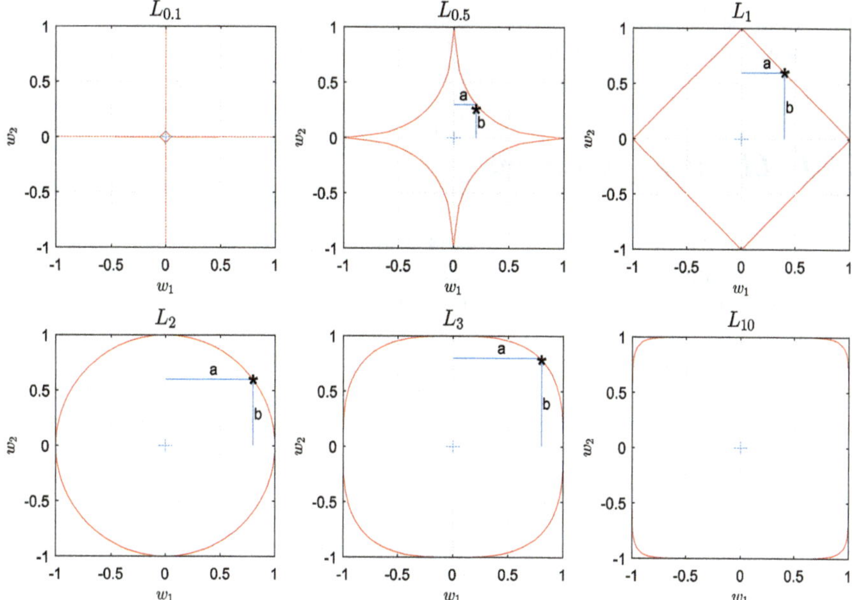

Fig. 2.2 Minkowski distance (except for $L_{0.1}$ and $L_{0.5}$ which violate the triangle inequality, all other distances are distance metrics) on a two-dimensional space $\boldsymbol{w}$. The red solid contour indicates one unit distance from the origin. For example, in the L_1 plot, the distance between the origin marked by "+" and "*" is $d(+, *) = a + b = 1$. Other examples include $L_{0.5}$: $d(+, *) = (a^{0.5} + b^{0.5})^{1/0.5} = 1$; L_2: $d(+, *) = \sqrt{a^2 + b^2} = 1$; and L_3: $d(+, *) = (a^3 + b^3)^{1/3} = 1$

The Minkowski distance is a metric for $p \geq 1$ since it satisfies all the four conditions listed above. However, the triangle inequality is not satisfied for $p < 1$. Hence, the Minkowski distance is not a metric for $p < 1$.

Figure 2.2 plots the Minkowski distance at several p-values. As illustrated in the plots of $L_{0.5}$, L_1, L_2, and L_3, the distances between the origin ('+') and a point on the unit contour ('*') are, respectively, given by $L_{0.5}$: $d(+, *) = (a^{0.5} + b^{0.5})^{1/0.5} = 1$, L_1: $d(+, *) = a + b = 1$, L_2: $d(+, *) = \sqrt{a^2 + b^2} = 1$ and L_3: $d(+, *) = (a^3 + b^3)^{1/3} = 1$. This illustration shows the warping of distance measures in these plots. Such warping produces different emphasis of dimension axes during parametric search.

2.3 Preprocessing Methods

The collected data can come in any of the above mentioned forms, namely, nominal, ordinal, interval, and ratio types. Moreover, multiple data sources can be collected in parallel and over a period of time. In other words, the data can come in large spatial dimensions and over time. In order to achieve consistency in data representation, the following techniques are often adopted for data preprocessing.

2.3.1 Cleaning or Cleansing

Data cleaning or cleansing is the process of detecting, correcting, and removal of corrupted, incomplete, erroneous, or inaccurate data samples from a collection. For text data, corrupted samples can be identified from the syntax error. However, for numerical data, only missing or out-of-range values can be readily identified. The following criteria could be useful for checking the data quality.

- Completeness: This is the degree to which all required measurements are available. For some applications, the experiments can be repeated to collect more measurements. For others, those data with incomplete measurement are frequently removed.
- Consistency: This is the degree to which the measurements are reproducible under similar conditions. For stationary data, the measurements should not drift. For non-stationary data, the drift can be modeled.
- Uniformity: This is the degree to which the measurements are made using the same unit of scale and representation. For example, in one system the weight measurement could be in kilogram but in another system the weight measurement could be in pounds. If these data are to be put together, they have to be represented under the same scale.
- Validity: This is the degree to which the measurements conforms to well-established constraints. For example, for data represented in percentage, the total sum of such measurements must be equal to 100 percent.

2.3.2 Alignment

In image processing, a frequent preprocessing step is to extract relevant portion for subsequent processing steps. For example, in face recognition, the face image excluding the hairs and wearing accessories is often extracted from the background image for comparison. Here, the alignment of face is often obtained based on the eyes, nose, and mouth locations where only regions around these regions are used for comparison.

2.3.3 Normalization

Min-max: When the bounds or range of each independent dimension of data are known, a common normalization technique is *min-max*. Alternatively, an estimation of the minimum and maximum values for a set of measurements can be made if the bounds are unknown. Given a set of raw measurements $\{x_i^{raw}\} \in [x^{min}, x^{max}]$, $i = 1, 2, \ldots, m$, the normalized measurements can be computed using

$$x_i = \frac{x_i^{raw} - x^{min}}{x^{max} - x^{min}}, \quad i = 1, 2, \ldots, m. \tag{2.8}$$

Standardization: When the population of measurements of each independent dimension of data is normally distributed where the parameters are known, the *standard score* or *z-score* is a popular choice. Mathematically, this normalization is obtained by subtracting its expected value $\mu = E[X]^1$ and then dividing the difference by its *standard deviation* $\sigma = \sqrt{\mathrm{Var}(X)}$:

$$x_i = \frac{x_i^{raw} - E[X]}{\sigma(X)}, \quad i = 1, 2, \ldots, m. \tag{2.9}$$

The exact values of $E[X]$ and $\sigma(X)$ may not be known in practice. Here, an estimation of the parameters can be made using a set of collected measurements:

$$\text{Sample mean}: \hat{\mu} = \frac{1}{m} \sum_{i=1}^{m} x_i, \tag{2.10}$$

$$\text{Sample variance}: \hat{\sigma} = \frac{1}{m} \sum_{i=1}^{m} (x_i - \hat{\mu})^2. \tag{2.11}$$

This normalization process is also known as *standardization* in statistics.

Median Absolute Deviation (MAD): Instead of using the mean value as the reference point in standard deviation, the *median value* is adopted as reference point in the *median absolute deviation*:

$$\mathrm{MAD} = \mathrm{median}(|x_i - \mathrm{median}(X)|), \tag{2.12}$$

where the normalization takes the form of

$$x_i = \frac{x^{raw} - \mathrm{median}(X)}{\mathrm{MAD}}. \tag{2.13}$$

---

[1] X is a generic notation for the random variable and x is a notation for its measured samples.

2.3.4 Other Transformations

Apart from the above normalization schemes, other types of data transformation may be adopted to translate or stretch the data. A simple transformation could be by means of translating the data toward a certain desired range of operation. Operations such as multiplying, adding, or subtracting a certain value uniformly to or from the measurement to get the data falling within a certain range provides a viable option to transform the data linearly.

For measurements which spread in exponential scale, taking the *logarithm* on such measurements (i.e., $\log(x_i)$) would be helpful to scale down the masking impact of large data values over small data values. Conversely, an exponential function (e.g., $\exp(x_i)$) can stretch the data when the measurements display a combination of flat and steep changing regions.

Some other nonlinear stretching include *sigmoid* and hyperbolic tanh functions:

$$x_i = \frac{1}{1 + e^{-h(x_i^{raw})}}, \tag{2.14}$$

$$x_i = \tanh(h(x_i^{raw})), \tag{2.15}$$

where the function $h(\cdot)$ can take any of the above normalization forms (such as (2.9)).

2.4 Bibliographic Notes

Several generalized forms of metric space are available by relaxing its axioms (Wikipedia 2024). A ***pseudometric*** on $\mathcal{X}$ is defined by excluding only the axiom of *identity of indiscernibles*, i.e., it is possible that $d(x, y) = 0$ for some distinct values $x \neq y$. A ***quasimetric*** is defined as a function that satisfies all axioms for a metric except the *symmetry* axiom. This metric is common in real-life applications when imbalance energy potential is considered. For example, the times between climbing up a hill and walking down the hill are different due to the difference in energy requirement to overcome the gravitational force. A ***semimetric***, on another hand, satisfies the first three axioms but not necessarily the triangle inequality. For ***metametrics***, all the axioms of metrics are satisfied except that the distance between identical points is not necessarily zero, i.e., $d(x, y) = 0$ implies $x = y$ but not vice versa. A classical book on set theory and metric spaces is Kaplansky (1972) and a classical book on related functional analysis is Kolmogorov & Fomin (1957), Kolmogorov and Fomin (1961). In Jain et al. (2005), several normalization techniques have been studied for multi-biometric scores fusion.

2.5 Exercises

2.1 There is no difference between information and data. True or False?

2.2 For categorical data such as the type of fruits, the one-hot encoding is preferred over an arbitrary number assignment such as one for orange, two for apple, three for mango, etc. True or False?

2.3 Which of the following represents the characteristics of ordinal data?

  (a) Ordinal data has a median
  (b) Value of interval is unknown
  (c) Measures non-numeric traits such as satisfaction, happiness, etc.
  (d) Establish a relative rank
  (e) None of these

2.4 Determine whether the following items belong to interval or ratio data?

  (a) The number of hours one spent studying each day
  (b) The score of an intelligence test
  (c) The temperatures in cities throughout the country that are listed in most newspapers
  (d) The birth weights of babies who were born at Samsung Hospital last week
  (e) Ages of students in a machine learning class
  (f) Number of pages in your statistic book
  (g) The annual salaries for all teachers in a town
  (h) The cost of a pair of shoes

2.5 Which of the following is incorrect regarding the properties of distance metrics?

  (a) The distance from $\mathbf{x}$ to $\mathbf{y}$ is the same as the distance from $\mathbf{y}$ to $\mathbf{x}$
  (b) The metric must satisfy the triangular inequality
  (c) The distance between two identical vectors, $\mathbf{x}$ and $\mathbf{y}$, is a non-zero value
  (d) The distance must never be negative
  (e) None of these

2.6 Which of the following distance measures calculates the distance between two binary vectors?

  (a) Euclidean distance
  (b) Manhattan distance
  (c) Minkowski distance
  (d) Hamming distance
  (e) None of these

2.7 What is the preferred way to handle missing or corrupted data in a small data set?

  (a) Drop the entire data sample that contains the corrupted features
  (b) Replace missing values with mean/median/mode
  (c) Replace missing features with zeros

(d) Do not change the corrupted features

(e) None of these

2.8 When the bounds of data are not known, the z-score standardization is preferred over min-max normalization. True or False?

2.9 You are given a set of data for supervised learning. A sample block of data looks like this:

1.0234,	0.3202,	25,	0.0081,	60033.81,	3
1.1356,	0.3308,	21,	0.0067,	69283.18,	-1
0.9988,	0.2526,	23,	0.0093,	66034.33,	2
1.1858,	0.3001,	22,	0.0077,	64037.35,	1
1.1533,	0.3853,	21,	0.0066,	62033.58,	-13
1.0755,	0.3102,	21,	0.0098,	60183.65,	1
1.0045,	0.2901,	22,	0.0065,	61093.98,	-2
1.1031,	0.3912,	21,	0.0088,	69033.23,	-1

Each row corresponds to a sample data measurement with 5 input features and 1 response. What kind of undesired effect can you anticipate if this set of raw data is used for learning?

(a) Those features with nominal values may create numerical instability

(b) Those features with very large values may overshadow those with very small values

(c) Those features with integer values may add bias to the system

(d) The sample with output response value -13 should be imputed

(e) None of these

2.10 Suppose you are given the following data regarding the monthly income of participants of a yoga learning group.

Participants	a	b	c	d	e	f	g	h	i
Income $	5000	3000	4000	4500	6000	3500	5000	800000	5500

Among mean, mode, and median, which is the most appropriate statistic to describe the data? Provide the value of this statistic based on the given table.

Chapter 3
Linear Parametric Models

Learning from data plays a key role in the fields of statistics, artificial intelligence, and many areas of science and technology. A typical task in such applications is to predict the future *outcome* based on a collection of historical *observations* or *measurements*. Under such a learning paradigm, the outcome is often called the *response* or the *output* of the system. The observations or measurements are also termed *features* or *inputs* to the system. The system is often modeled by a mathematical function which relates the features or inputs to the output. The task of learning is to find a mathematical model which predicts the responses of novel observations or measurements. In this chapter, we shall focus on the many ways of formulating such a mathematical model, particularly a linear one with respect to its parameters, assuming that it exists and describes the system relatively well. It is noted here that the linearity is only with respect to the parameters where nonlinearity can still live between the inputs and the outputs.

3.1 Systems of Linear Equations

3.1.1 Notations

Throughout the book, a scalar variable uses a non-boldface case letter such as x or X, and a vectored variable uses a boldface lowercase letter such as $\boldsymbol{x}$ or $\mathbf{x}$. For more specificity, $\mathbb{R}^d$ indicates that there are d number of real elements in the vector. Unless otherwise specified, a vector is taken to be a column vector by default. For a matrix, an uppercase letter in boldface such as $\mathbf{X}$ is denoted. The dimension of a matrix is indicated by specifying the row size followed by the column size. For example, $\mathbf{X} \in \mathbb{R}^{m \times d}$ indicates a matrix $\mathbf{X}$ of real numbers with m rows and d columns.

Consider a linear equation with d variables given by

© The Author(s), under exclusive license to Springer Nature Singapore Pte Ltd. 2025   25
K.-A. Toh et al., *Analytic Learning Methods for Pattern Recognition*,
https://doi.org/10.1007/978-981-96-2151-4_3

$$y = x_1 w_1 + x_2 w_2 + \cdots x_d w_d, \tag{3.1}$$

where $x_1, \ldots x_d$ are the measured variables, $w_1, \ldots w_d$ are fixed but unknown weighting parameters to be determined, and y is the measured system response. When the values of $x_1, \ldots x_d$ are varied, the system responses with corresponding y values. Suppose we have a collection of m variations of these $x_1, \ldots x_d$ values with corresponding y values, then we can write them as

$$
\begin{aligned}
x_{1,1} w_1 + x_{1,2} w_2 + \cdots + x_{1,d} w_d &= y_1 \\
x_{2,1} w_1 + x_{2,2} w_2 + \cdots + x_{2,d} w_d &= y_2 \\
&\;\;\vdots \\
x_{m,1} w_1 + x_{m,2} w_2 + \cdots + x_{m,d} w_d &= y_m.
\end{aligned}
\tag{3.2}
$$

Here we note that the first subscript of x indicates the sample size and the second subscript indicates the dimension index. This system corresponds to m simultaneous equations with d unknowns (i.e., the fixed but unknown weighting constants $w_1, \ldots w_d$) to be solved. Let

$$
\mathbf{X} = \begin{bmatrix}
x_{1,1} & x_{1,2} & \cdots & x_{1,d} \\
x_{2,1} & x_{2,2} & \cdots & x_{2,d} \\
\vdots & \vdots & \ddots & \vdots \\
x_{m,1} & x_{m,2} & \cdots & x_{m,d}
\end{bmatrix}, \quad
\mathbf{w} = \begin{bmatrix}
w_1 \\ w_2 \\ \vdots \\ w_d
\end{bmatrix}, \quad
\mathbf{y} = \begin{bmatrix}
y_1 \\ y_2 \\ \vdots \\ y_m
\end{bmatrix},
\tag{3.3}
$$

then the set of equations in (3.2) can be written in matrix-vector form:

$$\mathbf{X}\mathbf{w} = \mathbf{y}. \tag{3.4}$$

Depending on the sizes of m and d, different types of solutions are available for this system. We shall consider them separately in the following subsections.

3.1.2 Square or Even-Determined System

When the sample size is equal to the variable dimension, i.e., $m = d$, we have a square system (also known as even-determined system) with equal number of equations and unknowns. If $\mathbf{X}$ is invertible so that $\mathbf{X}^{-1}\mathbf{X} = \mathbf{I}$ with $\mathbf{I}$ being an identity matrix, then by pre-multiplying both sides of (3.4) by $\mathbf{X}^{-1}$, we have a unique solution given by

$$
\begin{aligned}
\mathbf{X}^{-1}\mathbf{X}\mathbf{w} &= \mathbf{X}^{-1}\mathbf{y} \\
\Rightarrow \quad \hat{\mathbf{w}} &= \mathbf{X}^{-1}\mathbf{y}.
\end{aligned}
\tag{3.5}
$$

Here, a hat has been placed on top of w to indicate that the solution $\hat{w}$ is a specific point in the space of w.

Example 3.1 Consider the following system of two equations with two unknowns:

$$w_1 + 2w_2 = 5 \tag{3.6}$$

$$3w_1 - 4w_2 = 6, \tag{3.7}$$

which can be packed into matrix-vector form $\mathbf{X}w = \mathbf{y}$ where

$$\mathbf{X} = \begin{bmatrix} 1 & 2 \\ 3 & -4 \end{bmatrix}, \quad w = \begin{bmatrix} w_1 \\ w_2 \end{bmatrix}, \quad \mathbf{y} = \begin{bmatrix} 5 \\ 6 \end{bmatrix}.$$

According to (3.5), and since $\mathbf{X}$ is invertible, we have

$$\hat{w} = \mathbf{X}^{-1}\mathbf{y} = \begin{bmatrix} 1 & 2 \\ 3 & -4 \end{bmatrix}^{-1} \begin{bmatrix} 5 \\ 6 \end{bmatrix} = \begin{bmatrix} 3.2 \\ 0.9 \end{bmatrix}. \tag{3.8}$$

The geometry of this system with its intersection solution point is shown in Fig. 3.1. $\qquad\square$

3.1.3 Over-Determined System

When the sample size is larger than the variable dimension, i.e., $m > d$, we have an over-determined system. This means that we have more equations than unknowns. In this case, there is no exact solution for this set of equations. However, an approximated

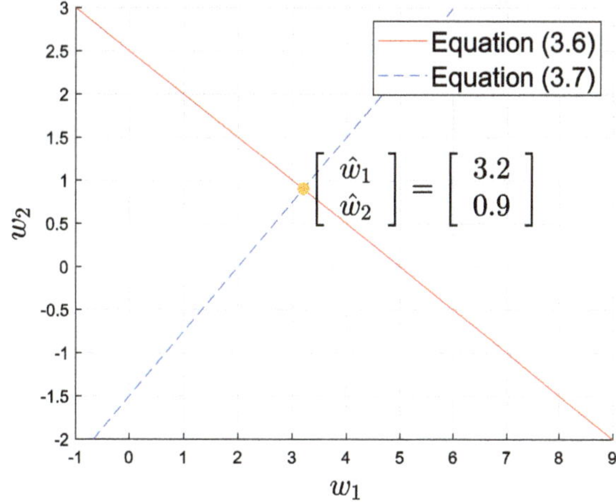

Fig. 3.1 An even-determined system with a unique solution

solution is yet available. If there exists a left-inverse of $\mathbf{X}$ such that $\mathbf{X}^{\dagger}\mathbf{X} = \mathbf{I}$, then by pre-multiplying both sides of (3.4) by the left-inverse $\mathbf{X}^{\dagger}$, we have, in general, an approximated solution given by

$$\mathbf{X}^{\dagger}\mathbf{X}\boldsymbol{w} = \mathbf{X}^{\dagger}\mathbf{y}$$
$$\Rightarrow \quad \hat{\boldsymbol{w}} = \mathbf{X}^{\dagger}\mathbf{y}, \tag{3.9}$$

where the left-inverse of $\mathbf{X}$ can be computed as $\mathbf{X}^{\dagger} = \left(\mathbf{X}^{T}\mathbf{X}\right)^{-1}\mathbf{X}^{T}$ if $\mathbf{X}^{T}\mathbf{X}$ is invertible. An exception is that when $\text{rank}(\mathbf{X}) = \text{rank}([\mathbf{X}, \mathbf{y}])$, the solution is exact.

Example 3.2 Consider the following system of three equations with two unknowns:

$$w_1 + 2w_2 = 1 \tag{3.10}$$
$$w_1 - 2w_2 = 2 \tag{3.11}$$
$$2w_1 = 1, \tag{3.12}$$

which can be packed into matrix-vector form $\mathbf{X}\boldsymbol{w} = \mathbf{y}$ where

$$\mathbf{X} = \begin{bmatrix} 1 & 2 \\ 1 & -2 \\ 2 & 0 \end{bmatrix}, \quad \boldsymbol{w} = \begin{bmatrix} w_1 \\ w_2 \end{bmatrix}, \quad \mathbf{y} = \begin{bmatrix} 1 \\ 2 \\ 1 \end{bmatrix}.$$

According to (3.9), and since $\mathbf{X}^{T}\mathbf{X}$ is invertible, we have

$$\hat{\boldsymbol{w}} = \left(\mathbf{X}^{T}\mathbf{X}\right)^{-1}\mathbf{X}^{T}\mathbf{y}$$
$$= \begin{bmatrix} 6 & 0 \\ 0 & 8 \end{bmatrix}^{-1} \begin{bmatrix} 1 & 1 & 2 \\ 2 & -2 & 0 \end{bmatrix} \begin{bmatrix} 1 \\ 2 \\ 1 \end{bmatrix} = \begin{bmatrix} 0.8333 \\ -0.2500 \end{bmatrix}, \tag{3.13}$$

which is an approximated solution. The geometry of this system with its approximated solution point is shown in Fig. 3.2. $\qquad\square$

Example 3.3 Consider the following system of three equations with two unknowns:

$$w_1 + 2w_2 = 1 \tag{3.14}$$
$$w_1 - 2w_2 = 2 \tag{3.15}$$
$$2w_1 = 3, \tag{3.16}$$

which can be packed into matrix-vector form $\mathbf{X}\boldsymbol{w} = \mathbf{y}$ where

$$\mathbf{X} = \begin{bmatrix} 1 & 2 \\ 1 & -2 \\ 2 & 0 \end{bmatrix}, \quad \boldsymbol{w} = \begin{bmatrix} w_1 \\ w_2 \end{bmatrix}, \quad \mathbf{y} = \begin{bmatrix} 1 \\ 2 \\ 3 \end{bmatrix}.$$

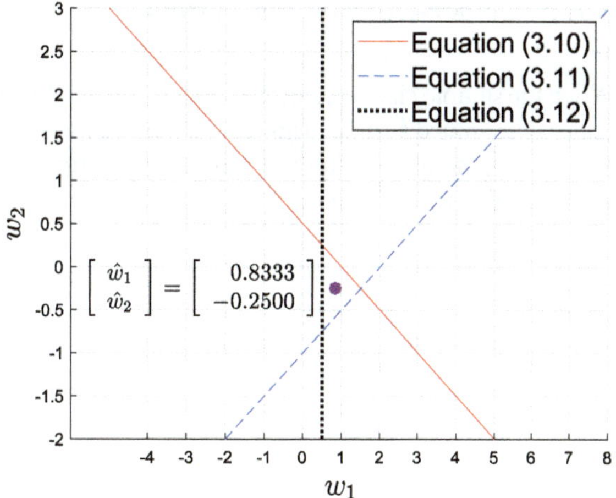

Fig. 3.2 An over-determined system with an approximated solution

Since $\mathbf{X}^T\mathbf{X}$ is invertible and $\text{rank}(\mathbf{X}) = \text{rank}([\mathbf{X}, \mathbf{y}])$, we have an exact solution as follows:

$$\hat{\boldsymbol{w}} = \left(\mathbf{X}^T\mathbf{X}\right)^{-1}\mathbf{X}^T\mathbf{y}$$

$$= \begin{bmatrix} 6 & 0 \\ 0 & 8 \end{bmatrix}^{-1} \begin{bmatrix} 1 & 1 & 2 \\ 2 & -2 & 0 \end{bmatrix} \begin{bmatrix} 1 \\ 2 \\ 3 \end{bmatrix} = \begin{bmatrix} 1.50 \\ -0.25 \end{bmatrix}. \tag{3.17}$$

The geometry of this system with an exact solution point is shown in Fig. 3.3. $\square$

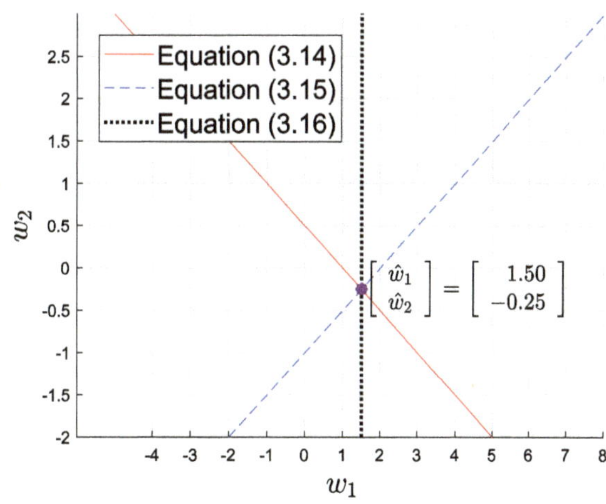

Fig. 3.3 An over-determined system with an exact solution

3.1.4 Under-Determined System

When the sample size is smaller than the variable dimension, i.e., $m < d$, we have an under-determined system. This means that we have fewer equations than unknowns. In this case, there are an infinite number of solutions for this set of equations. If there exists a right-inverse of $\mathbf{X}$ such that $\mathbf{X}\mathbf{X}^{\dagger} = \mathbf{I}$, then the d-vector

$$\hat{\boldsymbol{w}} = \mathbf{X}^{\dagger}\mathbf{y}, \tag{3.18}$$

which is one of the infinite solutions, satisfies Eq. (3.4), i.e.,

$$\begin{aligned}
\mathbf{X}\hat{\boldsymbol{w}} &= \mathbf{y} \\
\Rightarrow \quad \mathbf{X}\mathbf{X}^{\dagger}\mathbf{y} &= \mathbf{y} \\
\Rightarrow \quad \mathbf{y} &= \mathbf{y}.
\end{aligned} \tag{3.19}$$

The right-inverse of $\mathbf{X}$ can be computed as $\mathbf{X}^{\dagger} = \mathbf{X}^{T}\left(\mathbf{X}\mathbf{X}^{T}\right)^{-1}$ when $\mathbf{X}\mathbf{X}^{T}$ is invertible. The solution given by (3.18) which utilizes this right-inverse has a minimum norm among the norms of other solutions (see Sect. 6.1.2). An exception for this under-determined case is that the system has no solution if it is inconsistent or $\mathbf{X}\mathbf{X}^{T}$ is not invertible (such as having all parallel planes in the system of equations).

Example 3.4 Consider the following system of two equations with three unknowns:

$$w_1 + 2w_2 + 3w_3 = 1 \tag{3.20}$$
$$- w_1 - 2w_2 + 3w_3 = 2, \tag{3.21}$$

which can be packed into matrix-vector form $\mathbf{X}\boldsymbol{w} = \mathbf{y}$ where

$$\mathbf{X} = \begin{bmatrix} 1 & 2 & 3 \\ -1 & -2 & 3 \end{bmatrix}, \quad \boldsymbol{w} = \begin{bmatrix} w_1 \\ w_2 \\ w_3 \end{bmatrix}, \quad \mathbf{y} = \begin{bmatrix} 1 \\ 2 \end{bmatrix}.$$

According to (3.18), and since $\mathbf{X}\mathbf{X}^{T}$ is invertible, we have

$$\begin{aligned}
\hat{\boldsymbol{w}} &= \mathbf{X}^{T}\left(\mathbf{X}\mathbf{X}^{T}\right)^{-1}\mathbf{y} \\
&= \begin{bmatrix} 1 & -1 \\ 2 & -2 \\ 3 & 3 \end{bmatrix} \begin{bmatrix} 14 & 4 \\ 4 & 14 \end{bmatrix}^{-1} \begin{bmatrix} 1 \\ 2 \end{bmatrix} = \begin{bmatrix} -0.1 \\ -0.2 \\ 0.5 \end{bmatrix}.
\end{aligned} \tag{3.22}$$

The geometry of this under-determined system with the solution of minimum norm is shown in Fig. 3.4. In this plot, apart from the planes of Equations (3.20) and (3.21), the plane spanned by the row vectors of $\mathbf{X}$, namely, $[1, 2, 3]$ (marked as red stick) and $[-1, -2, 3]$ (marked as blue stick) is also shown. The infinite solutions of this

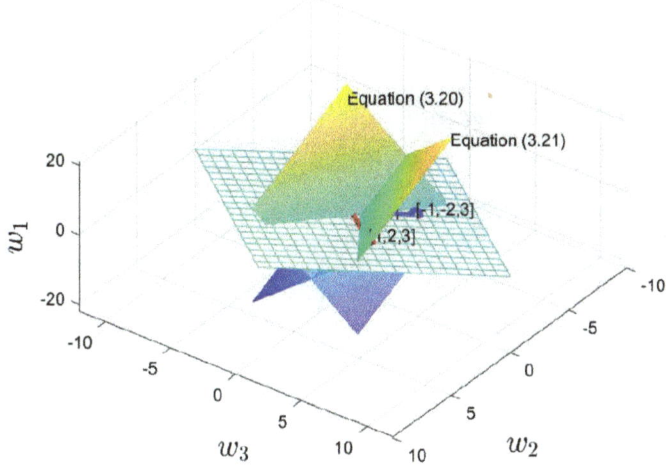

Fig. 3.4 An under-determined system with a unique minimum norm solution

under-determined system lie along the line of intersection between the two system equation planes. The solution of minimum norm lies on the intersection point of the spanned plane of the two row vectors with the infinite solution line. □

Example 3.5 Consider the following system of two equations with three unknowns:

$$w_1 + 2w_2 + 3w_3 = 1 \tag{3.23}$$
$$-w_1 - 2w_2 - 3w_3 = 8, \tag{3.24}$$

which can be packed into matrix-vector form $\mathbf{X}\boldsymbol{w} = \mathbf{y}$ where

$$\mathbf{X} = \begin{bmatrix} 1 & 2 & 3 \\ -1 & -2 & -3 \end{bmatrix}, \quad \boldsymbol{w} = \begin{bmatrix} w_1 \\ w_2 \\ w_3 \end{bmatrix}, \quad \mathbf{y} = \begin{bmatrix} 1 \\ 8 \end{bmatrix}.$$

This system is inconsistent because neither $\mathbf{X}^T\mathbf{X}$ nor $\mathbf{X}\mathbf{X}^T$ is invertible. The geometry of this system is shown in Fig. 3.5. □

3.2 Models with Linear Parameters

3.2.1 Linear Model

Next, we formulate the system of linear equations from a modeling perspective, allowing it to generalize across a broader range of applications. Suppose there

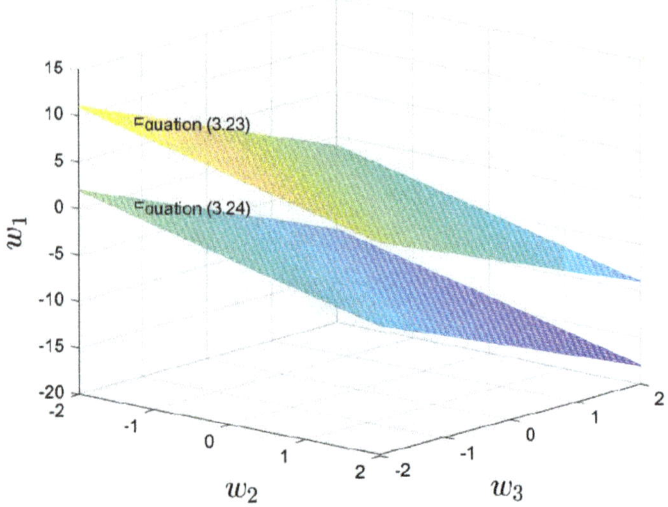

Fig. 3.5 An inconsistent under-determined system

are d number of input features where they are packed as a vector given by $x = [x_1, \ldots, x_d]^T$. Each feature vector can be indexed by a sample number $i \in \{1, \ldots, m\}$ for the pool of data, giving $x_i = [x_{i,1}, \ldots, x_{i,d}]^T$. These feature samples can be stacked in matrix form as

$$\mathbf{X} = \begin{bmatrix} x_1^T \\ \vdots \\ x_m^T \end{bmatrix} = [\mathbf{x}_1, \ldots, \mathbf{x}_d] = \overbrace{\begin{bmatrix} x_{1,1} & \cdots & x_{1,d} \\ \vdots & \ddots & \vdots \\ x_{m,1} & \cdots & x_{m,d} \end{bmatrix}}^{\begin{bmatrix} \mathbf{x}_1 & \cdots & \mathbf{x}_d \end{bmatrix}} \left. \right\} \begin{bmatrix} x_1^T \\ \vdots \\ x_m^T \end{bmatrix}. \tag{3.25}$$

Correspondingly, the system target response y can be indexed by the sample number giving $y_i, i \in \{1, \ldots, m\}$. Based on these notations, a linear parametric model

$$g(\boldsymbol{w}, \boldsymbol{x}_i) = \sum_{j=1}^{d} w_j x_{i,j} = \boldsymbol{w}^T \boldsymbol{x}_i, \quad i = 1, \ldots, m, \tag{3.26}$$

can be used to model the system target response or output as

$$y_i = g(\boldsymbol{w}, \boldsymbol{x}_i) + \epsilon_i = \sum_{j=1}^{d} w_j x_{i,j} + \epsilon_i, \quad i = 1, \ldots, m, \tag{3.27}$$

where ϵ_i is the modeling error. In a more compact manner, (3.27) can be rewritten as

$$\mathbf{y} = \mathbf{X}\mathbf{w} + \boldsymbol{\epsilon}, \tag{3.28}$$

where $\mathbf{y} = [y_1, \ldots, y_m]^T$, $\boldsymbol{\epsilon} = [\epsilon_1, \ldots, \epsilon_m]^T$, $\mathbf{X}$ is defined in (3.25) and $\mathbf{w} = [w_1, \ldots, w_d]^T$.

In practice, an augmented model is preferred since an additional degree of freedom is available to deal specifically with the overall data offset. If not otherwise mentioned, the notation $\mathbf{X}$ shall represent the regressor of the augmented form as follows:

$$\mathbf{X} = \begin{bmatrix} 1 & \boldsymbol{x}_1^T \\ \vdots & \vdots \\ 1 & \boldsymbol{x}_m^T \end{bmatrix} = [\mathbf{1}, \mathbf{x}_1, \ldots, \mathbf{x}_d] = \overbrace{\begin{bmatrix} 1 & x_{1,1} & \cdots & x_{1,d} \\ \vdots & \vdots & \ddots & \vdots \\ 1 & x_{m,1} & \cdots & x_{m,d} \end{bmatrix}}^{\left[\, \mathbf{x}_0\ \mathbf{x}_1\ \cdots\ \mathbf{x}_d \,\right]}. \tag{3.29}$$

For such an augmented model, the corresponding *model coefficients* or *weight parameters* are denoted as $\boldsymbol{w} = [w_0, w_1, \ldots, w_d]^T \in \mathbb{R}^{d+1}$.

3.2.2 Generalized Linear Model

(i) Transforming the input
The above linear model can be generalized to include transformed inputs. In other words, the inputs can be embedded under a nonlinear mapping with yet a linear relationship between the parameters and the embedment. Equation (3.26) can thus be generalized as

$$g(\boldsymbol{w}, \boldsymbol{x}_i) = \sum_{j=0}^{D} w_j p_j(\boldsymbol{x}_i) = \boldsymbol{w}^T \boldsymbol{p}(\boldsymbol{x}_i), \quad i = 1, \ldots, m, \tag{3.30}$$

where $p_j(\boldsymbol{x}_i)$, $j = 0, 1, \ldots, D$ are the transformed terms which can be linear or nonlinear with respect to $\boldsymbol{x}_i$. In general, the number of transformed terms D is different from the original feature size d. The vector $\boldsymbol{p}^T(\boldsymbol{x}_i) = [p_0(\boldsymbol{x}_i), p_1(\boldsymbol{x}_i), \ldots, p_D(\boldsymbol{x}_i)]$ is a packing of these transformed terms. By stacking multiple samples of an augmented model in matrix form given by

$$\mathbf{P} = \overbrace{\begin{bmatrix} 1 & p_1(\boldsymbol{x}_1) & \cdots & p_D(\boldsymbol{x}_1) \\ \vdots & \vdots & \ddots & \vdots \\ 1 & p_1(\boldsymbol{x}_m) & \cdots & p_D(\boldsymbol{x}_m) \end{bmatrix}}^{\left[\, \mathbf{p}_0(\boldsymbol{x})\ \mathbf{p}_1(\boldsymbol{x})\ \cdots\ \mathbf{p}_D(\boldsymbol{x}) \,\right]} \Bigg\} \begin{bmatrix} \boldsymbol{p}^T(\boldsymbol{x}_1) \\ \vdots \\ \boldsymbol{p}^T(\boldsymbol{x}_m) \end{bmatrix}, \tag{3.31}$$

a generalized linear system (with respect to (3.28)) can be written as

$$\mathbf{y} = \mathbf{P}\mathbf{w} + \boldsymbol{\epsilon}. \tag{3.32}$$

Here, we note that $\mathbf{P} \in \mathbb{R}^{m \times (D+1)}$ can be any arbitrary transformation matrix which maps the input feature space of $\mathbb{R}^d$ dimension to transformed feature space of $\mathbb{R}^{D+1}$ dimension. Popular choices for such transformation include polynomials, sigmoid function, Gaussian function, and random projection. In the following sections, the various transformations are grouped into representative categories and discussed in greater detail.

Example 3.6 Consider the following samples of two-dimensional input features (x_1, x_2):

Features	Sample 1	Sample 2	Sample 3	Sample 4	Sample 5	Sample 6
x_1:	−1	−3	1	2	5	8
x_2:	1	−1	6	8	2	3

which can be packed in matrix form with inclusion of an intercept/bias term as

$$\mathbf{X} = \begin{bmatrix} 1 & -1 & 1 \\ 1 & -3 & -1 \\ 1 & 1 & 6 \\ 1 & 2 & 8 \\ 1 & 5 & 2 \\ 1 & 8 & 3 \end{bmatrix}. \tag{3.33}$$

Suppose we have a transformation $p_1(\cdot) \triangleq e^{(\cdot)}$. Then the transformed features can be written in matrix form as

$$\mathbf{P} = \begin{bmatrix} 1 & e^{(-1)} & e^{(1)} \\ 1 & e^{(-3)} & e^{(-1)} \\ 1 & e^{(1)} & e^{(6)} \\ 1 & e^{(2)} & e^{(8)} \\ 1 & e^{(5)} & e^{(2)} \\ 1 & e^{(8)} & e^{(3)} \end{bmatrix}. \tag{3.34}$$

As another example, suppose we use a transformation given by $p_1(\cdot) \triangleq (\cdot)^2$ and $p_2(\cdot) \triangleq \sin(\cdot)$. Then the transformed features can be written in matrix form as

$$\mathbf{P} = \begin{bmatrix} 1 & (-1)^2 & (1)^2 & \sin(-1) & \sin(1) \\ 1 & (-3)^2 & (-1)^2 & \sin(-3) & \sin(-1) \\ 1 & (1)^2 & (6)^2 & \sin(1) & \sin(6) \\ 1 & (2)^2 & (8)^2 & \sin(2) & \sin(8) \\ 1 & (5)^2 & (2)^2 & \sin(5) & \sin(2) \\ 1 & (8)^2 & (3)^2 & \sin(8) & \sin(3) \end{bmatrix}. \tag{3.35}$$

$\square$

(ii) Transforming the output

For problems with binary outputs $y \in \{0, 1\}$, logistic regression models a transformed ratio of output probabilities $\Pr(Y = 1|X = x) = p$ and $\Pr(Y = 0|X = x) = 1 - p$ (X and Y are random variables here) as a linear function of input $x \in \mathbb{R}^d$. This transformation is the *logistic* (or *logit*) function, defined as $\log(p/(1 - p))$, which maps the probability ratio to the real line:

$$\log \frac{p}{1 - p} = w_0 + w^T x. \tag{3.36}$$

Based on this output transformation, the probability of the class-1 output ($\Pr(Y = 1|X = x) = p$) and the class-0 output ($\Pr(Y = 0|X = x) = 1 - p$) can be, respectively, expressed in terms of the input feature vector x as

$$p = \frac{1}{1 + e^{-(w_0 + w^T x)}}, \tag{3.37}$$

and

$$1 - p = \frac{e^{-(w_0 + w^T x)}}{1 + e^{-(w_0 + w^T x)}}. \tag{3.38}$$

For problems with multiple (N_c) number of categories, the logit model has the form

$$\log \frac{\Pr(Y = 1|X = x)}{\Pr(Y = N_c|X = x)} = w_1^0 + w_1^T x,$$
$$\log \frac{\Pr(Y = 2|X = x)}{\Pr(Y = N_c|X = x)} = w_2^0 + w_2^T x,$$
$$\vdots$$
$$\log \frac{\Pr(Y = N_c - 1|X = x)}{\Pr(Y = N_c|X = x)} = w_{(N_c-1)}^0 + w_{(N_c-1)}^T x, \tag{3.39}$$

where the probability for each category is

$$\Pr(Y = k|X = x) = \frac{e^{w_k^0 + w_k^T x}}{1 + \sum_{j=1}^{N_c-1} e^{w_j^0 + w_j^T x}}, \quad k = 1, \ldots, N_c - 1,$$
$$\Pr(Y = N_c|X = x) = \frac{1}{1 + \sum_{j=1}^{N_c-1} e^{w_j^0 + w_j^T x}}. \tag{3.40}$$

In the above notations, w_j^0 denotes the weight parameter of the bias term.

Since the transformation at the output space results in solving a nonlinear function with respect to the parameters of interest (parameter vector $\boldsymbol{w}$ and the bias term weight parameter w_0), we shall focus on transformation at the input space in the following sections.

3.3 Polynomials

3.3.1 Full Multivariate Polynomials

According to the Weierstrass approximation theory, the polynomials can approximate any continuous function on a closed and bounded interval to any degree of accuracy given sufficient polynomial terms. This implies that polynomial regression provides an effective way to describe complex nonlinear input-output relationships.

A two-dimensional polynomial model (with x_1 and x_2 as the input feature dimensions) of second order (the highest power of x_1 and x_2 terms is 2) without any bias term can be written as

$$g(x_1, x_2) = x_1 + x_2 + x_1 x_2 + x_1^2 + x_2^2. \tag{3.41}$$

Figure 3.6 shows a mesh plot of this polynomial (3.41) which portrays a typical quadratic curvature with positive definite second derivative. Figure 3.7 shows two plots of a third-order polynomial model with different coefficient signs. These figures demonstrate a higher mapping complexity of the third-order polynomials than that of the second-order polynomials.

The general multivariate polynomial model can be expressed as

$$g(\boldsymbol{w}, \boldsymbol{x}) = \sum_{j=0}^{D} w_j x_1^{r_1} x_2^{r_2} \cdots x_d^{r_d} = \sum_{j=0}^{D} w_j p_j(\boldsymbol{x}), \tag{3.42}$$

where each of the summation terms is call a *monomial* and the summation is taken over all nonnegative integers $r_1, \ldots, r_d$ for which $\sum_{i=1}^{d} r_i \leqslant r$ with r being the order of the model. $\boldsymbol{w} = [w_0, w_1, \ldots, w_D]^T$ is the parameter vector to be estimated and each $p_j(\cdot)$ represents a monomial term which is a product among the elements of the input vector $\boldsymbol{x} = [1, x_1, \ldots, x_d]^T$.

Example 3.7 Consider those samples of two-dimensional input features (x_1, x_2) from the previous example:

Features	Sample 1	Sample 2	Sample 3	Sample 4	Sample 5	Sample 6
x_1:	−1	−3	1	2	5	8
x_2:	1	−1	6	8	2	3

These data features can be expanded by a second-order polynomial model $\boldsymbol{p} = [1, x_1, x_2, x_1^2, x_1 x_2, x_2^2]$ and packed in matrix form as

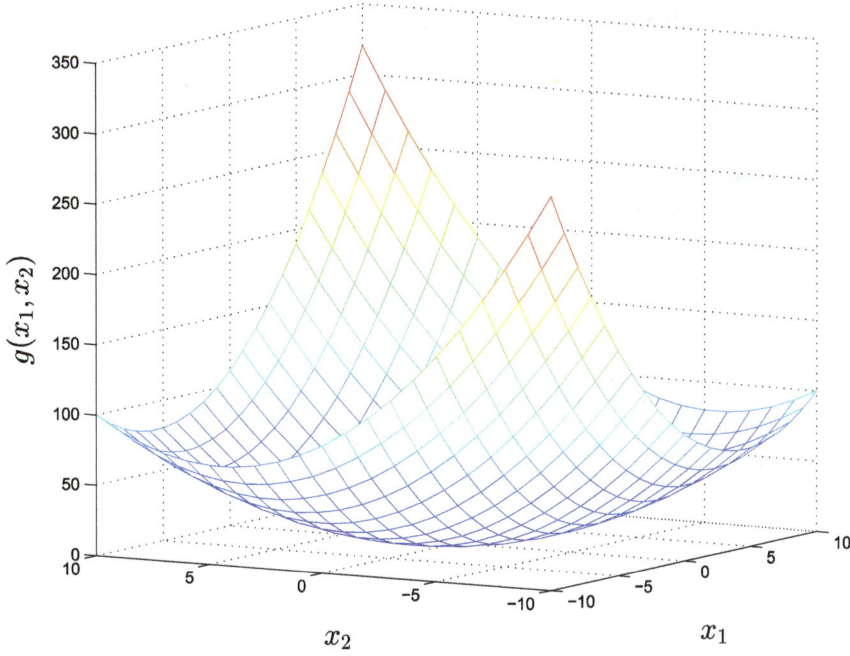

Fig. 3.6 A second-order polynomial model given by $g(x_1, x_2) = x_1 + x_2 + x_1 x_2 + x_1^2 + x_2^2$

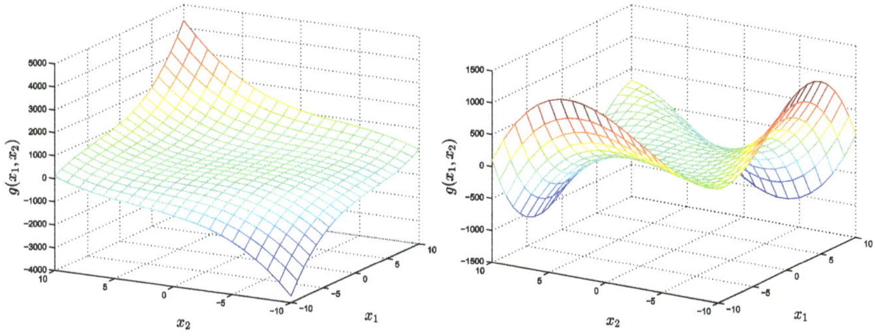

Fig. 3.7 Left panel: A third-order polynomial model given by $g(x_1, x_2) = x_1 + x_2 + x_1 x_2 + x_1^2 + x_2^2 + x_1 x_2^2 + x_2 x_1^2 + x_1^3 + x_2^3$; Right panel: a third-order polynomial model given by $g(x_1, x_2) = x_1 + x_2 + x_1 x_2 + x_1^2 + x_2^2 + x_1 x_2^2 + x_2 x_1^2 - x_1^3 - x_2^3$

$$\mathbf{P} = \begin{bmatrix} 1 & -1 & 1 & (-1)^2 & (-1) \times (1) & (1)^2 \\ 1 & -3 & -1 & (-3)^2 & (-3) \times (-1) & (-1)^2 \\ 1 & 1 & 6 & (1)^2 & (1) \times (6) & (6)^2 \\ 1 & 2 & 8 & (2)^2 & (2) \times (8) & (8)^2 \\ 1 & 5 & 2 & (5)^2 & (5) \times (2) & (2)^2 \\ 1 & 8 & 3 & (8)^2 & (8) \times (3) & (3)^2 \end{bmatrix}. \tag{3.43}$$

In Python, the full polynomials can be readily generated from the sklearn library function:

```python
import numpy as np
X = np.array([[-1, 1], [-3, -1], [1, 6], [2, 8], [5, 2], [8, 3]])
from sklearn.preprocessing import PolynomialFeatures
poly = PolynomialFeatures(2)
P = poly.fit_transform(X)
print(P)
[[ 1.  -1.   1.   1.  -1.   1.]
 [ 1.  -3.  -1.   9.   3.   1.]
 [ 1.   1.   6.   1.   6.  36.]
 [ 1.   2.   8.   4.  16.  64.]
 [ 1.   5.   2.  25.  10.   4.]
 [ 1.   8.   3.  64.  24.   9.]]
```

Each polynomial expansion term can be viewed as a new *axis*, *basis*, or *representation* of the data points. The left panel of Fig. 3.8 shows four data points of two colors (red and blue). When these data are viewed in their original axes (x and y), the two red dots are not linearly separable from the two blue dots (i.e., unable to separate the data using a straight line or a plane). This problem is known as the Exclusive-OR (XOR) problem where the response (output) is "one" (this response has been indicated by red coloring of the dots in the left panel of Fig. 3.8) when the two binary inputs are different (i.e., $x \neq y$), and "zero" (this response has been indicated by blue coloring of dots in the left panel of Fig. 3.8) when the two binary inputs are the same (i.e., $x = y$). However, when these data are projected onto the polynomial axis $z = xy$, we can see from the right panel of Fig. 3.8 that the projected blue dots (small blue dots) become linearly separable from the projected red dots (small red dots). As shown in Fig. 3.9, such a linearly separable property is also observed for translated

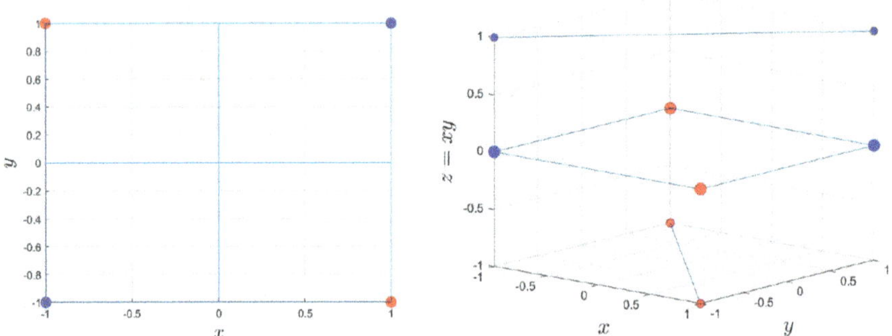

Fig. 3.8 Left panel: the Exclusive-OR (XOR) problem; Right panel: projected Exclusive-OR (XOR) points

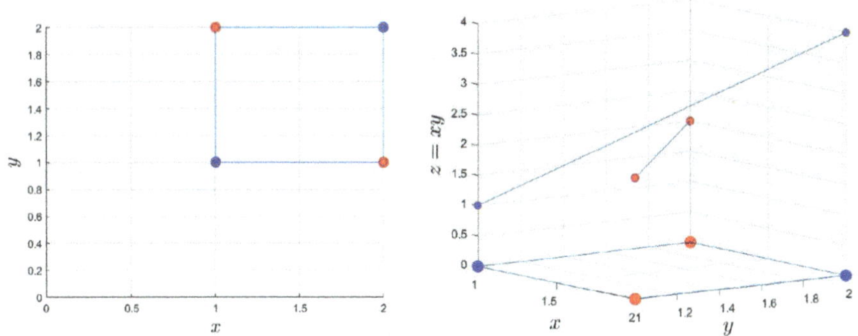

Fig. 3.9 Left panel: translated Exclusive-OR (XOR) problem; Right panel: projected translated Exclusive-OR (XOR) points

XOR points. These examples show that a projection onto the polynomial axes (such as xy, x^2, y^2, $\cdots$) may transform a set of data which is linearly non-separable into a linearly separable one.

However, according to the chosen order r, the total number of monomial terms in $g(\boldsymbol{w}, \boldsymbol{x})$ is given by $D + 1$ where

$$
D = \sum_{k=0}^{r} \frac{(k + d - 1)!}{k!(d - 1)!},
\tag{3.44}
$$

with $0! = 1$. This size of D can be approximately taken as an exponential growth of d^r with respect to the input dimension d and the model order r. For instance, a tenth-order model with 8 input features gives rise to a summation of 43758 monomial terms including the intercept. In view of this exponential growth of the number of parameters as well as the current limitation of memory and computing facility, the full multivariate polynomial model has yet to receive significant attention in applications.

3.3.2 Reduced Multivariate Polynomials

For regression applications, the above full multivariate polynomial model provides a versatile means for functional approximation. However, for classification applications, a model with full mapping capability may not be always necessary. This is because the main goal of a classifier is to find relevant decision boundary to segregate among data with different labels. And often enough, the complexity of such decision boundary cannot be too high where the generalization property may be lost. In other words, an overly complex model may result in overfitting of data and gives poor prediction for unseen data.

To significantly reduce the huge number of terms in the multivariate polynomials, a Reduced multivariate polynomial Model (RM) has been proposed:

$$g_{RM}(\boldsymbol{w}, \boldsymbol{x}) = \boldsymbol{w}^T p_{RM}(\boldsymbol{x})$$

$$= w_0 + \sum_{k=1}^{r} w_k \left(\sum_{j=1}^{l} x_j\right)^k + \sum_{k=1}^{r}\sum_{j=1}^{l} \beta_{k,j} x_j^k + \sum_{k=2}^{r} \left(\sum_{i=1}^{l} \gamma_{k,i} x_i\right)\left(\sum_{j=1}^{l} x_j\right)^{k-1} \tag{3.45}$$

$$l, r \geqslant 2,$$

where x_j, $j = 1, \ldots, l$ are the polynomial inputs, $\boldsymbol{w} = [w_0, w_k, \beta_{k,j}, \gamma_{k,i}]^T$, $j = 1, \ldots, l, k = 1, \ldots, r$ are the weighting coefficients to be estimated, and l, r correspond to input dimension, order of system, respectively. The number of terms in this model can be expressed as $K = 1 + r + l(2r - 1)$.

In view of some redundant terms in RM, a refined model is constructed as

$$g_{RM2}(\boldsymbol{w}, \boldsymbol{x}) = w_0 + \sum_{k=1}^{r}\sum_{j=1}^{l} \beta_{k,j} x_j^k + \sum_{k=2}^{r}\left(\sum_{i=1}^{l} \gamma_{k,i} x_i\right)\left(\sum_{j=1}^{l} x_j\right)^{k-1}, \tag{3.46}$$

where $l, r \geqslant 2$ and the total number of expansion terms is given by $K = 1 + l(2r - 1)$, which has r terms less than that in the original RM.

When there is no interaction among the input features, a power series model can be adopted as follows:

$$g_{PS}(\boldsymbol{w}, \boldsymbol{x}) = w_0 + \sum_{k=1}^{r}\sum_{j=1}^{d} \beta_{k,j} x_j^k. \tag{3.47}$$

Figure 3.10 shows the total number of monomial terms plotted over system order and input dimension for each of the full multivariate polynomial model (FMP), the reduced multivariate polynomial model (RM), and the power series model (PS). The FMP shows an exponential increase in its number of monomial terms whereas RM and PS models have linearly increasing number of terms.

Example 3.8 Consider those samples of two-dimensional input features (x_1, x_2) from the previous example:

Features	Sample 1	Sample 2	Sample 3	Sample 4	Sample 5	Sample 6
x_1:	−1	−3	1	2	5	8
x_2:	1	−1	6	8	2	3

These data features can be expanded by a Reduced Multivariate polynomial model by coding Eq. (3.46) in MATLAB/Octave as a function given by

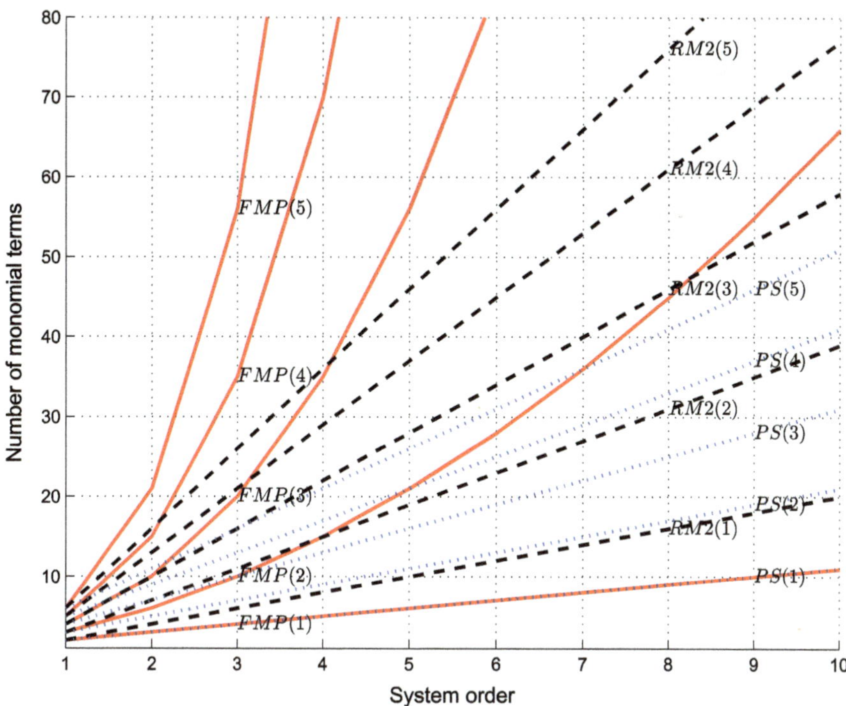

Fig. 3.10 Number of monomial terms versus system order and input dimension for each trend. The number within the parenthesis succeeding each model indicates the input dimension

```
function P = RM(X,order)
% The user assumes responsibility for determining appropriate use of the code,
% for consequences of its use, and for checking results against other reliable
% sources. No warrantee is given.
% Build regressor matrix P (mxK):
%     order = desired order of approximation,
%     X = input matrix (mxl), K = number of parameters to be est.
%     m = number of data samples, l = input dimension.
[m,l] = size(X); MM1=[]; MM3=[]; Msum=sum(X,2);
for i=1:order
    for k=1:l
        M1(:,k)=X(:,k).^i;
        if (i>1)
            M3(:,k)=X(:,k).*Msum.^(i-1);
        end
    end
    MM1=[MM1,M1];
    if (i>1)
```

```
    MM3=[MM3,M3];

  end

end

P = [ones(m,1),MM1,MM3];

return;
```

Then, the matrix expanded by the 2nd-order RM model can be generated using

```
X = [-1, 1; -3, -1; 1, 6; 2, 8; 5,2; 8,3]

P = RM(X,2)

P =

    1    -1     1     1     1     0     0
    1    -3    -1     9     1    12     4
    1     1     6     1    36     7    42
    1     2     8     4    64    20    80
    1     5     2    25     4    35    14
    1     8     3    64     9    88    33
```

$\square$

3.3.3   *Orthogonal Polynomials*

An orthogonal polynomial sequence refers to a family of polynomials with zero inner product between any two different polynomials within the sequence. Based on the distinctive property of linearly independent monomial terms, this family of polynomials has useful properties similar to that of the Fourier series which can be exploited for solving, expanding, and interpreting mathematical and physical problems.

For the univariate case, the inner product between two continuous real functions f and g defined on $x \in \mathbb{R}$ over an interval $[a, b]$ can be written as

$$\langle f, g \rangle = \int_a^b f(x)g(x)dx. \tag{3.48}$$

The Gram-Schmidt orthogonalization process can be applied to obtain a sequence of orthogonal polynomials as follows:

$$p_0(x) = 1,$$

$$p_1(x) = x - \frac{\langle x, p_0 \rangle}{\langle p_0, p_0 \rangle} p_0(x),$$

$$p_2(x) = x^2 - \frac{\langle x^2, p_0 \rangle}{\langle p_0, p_0 \rangle} p_0(x) - \frac{\langle x^2, p_1 \rangle}{\langle p_1, p_1 \rangle} p_1(x),$$

$$\vdots$$

$$p_n(x) = x^n - \frac{\langle x^n, p_0 \rangle}{\langle p_0, p_0 \rangle} p_0(x) \cdots - \frac{\langle x^n, p_{n-1} \rangle}{\langle p_{n-1}, p_{n-1} \rangle} p_{n-1}(x). \tag{3.49}$$

Several other examples of orthogonal polynomials which can be expressed explicitly are listed as follows:

Legendre polynomials:

$$p_n(x) = \frac{1}{2^n} \sum_{k=0}^{n} \binom{n}{k}^2 (x-1)^{n-k}(x+1)^k \tag{3.50}$$

Laguerre polynomials:

$$p_n(x) = \sum_{k=0}^{n} \binom{n}{k} \frac{(-1)^k}{k!} x^k. \tag{3.51}$$

Chebyshev polynomials:

$$p_n(x) = \frac{1}{2}\left[\left(x - \sqrt{x^2-1}\right)^n + \left(x + \sqrt{x^2-1}\right)^n \right]. \tag{3.52}$$

Hermite polynomials:

$$p_n(x) = n! \sum_{m=0}^{\lfloor \frac{n}{2} \rfloor} \frac{(-1)^m}{m!(n-2m)!}(2x)^{n-2m}. \tag{3.53}$$

The full multivariate orthogonal polynomials remain a research topic and yet to find its way near common applications.

3.4 Shallow Feedforward Networks

3.4.1 Perceptron Network

The perceptron is commonly referred to as a model for supervised binary classification. A typical perceptron neuron unit uses the hard-limit transfer function on a linear combination of inputs:

$$f(\boldsymbol{w}, \boldsymbol{x}) = L(\boldsymbol{w}^T \boldsymbol{x} + w_0), \tag{3.54}$$

where $L(\cdot)$ can be either a unit **step function** given by

$$L(x) = u(x) = \begin{cases} 1 & \text{if } x \geqslant 0 \\ 0 & \text{otherwise} \end{cases}, \tag{3.55}$$

or a **signum function** given by

$$L(x) = \text{sgn}(x) = \begin{cases} 1 & \text{if } x > 0 \\ 0 & \text{if } x = 0 \\ -1 & \text{if } x > 0 \end{cases}. \tag{3.56}$$

Alternatively, a **ramp function** can be adopted to enable piecewise differentiation:

$$L(x) = \text{ramp}(x) = \begin{cases} 1, & x \geqslant 1 \\ x, & -1 < x < 1 \\ -1, & x \leqslant -1 \end{cases}. \tag{3.57}$$

Figure 3.11 shows the plots for step, signum, and ramp functions which are plotted over $-2 \leqslant x \leqslant 2$.

For a series of interconnected perceptron neurons, a compact representation can be written as

$$f(\mathbf{W}, x) = L(\mathbf{W}x), \tag{3.58}$$

when an augmented input vector $x = [1, x_1, \ldots, x_d]^T$ is adopted with the bias term being absorbed into the weight matrix:

$$\mathbf{W} = \begin{bmatrix} w_{1,0} & w_{1,1} & \cdots & w_{1,d} \\ \vdots & & & \vdots \\ w_{h,0} & w_{h,1} & \cdots & w_{h,d} \end{bmatrix}. \tag{3.59}$$

This is known as a single layer perceptron network with multiple outputs.

Based on the above basic perceptron unit, a similar perceptron structure can be applied to itself compositely as

$$g(\mathbf{W}_o, \mathbf{W}_h, x) = L(\mathbf{W}_o f) = L(\mathbf{W}_o L(\mathbf{W}_h x)), \tag{3.60}$$

where $\mathbf{W}_o \in \mathbb{R}^{o \times h}$ and $\mathbf{W}_h \in \mathbb{R}^{h \times (d+1)}$, respectively, denote the output weights and the hidden neuron weights. This results in a two-layer perceptron network.

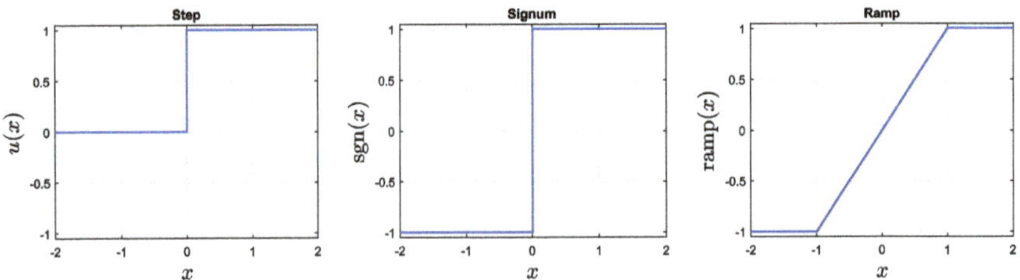

Fig. 3.11   Step, signum, and ramp functions

The above recursive composition of the perceptron structure can be repeated to form a multilayer perceptron network with complex mapping capability. However, due to the highly nonlinear structure, learning of the weight parameters cannot be analytic. Nevertheless, when a linear output layer with the hidden layer weights predetermined, an analytic learning of the output weights can be achieved:

$$g(\mathbf{W}_o, \mathbf{W}_h, \mathbf{x}) = \mathbf{W}_o f = \mathbf{W}_o L(\mathbf{W}_h \mathbf{x}). \tag{3.61}$$

For single output unit, we can write in the familiar linear projection form as

$$g(\mathbf{w}, \mathbf{W}_h, \mathbf{x}) = \mathbf{w}^T f = \mathbf{w}^T L(\mathbf{W}_h \mathbf{x}). \tag{3.62}$$

To allow for smooth transition and differentiability, the hard-limit function is often replaced by a smooth function. The following functions are popular choices in the neural network literature.

Rectified Linear Unit (ReLU) function: This function is a popular choice in deep learning because of its implementation simplicity in terms of both its original form and its derivative form despite its non-analytic nature. The *rectifier* or ReLU function can be written as

$$L(x) = \max(0, x), \tag{3.63}$$

which also means taking only the positive part of the real variable x, i.e., $L(x) = x^+$. This function is also called the *ramp* function and is analogous to the *half-wave rectification* in electrical engineering, thus the name.

Softplus function: This function is a smooth approximation to the ReLU function given by

$$L(x) = \log(1 + e^x), \tag{3.64}$$

where its derivative is the logistic function given by $1/(1 + e^{-x})$.

Softmax function: This is a generalization of the logistic function that squashes a d-dimensional input x of real values into the range $(0, 1)$ that adds up to 1. The function can be written as

$$L(x_i) = \frac{e^{x_i}}{\sum_{i=1}^{d} e^{x_i}}, \quad i = 1, 2, \ldots, d. \tag{3.65}$$

Sigmoid/logistic function: A frequent choice for a soft-limit function is the sigmoid function given by

$$L(x) = \sigma(x) = \frac{1}{1 + e^{-ax}}, \tag{3.66}$$

where $\sigma \in (0, 1)$ and a controls the slope of transition. Without explicitly expressing the predetermined hidden weights, the sigmoidal network with single output can then be written as

$$g(\boldsymbol{w}, \boldsymbol{x}) = \boldsymbol{w}^T \sigma(\boldsymbol{x}), \tag{3.67}$$

where $\boldsymbol{w} = [w_0, w_1, \ldots, w_h]^T$ and $\sigma(\boldsymbol{x}) = [1, \sigma_1(\boldsymbol{w}_1^T \boldsymbol{x}), \ldots \sigma_h(\boldsymbol{w}_h^T \boldsymbol{x})]^T$ and h is the number of sigmoidal units.

Logit function: This function is the inverse of the sigmoid/logistic function commonly used in statistics. The function can be written as

$$L(x) = \log\left(\frac{x}{1-x}\right), \quad x \in [0, 1], \tag{3.68}$$

where the base of the logarithm can be arbitrarily chosen to be larger than 1. The natural logarithm with base e is the one most often used.

Arctan function: Instead of having a range in $(0, 1)$, the inverse tangent (also known as arctan) function gives a range with arctan $\in (-\pi/2, +\pi/2)$:

$$\arctan(x) = \tan^{-1}(ax), \tag{3.69}$$

where a controls the slope of the activation region. A normalization to the range of $(-1, 1)$ can be performed when adopting it to replace the above hard-limit function:

$$L(x) = \phi(x) = \frac{2}{\pi}\arctan(x) = \frac{2}{\pi}\tan^{-1}(ax). \tag{3.70}$$

Consider h number of the above activation functions given by

$$\phi_i(\boldsymbol{a}^T \boldsymbol{x}) = \phi_i(a_0 \cdot 1 + a_1 x_{1,i} + \cdots + a_d x_{d,i}), \quad i = 1, \ldots, h, \tag{3.71}$$

a linear prediction model can be written as

$$g(\boldsymbol{w}, \boldsymbol{x}) = \boldsymbol{w}^T \phi(\boldsymbol{x}), \tag{3.72}$$

where $\boldsymbol{w} = [w_0, w_1, \ldots, w_h]^T$ and $\phi(\boldsymbol{x}) = [1, \phi_1(\boldsymbol{a}_1^T \boldsymbol{x}), \ldots, \phi_h(\boldsymbol{a}_h^T \boldsymbol{x})]^T$.

Radial Basis function: A radial basis function is a real-valued function whose value on a point depends on the *distance* between this point and a reference *center* point. Consider a point $\boldsymbol{x} \in \mathbb{R}^d$ and a center $\boldsymbol{c} \in \mathbb{R}^d$, the distance $r \in \mathbb{R}$ between $\boldsymbol{x}$ and $\boldsymbol{c}$ in Euclidean space can be expressed as

$$r = \|\boldsymbol{x} - \boldsymbol{c}\|_2 = \sqrt{(\boldsymbol{x} - \boldsymbol{c})^T (\boldsymbol{x} - \boldsymbol{c})}. \tag{3.73}$$

The popular Gaussian-based radial basis function can be written as

$$\phi(r) = \frac{1}{\rho\sqrt{2\pi}} e^{-\frac{1}{2}(r/\rho)^2}, \tag{3.74}$$

where ρ controls the width (spread) of the Gaussian function. Other types of radial basis functions include

$$\phi(r) = \sqrt{1 + r/\rho}, \tag{3.75}$$

$$\phi(r) = \frac{1}{\sqrt{1 + r/\rho}}, \tag{3.76}$$

and

$$\phi(r) = \frac{1}{1 + r/\rho}. \tag{3.77}$$

By packing a set containing h number of radial basis functions with different preset internal variables (ρ, c) in $\phi = [1, \phi_1, \ldots, \phi_h]^T$, a linear model can be written as

$$g(w, x) = w^T \phi(x). \tag{3.78}$$

Figure 3.12 shows the graphical plots for ReLU, softplus, logit, sigmoid, arctan, and Gaussian radial basis functions.

3.4.2 Reciprocal Sigmoid Network

The successful application of the sigmoidal network can be attributed to its approximation capability and the ease of network to converge to a certain local solution. Adequate use of the sigmoid-based models, however, is complicated by the effort required to fix the various learning hyper-parameters (e.g., number of layers, number of neurons in each layer, momentum, learning rate, etc.) in order that the learning converges to an acceptable local solution with good predictivity. Here, based on observations regarding some properties of the sigmoid function, a linear model can be constructed, which carries much of the approximation capability as well as relieving much of the effort to search for appropriate local solutions among the various combinations of hyper-parameters. In the following derivations, we provide a sketch to show that $\sigma(ax)$ can be expressed as a linear combination of products containing the basic form of $\sigma(x)$ term itself.

Let $x_1 = x(a - 1)$ and $x_2 = x$, then $\sigma(ax)$ can be written as $\sigma(x(a - 1) + x) = \sigma(x_1 + x_2)$. Next, consider the basic form of sigmoid function given by

$$\sigma(x) = \frac{1}{1 + e^{-x}}, \tag{3.79}$$

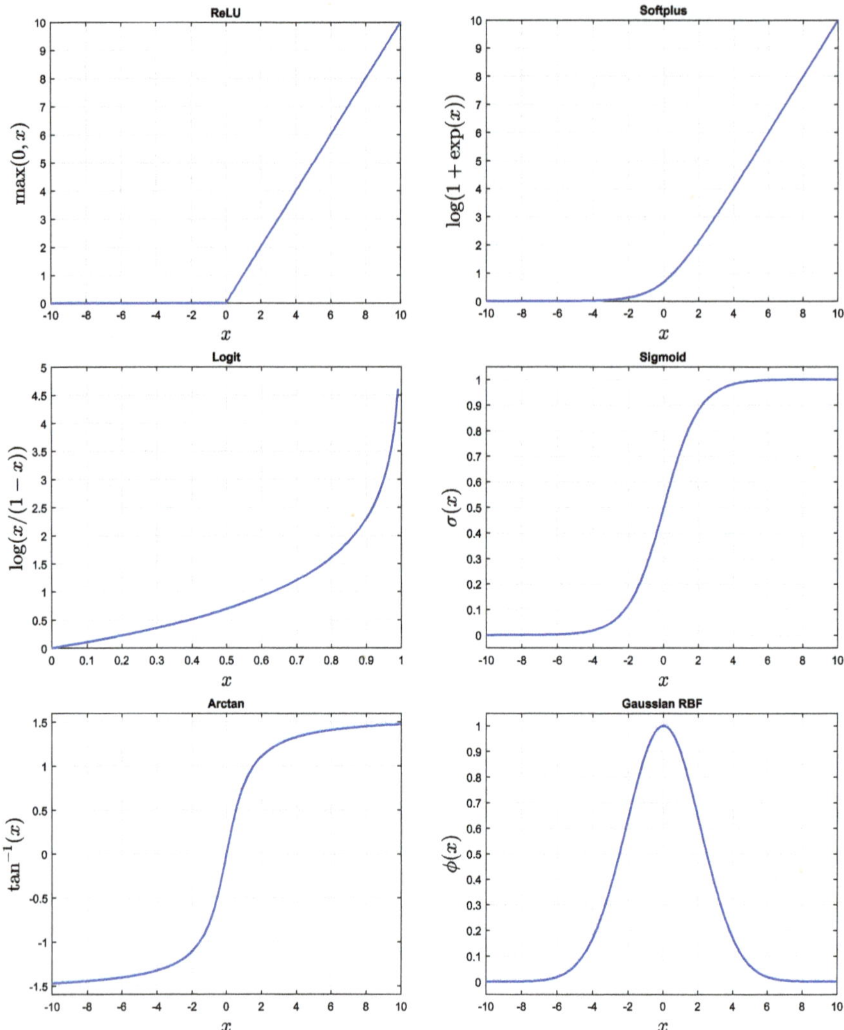

Fig. 3.12 ReLU, softplus, logit, sigmoid, arctan, and Gaussian radial basis ($\rho = 3$ and $c = 0$ in (3.74)) functions

and rearrange it as

$$e^{-x} = (1 - \sigma(x))/\sigma(x).\tag{3.80}$$

Adoption of (3.80) for e^{-x_1} and e^{-x_2} in $\sigma(x_1 + x_2)$ gives

$$\sigma(x_1 + x_2) = \frac{1}{1 + e^{-x_1}e^{-x_2}}$$

$$= \frac{\sigma(x_1)\sigma(x_2)}{\sigma(x_1)\sigma(x_2) + (1 - \sigma(x_1))(1 - \sigma(x_2))}. \tag{3.81}$$

The numerator and the denominator are next divided by $\sigma(x_1)\sigma(x_2)$ such that (3.81) can be rewritten as

$$\sigma(x_1 + x_2) = \frac{1}{1 + \omega}, \tag{3.82}$$

where

$$\omega = \frac{(1 - \sigma(x_1))(1 - \sigma(x_2))}{\sigma(x_1)\sigma(x_2)}$$

$$= 1 - (\sigma(x_1))^{-1} - (\sigma(x_2))^{-1} + (\sigma(x_1)\sigma(x_2))^{-1}. \tag{3.83}$$

By means of Binomial expansion, (3.82) can be expressed as

$$\sigma(x_1 + x_2) = (1 + \omega)^{-1} = 1 - \omega + \omega^2 - \omega^3 + \omega^4 - \cdots, \tag{3.84}$$

which converges for $|\omega| < 1$. Equations (3.84) and (3.83) show that $\sigma(x_1 + x_2)$ can be expressed in terms of products and powers of $(\sigma(x_1))^{-1}$ and $(\sigma(x_2))^{-1}$ when $|\omega| < 1$. Putting them together, we have

$$\sigma(x_1 + x_2) = w_0 + w_1(\sigma(x_1))^{-1} + w_2(\sigma(x_2))^{-1} + w_3(\sigma(x_1)\sigma(x_2))^{-1}$$
$$+ w_4(\sigma(x_1))^{-2} + w_5(\sigma(x_2))^{-2} + w_6(\sigma(x_1)\sigma(x_2))^{-2} + \cdots, \tag{3.85}$$

where $w_i, i = 0, 1, 2, \ldots$ are the weighting coefficients. In addition, based on the assumption that $\sigma(x_1 + x_2) = \sigma(ax)$ for $x_1 = x(a - 1) = bx$ and $x_2 = x$, we can write

$$\sigma(ax) = w_0 + w_1(\sigma(bx))^{-1} + w_2(\sigma(x))^{-1} + w_3(\sigma(bx)\sigma(x))^{-1}$$
$$+ w_4(\sigma(bx))^{-2} + w_5(\sigma(x))^{-2} + w_6(\sigma(bx)\sigma(x))^{-2} + \cdots. \tag{3.86}$$

Since $\sigma(bx)$ can be further decomposed, we see that $\sigma(ax)$ can be approximated by a series of powered products of $(\sigma(x))^{-1}$.

This expansion shows that the reciprocal of sigmoid functions $(\sigma(x))^{-1}$ can be linearly combined by means of product and powered basis terms, to achieve an equivalent "shifting" (e.g., $x_1 + x_2$) and "scaling" (e.g., ax) effects for a sigmoid function containing nonlinear "shifting" and "scaling" operation on the parameters (i.e., a and b in $\sigma(a(x - b))$). This implies that any scaled or shifted sigmoid function might be expanded as a series containing linear combination of product and power terms of the reciprocal sigmoid function. A collection of these series, which represent differently scaled and shifted sigmoids, can thus be used for function approximation. Moreover, as the reciprocal of the sigmoid function (3.79) is $1 + e^{-x}$, there is no need to perform a reciprocal inversion on the term $(1 + e^{-x})$. This translates into a gain of computational simplicity.

By packing a set containing h number of product terms of reciprocal sigmoid functions $\phi_j = (1 + e^{-x_i})$, $j = 1, \ldots, h$ (where i indexes the feature dimensions $i = 1, \ldots, d$) as $\boldsymbol{\phi} = [1, \phi_1, \ldots, \phi_h]^T$, and then embed $\boldsymbol{\phi}$ within a polynomial expansion (or using a reduced polynomial to avoid explosion of parameter size), a linear parametric model with input-output mapping nonlinearity can be written as

$$g(\boldsymbol{w}, \boldsymbol{x}) = \boldsymbol{w}^T \boldsymbol{p}(\boldsymbol{\phi}(\boldsymbol{x})). \tag{3.87}$$

3.4.3 Hyperbolic Function Network

Following similar reasoning as above, the hyperbolic functions can also be considered as a candidate for nonlinear input-output mapping and being expressed as a generalized linear model. Several examples of "scaling" and "shifting" expressions are observed as follows:

$$\left. \begin{array}{l} cosh(2x) = 2cosh^2(x) - 1 \\ cosh(3x) = 3cosh^3(x) - 3cosh(x) \\ cosh(4x) = 8cosh^4(x) - 8cosh^2(x) + 1 \end{array} \right\} \tag{3.88}$$

$$\left. \begin{array}{l} sinh(2x) = 2sinh(x)cosh(x) \\ sinh(3x) = 3sinh(x) + 4sinh^3(x) \\ sinh(4x) = 8sinh^3(x)cosh(x) + 4sinh(x)cosh(x) \end{array} \right\} \tag{3.89}$$

$$\left. \begin{array}{l} tanh(2x) = \frac{2tanh(x)}{1+tanh^2(x)} \\ tanh(3x) = \frac{3tanh(x)+tanh^3(x)}{1+3tanh^2(x)} \\ tanh(4x) = \frac{4tanh(x)+4tanh^3(x)}{1+6tanh^2(x)+tanh^4(x)} \end{array} \right\} \tag{3.90}$$

$$cosh(x \pm y) = cosh(x)cosh(y) \pm sinh(x)sinh(y), \tag{3.91}$$

$$sinh(x \pm y) = sinh(x)cosh(y) \pm cosh(x)sinh(y), \tag{3.92}$$

$$tanh(x \pm y) = \frac{tanh(x) \pm tanh(y)}{1 \pm tanh(x)tanh(y)}. \tag{3.93}$$

Figure 3.13 shows a plot of the fundamental hyperbolic functions on a single input dimension. Here, it is observed that tanh behaves like an arctan function or a sigmoid with different scaling and offset. The sinh function shows a "gentle" activation near the origin and a "steep" activation away from the origin. The cosh function shows a unit offset for the output. This means that the origin is not included where the output will not give a zero value even if the input is zero. This may be undesirable for mapping of input near the origin. An inclusion of an offset of "-1" into the cosh function provides a resolution to this issue.

Let

$$\phi_{sinh}(\boldsymbol{x}) = sinh(\boldsymbol{x}) \tag{3.94}$$

$$\phi_{cosh}(\boldsymbol{x}) = cosh(\boldsymbol{x}) - 1 \tag{3.95}$$

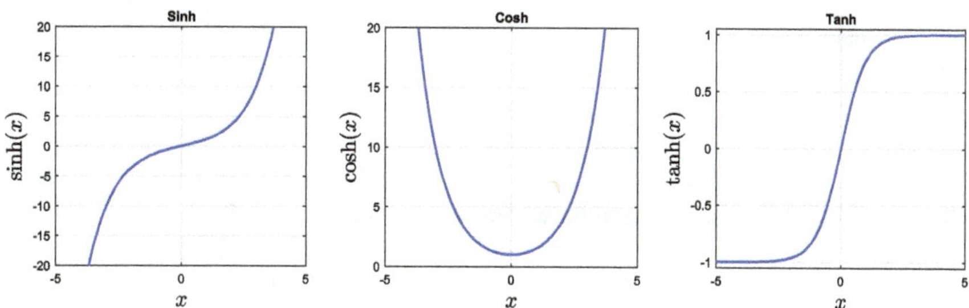

Fig. 3.13 Hyperbolic functions: $\sinh(x)$, $\cosh(x)$, and $\tanh(x)$

$$\phi_{tanh}(\boldsymbol{x}) = \tanh(\boldsymbol{x}), \tag{3.96}$$

then, a linear combination of the product terms of hyperbolic functions forms a generalized linear model:

$$g_{\sinh}(\boldsymbol{w}, \boldsymbol{x}) = \boldsymbol{w}^T \boldsymbol{p}(\phi_{sinh}(\boldsymbol{x})) \tag{3.97}$$
$$g_{\cosh}(\boldsymbol{w}, \boldsymbol{x}) = \boldsymbol{w}^T \boldsymbol{p}(\phi_{cosh}(\boldsymbol{x})) \tag{3.98}$$
$$g_{\tanh}(\boldsymbol{w}, \boldsymbol{x}) = \boldsymbol{w}^T \boldsymbol{p}(\phi_{tanh}(\boldsymbol{x})), \tag{3.99}$$

where $\boldsymbol{p}(\phi(\boldsymbol{x}))$ represents a polynomial-based combination of the transformed features $\phi(\boldsymbol{x})$. Alternatively, a linear combination of a mixture of these functions can be constructed:

$$\begin{aligned}
g_{hyperbolic}&(\boldsymbol{w}, \boldsymbol{\beta}, \boldsymbol{\gamma}, \boldsymbol{x}) \\
&= \boldsymbol{w}^T \boldsymbol{p}(\phi_{sinh}(\boldsymbol{x})) + \boldsymbol{\beta}^T \boldsymbol{p}(\phi_{cosh}(\boldsymbol{x})) + \boldsymbol{\gamma}^T \boldsymbol{p}(\phi_{tanh}(\boldsymbol{x})).
\end{aligned} \tag{3.100}$$

3.5 Projection Models

3.5.1 Random Projection

A visualization of a low-dimensional projection is the casting of shadows onto a plane. The projection of an object on the plane is related to its shadow on the plane. The shadow of a three-dimensional sphere is a two-dimensional circle.

In vector space, an *inner product* or *dot product* between two vectors $\boldsymbol{x} \in \mathbb{R}^d$ and $\boldsymbol{y} \in \mathbb{R}^d$ given by

$$\langle \boldsymbol{x}, \boldsymbol{y} \rangle = \boldsymbol{x}^T \boldsymbol{y} = \boldsymbol{x} \cdot \boldsymbol{y} = \sum_{i=1}^{d} x_i y_i, \tag{3.101}$$

can be considered a projection, either of x onto y or otherwise. In $\mathbb{R}^2$, the geometry of the dot product between the two vectors is the shadow of one vector on another:

$$x \cdot y = \cos(\angle(x, y)) \cdot |x| \cdot |y|, \tag{3.102}$$

where $\angle(x, y)$ denotes the angle between the two vectors x and y.

Based on this perspective, the product between a matrix and a vector can be considered a projection. This includes all the above-mentioned models, such as polynomials, sigmoid function, radial basis functions, and hyperbolic functions, where a linear projection between the parameter vector w and the input transformation model vector p can be observed.

Consider a model vector obtained by the following projection of input data x onto the matrix $\mathbf{R}$:

$$q = \mathbf{R}x, \tag{3.103}$$

where $\mathbf{R} \in \mathbb{R}^{D \times d}$ is a matrix of random numbers and $x \in \mathbb{R}^d$. Then, the input transformation model $q \in \mathbb{R}^D$ is obtained by random projection. To add nonlinearity to the transformation, we can have

$$p(x) = \boldsymbol{\phi}(q) = [1, \phi(q_1), \ldots, \phi(q_D)]^T = [1, \phi(r_1^T x), \ldots, \phi(r_D^T x)]^T, \tag{3.104}$$

where r_i^T, $i = 1, \ldots, D$ denote the row vectors of $\mathbf{R}$ and ϕ can be any nonlinear transformation, such as unit step, sigmoid, or exponential functions discussed in previous subsections. In order to include the linear mapping property in the model, the linear term x can be included as follows:

$$p(x) = [1, x^T, \phi(r_1^T x), \ldots, \phi(r_D^T x)]^T. \tag{3.105}$$

In general, different nonlinear functions ϕ_i can be included in a mixture for transforming each q_i, $i = 1, \ldots, D$. However, for simplicity reason, a single type of nonlinear function is often considered for ϕ. Here, two hyper-parameters are important for deriving useful features in applications. These hyper-parameters are the size of projection D and the nonlinear parameter within ϕ such as the slope of sigmoid function or the power of exponential function.

3.5.2 Multi-scale Projection

Apart from using random numbers, a certain structure could also be imposed on $\mathbf{R}$ so long that it does not contain only vectors of constant elements.

Consider the following non-random projection matrix

$$\mathbf{R}(s_j) = \begin{bmatrix} \varphi_1(s_j) & \varphi_2(s_j) & \cdots & \varphi_{d-1}(s_j) & \varphi_d(s_j) \\ \varphi_2(s_j) & \varphi_3(s_j) & \cdots & \varphi_d(s_j) & \varphi_1(s_j) \\ \vdots & \vdots & \ddots & \vdots & \vdots \\ \varphi_{d-1}(s_j) & \varphi_d(s_j) & & \varphi_{d-3}(s_j) & \varphi_{d-2}(s_j) \\ \varphi_d(s_j) & \varphi_1(s_j) & \cdots & \varphi_{d-2}(s_j) & \varphi_{d-1}(s_j) \end{bmatrix} \tag{3.106}$$

where each row is formed by rotating its elements. Each element in each row of the matrix is a differently scaled version of the seed element $\varphi_1(s_j)$ where

$$\varphi_k(s_j) = (1/s_j)^k, \tag{3.107}$$

for arbitrary jth scaling factor $s_j \in \mathbb{R}$. If s_j admits negative values, then care must be taken for admissible power k in order to avoid complex numbers.

Based on the projection matrix given by (3.106), a linear projection model can be constructed as

$$p(x) = [1, x^T, \phi^T], \tag{3.108}$$

where

$$\phi = \phi(\mathbf{R}x) = \begin{pmatrix} \phi(r_1^T x) \\ \vdots \\ \phi(r_D^T x) \end{pmatrix}. \tag{3.109}$$

3.6 Kernel Models

The word "kernel" has several meanings in mathematics. In machine learning, kernel methods refer to a class of algorithms for pattern analysis. An application example is the embedding of features in a *support vector machine* (SVM) which relies on the supporting vectors at data distribution boundary to make classification decisions. The key idea of kernel method is to represent the raw data based on a *similarity function* called *kernel function* over *pairs of data points* whereas an explicit feature map, such as those projection and nonlinear transformation models $p(x)$ mentioned in above subsections, need not be known. In other words, the estimation is made from data instances utilizing correlation (inner product) among data points with implicit parameters instead of estimating a fixed set of explicit data feature coordinates themselves. Hence, kernel methods are also known as instance-based learners. Primal ridge regression in Sect. 5.1.3 directly estimates the data feature parameter w, while dual ridge regression in Sect. 5.1.4 estimates the instance-related parameters α in the sample space where the primal feature parameter w becomes its projected image, i.e., $w = \mathbf{P}^T \alpha$.

3.6.1 Linear Kernel

Mathematically, a kernel function $k(\boldsymbol{x}_1, \boldsymbol{x}_2)$ maps two data points $\boldsymbol{x}_1 \in \mathcal{X}$ and $\boldsymbol{x}_2 \in \mathcal{X}$ from their joint input space $\mathcal{X} \times \mathcal{X}$ onto another space $\mathcal{K}$, i.e., $k : \mathcal{X} \times \mathcal{X} \mapsto \mathcal{K}$. The kernel function is said to satisfy Mercer's condition if the kernel matrix for a set of data points is positive semidefinite. In our applications, it is frequent to find $\mathcal{X} \subseteq \mathbb{R}^d$ and $\mathcal{K} \subseteq \mathbb{R}$, then $k(\boldsymbol{x}_1, \boldsymbol{x}_2)$ can be expressed as an inner product between the two points:

$$k(\boldsymbol{x}_1, \boldsymbol{x}_2) = \langle \boldsymbol{x}_1, \boldsymbol{x}_2 \rangle. \tag{3.110}$$

According to Wahba's representer theorem, the minimizer of any arbitrary cost function plus a strict monotonic increasing transformation of the minimizer under a normed RKHS admits a representation of the form as a linear combination of kernelized training data points. In other words, the model output g of any data point $\boldsymbol{x}$ can be expressed in terms of the training data $\boldsymbol{x}_i$, $i = 1, \ldots, m$ in inner product form given by

$$
\begin{aligned}
g(\boldsymbol{w}, \boldsymbol{x}) &= \sum_{i=1}^{m} \alpha_i \langle \boldsymbol{x}_i, \boldsymbol{x} \rangle \\
&= \sum_{i=1}^{m} \alpha_i k(\boldsymbol{x}_i, \boldsymbol{x}),
\end{aligned}
\tag{3.111}
$$

when the kernel function satisfies Mercer's condition. By defining a mapping given by

$$\boldsymbol{w} = \sum_{i=1}^{m} \alpha_i \boldsymbol{x}_i = \sum_{i=1}^{m} \alpha_i \begin{bmatrix} x_1 \\ \vdots \\ x_d \end{bmatrix}_i, \tag{3.112}$$

where $\boldsymbol{x}_i$, $i = 1, \ldots, m$ are the training samples, we can relate the model in the *primal* parametric space ($\boldsymbol{w} \in \mathbb{R}^d$ or $\boldsymbol{w} \in \mathbb{R}^{d+1}$ when the bias term is included) to the model expressed in the *dual* parametric space ($\boldsymbol{\alpha} \in \mathbb{R}^m$):

$$
\begin{aligned}
g(\boldsymbol{w}, \boldsymbol{x}) &= \boldsymbol{w}^T \boldsymbol{x} \\
&= \left(\sum_{i=1}^{m} \alpha_i \boldsymbol{x}_i \right)^T \boldsymbol{x} \\
&= \sum_{i=1}^{m} \alpha_i \boldsymbol{x}_i^T \boldsymbol{x} \\
&= \sum_{i=1}^{m} \alpha_i k(\boldsymbol{x}_i, \boldsymbol{x})
\end{aligned}
$$

$$= \boldsymbol{\alpha}^T k(\boldsymbol{x})$$
$$= g(\boldsymbol{\alpha}, \boldsymbol{x}), \tag{3.113}$$

where

$$\boldsymbol{\alpha} = [\alpha_1, \ldots, \alpha_m]^T, \tag{3.114}$$

and

$$k(\boldsymbol{x}) = [k(\boldsymbol{x}_1, \boldsymbol{x}), \ldots, k(\boldsymbol{x}_m, \boldsymbol{x})]^T. \tag{3.115}$$

Here, we see a change of parameters based on the input feature space $\boldsymbol{w} \in \mathbb{R}^d$ or $\mathbb{R}^{d+1}$ to that based on the data sample space $\boldsymbol{\alpha} \in \mathbb{R}^m$ in (3.113) since $\boldsymbol{\alpha} = [\alpha_1, \ldots, \alpha_m]^T$ where m is the training sample size. In other words, $\boldsymbol{\alpha}$ is the new parameters to be estimated in (3.113) under the kernel paradigm.

Collectively, for m number of training points $\mathbf{X} = [\boldsymbol{x}_1, \ldots, \boldsymbol{x}_m]^T \in \mathbb{R}^{m \times d}$ and target $\mathbf{y} = [y_1, \ldots, y_m]^T \in \mathbb{R}^m$, we can represent the set of training prediction by

$$\begin{aligned}
\mathbf{y} &= \mathbf{g}(\boldsymbol{w}, \boldsymbol{x}) + \boldsymbol{\epsilon} \\
&= \mathbf{X}\boldsymbol{w} + \boldsymbol{\epsilon} \\
&= \begin{bmatrix} \boldsymbol{w}^T \boldsymbol{x}_1 \\ \vdots \\ \boldsymbol{w}^T \boldsymbol{x}_m \end{bmatrix} + \begin{bmatrix} \epsilon_1 \\ \vdots \\ \epsilon_m \end{bmatrix} \\
&= \begin{bmatrix} \sum_{i=1}^m \alpha_i k(\boldsymbol{x}_i, \boldsymbol{x}_1) \\ \vdots \\ \sum_{i=1}^m \alpha_i k(\boldsymbol{x}_i, \boldsymbol{x}_m) \end{bmatrix} + \boldsymbol{\epsilon} \\
&= \mathbf{X}\mathbf{X}^T \boldsymbol{\alpha} + \boldsymbol{\epsilon} \\
&= \mathbf{K}\boldsymbol{\alpha} + \boldsymbol{\epsilon}, \tag{3.116}
\end{aligned}$$

where

$$\mathbf{K} = \mathbf{X}\mathbf{X}^T, \tag{3.117}$$
$$\boldsymbol{w} = \mathbf{X}^T \boldsymbol{\alpha}, \tag{3.118}$$

with $\boldsymbol{\alpha} \in \mathbb{R}^m$ and when $\mathbf{K} \in \mathbb{R}^{m \times m}$ is positive semidefinite (Mercer's condition).

3.6.2 Polynomial Kernel

The main advantage of adopting a kernel model is its capability of embedding non-linearity and dimensionality under the data sample space without explicit knowledge of the feature mapping coordinates. For example, an rth-order model adopting the polynomial kernel can be written as

$$g(\boldsymbol{w}, \boldsymbol{x}) = \sum_{i=1}^{m} \alpha_i k(\boldsymbol{x}_i, \boldsymbol{x}) \tag{3.119}$$

with

$$k(\boldsymbol{x}_i, \boldsymbol{x}) = \langle \boldsymbol{\phi}(\boldsymbol{x}_i), \boldsymbol{\phi}(\boldsymbol{x}) \rangle = (a + b\langle \boldsymbol{x}_i, \boldsymbol{x} \rangle)^r, \tag{3.120}$$

where $a, b \in \mathbb{R}$ and an explicit expression for ϕ need not be known in practice.

An example for the unknown ϕ in a second-order polynomial model can be identified as

$$\boldsymbol{\phi}(\boldsymbol{x}) = [1, \sqrt{2}x_1, \sqrt{2}x_2, x_1^2, x_2^2, \sqrt{2}x_1 x_2]^T, \tag{3.121}$$

where

$$\begin{aligned} k(\boldsymbol{x}_i, \boldsymbol{x}) &= \langle \boldsymbol{\phi}(\boldsymbol{x}_i), \boldsymbol{\phi}(\boldsymbol{x}) \rangle \\ &= (1 + \langle \boldsymbol{x}_i, \boldsymbol{x} \rangle)^2 \\ &= 1 + 2x_{i1}x_1 + 2x_{i2}x_2 + (x_{i1}x_1)^2 + (x_{i2}x_2)^2 + 2x_{i1}x_1 x_{i2}x_2. \end{aligned} \tag{3.122}$$

Collectively, for m number of training points $\mathbf{X} = [\boldsymbol{x}_1, \ldots, \boldsymbol{x}_m]^T \in \mathbb{R}^{m \times d}$, the polynomial kernel for training can be expressed as

$$\mathbf{K} = (1 + \mathbf{X}\mathbf{X}^T)^r, \tag{3.123}$$

where r denotes the polynomial order which is taken elementwise for the terms within. In this case, the polynomial embedding (3.122) need not be explicit. For use in the testing procedure, the test kernel matrix can be obtained as

$$\mathbf{K}_t = (1 + \mathbf{X}_t\mathbf{X}^T)^r, \tag{3.124}$$

where $\mathbf{X} \in \mathbb{R}^{m \times d}$ and $\mathbf{X}_t \in \mathbb{R}^{n \times d}$ denote the training and test data matrices, respectively.

3.6.3 Gaussian Kernel

A model adopting the Gaussian kernel can be written as

$$g(\boldsymbol{w}, \boldsymbol{x}) = \sum_{i=1}^{m} \alpha_i k(\boldsymbol{x}_i, \boldsymbol{x}), \tag{3.125}$$

with

$$k(\boldsymbol{x}_i, \boldsymbol{x}) = e^{-\|\boldsymbol{x} - \boldsymbol{x}_i\|^2 / 2\sigma^2}, \tag{3.126}$$

or

$$k(\boldsymbol{x}_i, \boldsymbol{x}) = e^{-\gamma \|\boldsymbol{x}-\boldsymbol{x}_i\|^2}, \tag{3.127}$$

where $\sigma, \gamma \in \mathbb{R}$ correspond to the width of Gaussian function. When the square of the norm term is left out, the kernel is called an exponential kernel:

$$k(\boldsymbol{x}_i, \boldsymbol{x}) = e^{-\|\boldsymbol{x}-\boldsymbol{x}_i\|/2\sigma^2}. \tag{3.128}$$

For this type of kernel which utilizes the difference between two sample points instead of the inner product, the kernel matrix construction needs to enumerate the distance for each pair of sample points. For example, the test kernel can be constructed using the test matrix $\mathbf{X}_t \in \mathbb{R}^{n \times d}$ and the training matrix $\mathbf{X} \in \mathbb{R}^{m \times d}$:

```
% Test kernel (based on looping)

for i=1:size(Xt,1)

    for j=1:size(X,1)

        Kt(i,j) = exp(-norm(Xt(i,:)-X(j,:))^2/(2*sigma^2));

    end

end
```

3.6.4 Asymmetric Kernels

Consider a nonlinear model (nonlinearity with respect to the input matrix $\mathbf{X} \in \mathbb{R}^{m \times d}$) given by

$$g(\boldsymbol{w}, \boldsymbol{x}) = \phi(\mathbf{X})\boldsymbol{w}, \tag{3.129}$$

where $\phi(\cdot)$ is a nonlinear function such as the exponential or sigmoid function and $\boldsymbol{w} \in \mathbb{R}^d$ is the parameter vector. Since we have $\boldsymbol{w} = \mathbf{X}^T \boldsymbol{\alpha}$ from (3.118), Eq. (3.129) can be rewritten as

$$g(\boldsymbol{w}, \boldsymbol{x}) = \phi(\mathbf{X})\mathbf{X}^T \boldsymbol{\alpha}, \tag{3.130}$$

where $\boldsymbol{\alpha} \in \mathbb{R}^m$ is the kernel parameter. Based on (3.130), we can interpret the nonlinear function in the form of (3.129) as an asymmetric kernel with kernel matrix given by $\mathbf{K}_{asym} = k(\phi(\mathbf{X}), \mathbf{X}) = \phi(\mathbf{X})\mathbf{X}^T$.

3.6.5 Creating More Kernels

Apart from the basic forms in above, more kernels can be created by a linear combination or product among kernels. Some examples are given below for immediate reference:

$$k(\boldsymbol{x}_i, \boldsymbol{x}) = k_1(\boldsymbol{x}_i, \boldsymbol{x}) + k_2(\boldsymbol{x}_i, \boldsymbol{x}), \tag{3.131}$$

$$k(\boldsymbol{x}_i, \boldsymbol{x}) = ak(\boldsymbol{x}_i, \boldsymbol{x}), \quad a \in \mathbb{R}^+, \tag{3.132}$$

$$k(\boldsymbol{x}_i, \boldsymbol{x}) = k_1(\boldsymbol{x}_i, \boldsymbol{x})k_2(\boldsymbol{x}_i, \boldsymbol{x}), \tag{3.133}$$

$$k(\boldsymbol{x}_i, \boldsymbol{x}) = f(\boldsymbol{x}_i)f(\boldsymbol{x}). \tag{3.134}$$

3.7 Bibliographic Notes

Polynomials

The ancient Chinese, Sumerian, Babylonians, and Egyptians were among the earliest known civilization to deal with polynomial equations (Pan, 1997). Through sketches and symbols, they were able to handle equations such as $x^2 + y^2 = z^2$ which is known as the Pythagoras theorem (Posamentier, 2010; Maor, 2010) today. Prior to development of elegant mathematical notations in the fifteenth century, equations were often expressed in words. For example, an algebraic problem from the Chinese Arithmetic in Nine Sections (九章算术), circa 200 BCE, begins with "Three sheafs of good crop, two sheafs of mediocre crop, and one sheaf of bad crop are sold for 29 dou" which is equivalent to $3x + 2y + z = 29$ (Wikipedia, 2015b).

According to Weierstrass's approximation theorem (1885), "Any function which is continuous in the interval $a \leqslant x \leqslant b$ may be approximated uniformly by polynomials in this interval" (Courant & Hilbert (1953), p. 65). This theorem was later generalized by Marshall H. Stone to an arbitrary compact Hausdorff space and approximation with elements from more general subalgebras with a simplified proof in 1948 (now known as Stone-Weierstrass theorem). These results paved the theoretical ground for use of polynomials particularly in function approximation. However in practice, adoption of a multivariate polynomial model often encounters huge number of expansion terms for high-dimensional inputs and high-order systems.

In order to address the issue of exponential growth of polynomial terms with respect to input dimension and system order, a linearly growing reduced multivariate polynomial model was proposed in Toh (2003b), Toh et al. (2004b), Toh et al. (2004a) for pattern classification. This model was subsequently extended to hyperbolic basis functions for multimodal biometrics data fusion (Toh and Yau, 2004) and being refined with a reciprocal sigmoid realization in Toh (2006b) to simplify the sigmoidal computation. As for orthogonal polynomials, two implementations can be found in Sun et al. (2015) for binary pattern classification.

Neural Networks

In the neural network literature, it was suggested (Hecht-Nielsen, 1987, 1990; Kůrková, 1991, 1992; Sprecher, 1993) that Kolmogorov's theorem on realization of arbitrary multivariate functions provides theoretical support regarding the function approximation capability of neural networks. However, Girosi and Poggio (1989),

Lin and Unbehauen (1993) argued that the interpretation of Kolmogorov's theorem as irrelevant to neural networks.

Theorem 3.1 (Kolmogorov and Fomin 1957) (see, e.g., Hassoun (1995))
Any continuous real-valued functions $f(x_1, x_2, \ldots, x_n)$ defined on $[0, 1]_n$, $n \geqslant 2$ can be represented in the form:

$$f(x_1, x_2, \ldots, x_n) = \sum_{j=1}^{2n+1} g_j \left(\sum_{i=1}^{n} \phi_{ij}(x_i) \right), \tag{3.135}$$

where the g_j's are properly chosen continuous functions of one variable, and the ϕ_{ij}'s are continuous monotonically increasing functions independent of f.

Independently, the *universal approximation theorem* for neural networks was shown by Cybenko (1989), Hornik et al. (1989), Funahashi (1989) which in simple terms reads as "a feedforward network with a single hidden layer containing a finite number of neurons, can approximate continuous functions on compact subsets of $\mathbb{R}^n$, under mild assumptions on the activation function" (see, e.g., Wikipedia (2015c)). In Poggio and Girosi (1990), a class of three-layer networks called regularization networks was developed for function approximation under practical considerations. Other developments related to approximation capability of neural network include (Hecht-Nielsen, 1987, 1989; Cotter, 1990; Hornik et al., 1990; Park and Sandberg, 1991; Schürmann, 1996). Since then, there was a surge of developments in the field, particularly the deep learning networks which were found useful for complex and big data learning tasks. The interested readers are referred to Bishop (1995), Haykin (2009) for an overview of neural network developments and (Goodfellow et al., 2016) for techniques in deep learning.

Theorem 3.2 (Cybenko, 1989)
Let σ be any continuous discriminatory function. Then finite sums of the form

$$G(x) = \sum_{j=1}^{N} w_j \sigma(y_j^T x + \theta_j) \tag{3.136}$$

are dense in $C(I_n)$. In other words, given any $f \in C(I_n)$ and $\varepsilon > 0$, there is a sum, $G(x)$, of the above form, for which

$$|G(x) - f(x)| < \varepsilon \quad \text{for all} \quad x \in I_n. \tag{3.137}$$

In view of the computational limitation at the time, the backpropagation learning algorithm facilitated estimation of network weights layer by layer with an option to process sample by sample (Werbos, 1974, 1990; Rumelhart et al., 1986). Since then, attributed to the advancement of computer technology, network learning became feasible in personal computers where the development had been seen through various nonlinear optimization methods (Brent, 1991; Battiti, 1992; Barnard, 1992; Toh, 2003a). Under current context, while the mainstream was geared toward utilizing deep learning to explore complex feature representation and extraction (LeCun et al., 2015), some explored the well-known linear estimation model with large quantity of preset random hidden weights (Jaeger & Haas, 2004; Huang et al., 2006; Toh, 2008; Widrow et al., 2013).

Random Projection

Random projection for dimensionality reduction can be traced to the celebrated Johnson-Lindenstrauss (JL) lemma (see Lemma 3.1) which in words states that "a small set of points in a high-dimensional space can be embedded into a space of much lower dimension in such a way that distances between the points are nearly preserved" Wikipedia (2015a).

Lemma 3.1 (Johnson-Lindenstrauss, see, e.g., Vempala (2004)) *A set of n points $u_1, \ldots, u_n$ in $\mathbb{R}^d$ can be projected down to $v_1, \ldots, v_n$ in $\mathbb{R}^k$ such that all pairwise distances are preserved:*

$$(1 - \epsilon)\|u_i - u_j\|^2 \leqslant \|v_i - v_j\|^2 \leqslant (1 + \epsilon)\|u_i - u_j\|^2 \tag{3.138}$$

if

$$k > \frac{9\ln n}{\epsilon^2 - \epsilon^3}, \quad \text{and } 0 \leqslant \epsilon \leqslant 1/2. \tag{3.139}$$

Consider a random projection given by

$$v = \frac{1}{\sqrt{k}}\mathbf{R}u, \tag{3.140}$$

where $\mathbf{R}$ is a random $k \times d$ matrix such that its elements are drawn identically and independently distributed from a normal distribution with mean 0 and variance 1. A single hidden layer feedforward neural network can be constructed by passing a randomly projected input vector (e.g., $\mathbf{R}x$ in (3.103)) through a nonlinear activation function (e.g., ϕ in (3.74)) where the yielded vector elements (e.g., elements of $p(x)$ in (3.104)) are weighted (using w) and summed to form the network output (i.e., $g(w, x) = w^T p(x)$). When the random projection matrix is generated and fixed thereon, the output weight elements (i.e., elements of w) become a linear set of parameters to be adjusted during training. By exploiting various nonlinear functions and formation of network parameters, several variants of learning algorithms can be found: radial basis function network (Broomhead & Lowe, 1988; Lowe, 1989), random vector functional-link net (Pao & Takefuji, 1992; Pao et al., 1994), feedforward neural networks with random weights (Schmidt et al., 1992; Toh, 2008), echo state networks (Jaeger & Haas, 2004; Jaeger, 2002), extreme learning machines (Huang et al., 2000, 2006), and no-prop algorithm (Widrow et al., 2013).

Kernel Methods

The history of studying the conditions under which a symmetric and continuous function is definite can be traced back to Mercer (1909) where the determinant of matrix was employed for analysis. According to Vert et al. (2004), Aronszajn (1950) and Parzen (1962) were among the first to employ methods of positive definite kernels in statistics. Subsequently, Aizerman et al. (1964) used positive definite kernels in a way close to what we call a *kernel trick*. However, it was not until 30 years later that the feature space view (Kimeldorf & Wahba, 1971; Schölkopf et al., 2001) was adopted to design new algorithms such as Support Vector Machines (SVM) (Boser et al., 1992; Vapnik, 1998, 1999) by Bernhard E. Boser, Isabelle M. Guyon and Vladimir N. Vapnik with Vladimir N. Vapnik and Alexey Y. Chervonenkis (Vapnik & Chervonenkis, 1991; Chervonenkis, 2001) being the developers of the background Vapnik-Chervonenkis theory of statistical learning. Some early tutorials of the subject can be found in Burges (1998), Osuna et al. (1997). The book by Schölkopf & Smola (2001) provides a comprehensive account on learning with kernels. In He et al. (2023), a discussion on relaxing the symmetry condition can be found. Mercer's theorem and the representer theorem from Schölkopf & Smola (2001) are reproduced below for immediate reference.

Mercer's Theorem (Mercer, 1909; Schölkopf & Smola, 2001): Suppose $k \in L_\infty(\mathcal{X}^2)$ is a symmetric real-valued function such that the integral operator

$$T_k : L_2(\mathcal{X}) \to L_2(\mathcal{X})$$

$$(T_k f)(\boldsymbol{x}) := \int_{\mathcal{X}} k(\boldsymbol{x}, \boldsymbol{x}') f(\boldsymbol{x}') d\mu(\boldsymbol{x}') \tag{3.141}$$

is positive definite; that is, for all $f \in L_2(\mathcal{X})$, we have

$$\int_{\mathcal{X}^2} k(\boldsymbol{x}, \boldsymbol{x}') f(\boldsymbol{x}) f(\boldsymbol{x}') d\mu(\boldsymbol{x}) d\mu(\boldsymbol{x}') \geqslant 0. \tag{3.142}$$

Let $\psi_j \in L_2(\mathcal{X})$ be the normalized orthogonal eigenfunctions of T_k associated with the eigenvalues $\lambda_j > 0$, sorted in non-increasing order. Then

1. $(\lambda_j)_j \in \ell_1$,
2. $k(\boldsymbol{x}, \boldsymbol{x}') = \sum_{j=1}^{N_{\mathcal{H}}} \lambda_j \psi(\boldsymbol{x}) \psi(\boldsymbol{x}')$ holds for almost all $(\boldsymbol{x}, \boldsymbol{x}')$. Either $N_{\mathcal{H}} \in \mathbb{N}$, or $N_{\mathcal{H}} = \infty$; in the latter case, the series converges absolutely and uniformly for almost all $(\boldsymbol{x}, \boldsymbol{x}')$.

Representer Theorem (Kimeldorf & Wahba, 1971; Schölkopf & Smola, 2001): Denote by $\Omega : [0, \infty) \to \mathbb{R}$ a strictly monotonic increasing function, by $\mathcal{X}$ a set, and by $c : (\mathcal{X}, \mathbb{R}^2)^m \to \mathbb{R} \cup \{\infty\}$ an arbitrary loss function. Then each minimizer $f \in \mathcal{H}$ of the regularized risk

$$c((\boldsymbol{x}_1, \boldsymbol{y}_1, f(\boldsymbol{x}_1)), \dots, (\boldsymbol{x}_m, \boldsymbol{y}_m, f(\boldsymbol{x}_m))) + \Omega(\|f\|_{\mathcal{H}}) \tag{3.143}$$

admits a representation of the form:

$$f(x) = \sum_{i=1}^{m} \alpha_i k(x_i, x).$$

(3.144)

3.8   Exercises

3.1 Given $\mathbf{Xw} = \mathbf{y}$ where $\mathbf{X} = \begin{bmatrix} 1 & 1 \\ 5 & 6 \end{bmatrix}$, $\mathbf{y} = \begin{bmatrix} 0 \\ 1 \end{bmatrix}$.

   (a) What kind of system is this? (even-, over- or under-determined?)
   (b) Is $\mathbf{X}$ invertible? Why?
   (c) Solve for $\mathbf{w}$ if it is solvable

3.2 The linear system given by $\begin{bmatrix} 1 & 2 \\ 3 & 4 \\ 5 & 6 \end{bmatrix} \begin{bmatrix} w_1 \\ w_2 \end{bmatrix} = \begin{bmatrix} 1 \\ 2 \\ 3 \end{bmatrix}$ is over-determined and thus
has no exact solution. True or False?

3.3 Consider the following system:

$$w_1 + w_2 = 0$$
$$w_1 - w_2 + w_3 - w_4 = 1$$
$$w_3 + w_4 = 2.$$

What kind of system is this? Can the system be solved? Select one or more
from below:

   (a) The system is even-determined and can be solved with an exact solution
   (b) The system is over-determined and can be solved with an approximated
       solution
   (c) The system is over-determined and it does not have an exact solution
   (d) The system is under-determined and can be solved with an exact least norm
       solution
   (e) The system is under-determined and it has infinite number of solutions

3.4 Identify those polynomial models shown below which include at least a second-
order term:

   (a) $1 + x_1 + x_2 + x_1 x_2 - x_1^2 - x_2^2$
   (b) $1 + x_1 + x_2 + x_3 + x_4 + x_5$
   (c) $1 + x_1 + x_2 + x_1 x_2$
   (d) $1 + x_1 + x_2 + x_1(x_1 - x_2)$
   (e) None of these

3.5 Given the input feature x and the corresponding response y as follows:

x	-10	-7	-3	-1	4	8
y	10	11	8	6	3	4

    (a) Perform a third-order polynomial regression and plot the result of line fitting

    (b) Given a test point $x = -5$ predict y using the learned polynomial model

3.6 This question is also about polynomial models. Which of the following models has discriminative feature for the data given by

$$\begin{bmatrix} x_1 \\ x_2 \end{bmatrix} : \begin{bmatrix} -1 \\ -1 \end{bmatrix}, \begin{bmatrix} +1 \\ +1 \end{bmatrix}, \begin{bmatrix} -1 \\ +1 \end{bmatrix}, \begin{bmatrix} +1 \\ -1 \end{bmatrix}$$

with corresponding target values $y \in \{+1, +1, -1, -1\}$?

    (a) $g(\mathbf{x}) = 1 + w_1 x_1 + w_2 x_2$

    (b) $g(\mathbf{x}) = 1 + w_1 x_1 + w_2 x_2 + w_3 x_1 x_2$

    (c) $g(\mathbf{x}) = 1 + w_1 x_1 + w_2 x_2 + w_3 x_1^2$

    (d) $g(\mathbf{x}) = 1 + w_1 x_1 + w_2 x_2 + w_3 x_2^2$

    (d) $g(\mathbf{x}) = 1 + w_1 x_1 + w_2 x_2 + w_3 x 1^2 + w_4 x_2^2$

    (e) None of these.

3.7 This question is also about multivariate polynomials.

    (a) Write down the expression for a third-order polynomial model having a three-dimensional input

    (b) Write down the $\mathbf{P}$ matrix for this polynomial given

$$\mathbf{X} = \begin{bmatrix} x_{11} \ x_{12} \ x_{13} \\ x_{21} \ x_{22} \ x_{23} \end{bmatrix} = \begin{bmatrix} 1 & 0 & 1 \\ 1 & -1 & 1 \end{bmatrix}$$

    (c) Given $\mathbf{y} = \begin{bmatrix} 0 \\ 1 \end{bmatrix}$, can a unique solution be obtained for this under-determined system of $\mathbf{P}$ matrix?

3.8 This question is about the reduced multivariate polynomials.

    (a) Write down the expression for a third-order RM model having a three-dimensional input

    (b) Write down the $\mathbf{P}$ matrix for this polynomial given

$$\mathbf{X} = \begin{bmatrix} x_{11} \ x_{12} \ x_{13} \\ x_{21} \ x_{22} \ x_{23} \end{bmatrix} = \begin{bmatrix} 1 & 0 & 1 \\ 1 & -1 & 1 \end{bmatrix}$$

    (c) Given $\mathbf{y} = \begin{bmatrix} 0 \\ 1 \end{bmatrix}$, can a unique solution be obtained for this under-determined system of $\mathbf{P}$ matrix?

3.9 The two-layer perceptron network with linear output layer can be written in two ways, namely,

$$g(\mathbf{W}_o, \mathbf{W}_h, \mathbf{x}) = \mathbf{W}_o L(\mathbf{W}_h \mathbf{x}), \quad \mathbf{W}_h \in \mathbb{R}^{h \times d}, \ \mathbf{W}_o \in \mathbb{R}^{o \times h},$$

and

$$g(\mathbf{W}_o, \mathbf{W}_h, \mathbf{x}) = L(\mathbf{x}^T \mathbf{W}_h)\mathbf{W}_o, \quad \mathbf{W}_h \in \mathbb{R}^{d \times h}, \ \mathbf{W}_o \in \mathbb{R}^{h \times o},$$

where $L(\cdot) = \max(0, \cdot)$ for $\mathbf{x} \in \mathbb{R}^{d \times 1}$. Let's use the second form for our computation and assume $\mathbf{W}_h$ has been fixed at

$$\mathbf{W}_h = \begin{bmatrix} 2 & 2 & 3 \\ 3 & 0 & 2 \end{bmatrix}.$$

Suppose you are given several training samples $\mathbf{x}$ with corresponding target y:

$$\mathbf{x} : \left\{ \begin{bmatrix} 1 \\ 0 \end{bmatrix}, \begin{bmatrix} 0 \\ 1 \end{bmatrix}, \begin{bmatrix} 0 \\ 0 \end{bmatrix}, \begin{bmatrix} 1 \\ 1 \end{bmatrix} \right\}, \quad y : \{0, 0, 1, 1\}.$$

Can the output weights $\mathbf{W}_o$ be solved deterministically?

3.10 Code a function to build a regressor matrix for the Reduced Multivariate Polynomials Reciprocal Sigmoid features with MATLAB/Octave.

Chapter 4
Learning Score Functions

The previous chapter has introduced several learning models with particular focus on linear parametric form. In this chapter, these learning models will be utilized in a *score metric* to form the *learning objective function*. The score metric is also known as the *loss function* which measures the error of each learning instant. The sum of the learning scores or error losses over a set of training samples forms the learning objective or *cost* function, which is also known as the *learning criterion function*. It can be expressed as $J(\boldsymbol{w})$, where $\boldsymbol{w}$ is the explicit parameter vector to be learned. The objective of learning is to minimize $J(\boldsymbol{w})$ by adjusting $\boldsymbol{w}$:

$$\arg\min_{\boldsymbol{w}} J(\boldsymbol{w}) = \arg\min_{\boldsymbol{w}} \sum_{i=1}^{m} L(g(\boldsymbol{x}_i, \boldsymbol{w}), y_i) + \lambda R(\boldsymbol{w}),$$

where

$$
\begin{aligned}
J(\boldsymbol{w}) &: \text{Learning cost function (optimization criterion)} \\
L(\cdot) &: \text{Loss function (also known as Score function)} \\
g(\boldsymbol{x}_i, \boldsymbol{w}) &: \text{Learning model} \\
\boldsymbol{x}_i &: \text{Learning input vector} \\
\boldsymbol{w} &: \text{Learning parameter vector} \\
y_i &: \text{Learning target} \\
m &: \text{Sample size} \\
R(\boldsymbol{w}) &: \text{Regularization function} \\
\lambda &: \text{Regularization parameter}
\end{aligned}
$$

Three basic types of learning objective or criterion function, namely, the *regression accuracy*, the *classification accuracy*, and the *ranking and operating characteristics* will be presented in this chapter.

© The Author(s), under exclusive license to Springer Nature Singapore Pte Ltd. 2025   65
K.-A. Toh et al., *Analytic Learning Methods for Pattern Recognition*,
https://doi.org/10.1007/978-981-96-2151-4_4

Under the regression accuracy criterion, the *squared error loss* term is frequently utilized to formulate the *sum of squared errors* or *mean squared error* minimization learning criterion. For the classification accuracy criterion, there are several variants with slightly different interpretation. These variants include *correct classification rate, precision, F-measure, misclassification error rate, total error rate*, and *equal error rate*. For criterion related to ranking and operating characteristics, the *receiver operating characteristic* (ROC) curve and the *area under the ROC curve* are formulated. For all the above three types of learning objectives or criteria, a *regularization* term is often added to the *loss* term to stabilize the learning for good prediction.

4.1 Regression Accuracy

The regression accuracy is often expressed in terms of the *mean squared error* (MSE) defined as

$$\text{MSE} : J(\boldsymbol{w}) = \frac{1}{m} \sum_{i=1}^{m} [g(\boldsymbol{x}_i, \boldsymbol{w}) - y_i]^2. \tag{4.1}$$

This criterion measures the mean of the squared errors over the set of samples utilized. The learning score metric or loss function adopted here is the error distance given by $[g(\boldsymbol{x}_i, \boldsymbol{w}) - y_i]^2$. In terms of the minimizer solution, the scaling factor $1/m$ is immaterial where the *sum of squared errors* (SSE) (4.2) can also be used for optimization.

$$\text{SSE} : J(\boldsymbol{w}) = \sum_{i=1}^{m} [g(\boldsymbol{x}_i, \boldsymbol{w}) - y_i]^2. \tag{4.2}$$

Example 4.1 Suppose we have m samples of $\boldsymbol{x} = [x_1, \ x_2, \ldots, \ x_d]^T$ with corresponding target $y \in \mathbb{R}$, then these samples can be indexed as $\boldsymbol{x}_i$ and y_i with $i = 1, \ldots, m$. With inclusion of intercept or bias, the input samples $\boldsymbol{x}_i, i = 1, \ldots, m$ can be packed in matrix form as

$$\mathbf{X} = \begin{bmatrix} 1 & x_{1,1} & x_{1,2} & \cdots & x_{1,d} \\ 1 & x_{2,1} & x_{2,2} & \cdots & x_{2,d} \\ \vdots & & & & \\ 1 & x_{m,1} & x_{m,2} & \cdots & x_{m,d} \end{bmatrix}. \tag{4.3}$$

Correspondingly, the target samples $y_i, i = 1, \ldots, m$ can be packed as a column vector $\mathbf{y} = [y_1, y_2, \ldots, y_m]^T$. The outputs given by the learning model $g(\boldsymbol{x}_i, \boldsymbol{w}), i = 1, \ldots, m$ can also be packed as $\boldsymbol{g} = [g(\boldsymbol{x}_1, \hat{\boldsymbol{w}}), g(\boldsymbol{x}_2, \hat{\boldsymbol{w}}), \ldots, g(\boldsymbol{x}_m, \hat{\boldsymbol{w}})]^T$ for a certain parameter value given by $\hat{\boldsymbol{w}}$. To facilitate finding the best value of $\boldsymbol{w}$ which minimizes the error, the SSE objective (which is expressed using vector notation) can be written as

$$\text{SSE} : J(\boldsymbol{w}) = (\boldsymbol{g} - \mathbf{y})^T (\boldsymbol{g} - \mathbf{y}). \tag{4.4}$$

If $g(x, w)$ is a linear model given by $w^T x$, then the SSE can be written as

$$\text{SSE} : J(w) = (Xw - y)^T (Xw - y). \tag{4.5}$$

More generally, if $g(x, w)$ is a linear model with transformed features, the SSE can be written as

$$\text{SSE} : J(w) = (Pw - y)^T (Pw - y), \tag{4.6}$$

where P is a stacking of transformed sample row vectors $p(x_i) = [1, p(x_1),$ $p(x_2), \ldots, p(x_m)]$ (see, e.g., (3.43) in Example 3.7). These linear formulations in matrix-vector form allow utilization of matrix algebra for solving the minimization problem. $\qquad\square$

4.2 Classification Accuracy

The classification accuracy is different from the regression accuracy in the sense that the former is a discrete counting measure whereas the latter is a distance measure of continuous value. Before going into the several variants of the classification accuracy, we shall begin with the types of classification outcomes. For binary classification of labels "0" and "1", there are four types of classification outcomes, namely, class-0 correctly classified as class-0, class-1 correctly classified as class-1, class-0 incorrectly classified as class-1, and class-1 incorrectly classified as class-0. These four types of classification outcomes form the confusion matrix. In order to quantify each classified sample into one of these four types of outcomes, a classification score function or loss function plays the role of decision-making. These classification loss functions shall be discussed separately below.

4.2.1 Confusion Matrix

Given a training set consisting of m examples $(x_i, y_i), i = 1, 2, \ldots, m$, where $x_i \in \mathbb{R}^d$ denotes the ith feature sample, and $y_i \in \{0, 1\}$ denotes the corresponding category indicator or target label. The value y_i can also be viewed as the class associated with x_i.

For the binary classification problem, our goal is to determine a predictor g based on the models covered in the previous chapter and a threshold τ, utilizing the given training set so that a correct class label prediction can be obtained for unseen data $(x_j, y_j), j = 1, 2, \ldots, n$. An ideal classifier g° is such that $cls(g^\circ(x_j)) = y_j$ for all $j = 1, 2, \ldots, n$, which relates each x_j to its truth target label y_j via the following labeling or scoring function:

$$cls(g(x_j)) = \begin{cases} 1 \text{ if } g(x_j) \geq \tau \\ 0 \text{ if } g(x_j) < \tau \end{cases}, \quad j = 1, 2, \ldots, n. \tag{4.7}$$

Table 4.1 Confusion matrix for binary problems

	Ground truth		
Prediction	P	N	
$\hat{P}$	TP	FP	$\rightarrow$ Positive predictive value (Precision) TP/(TP + FP)
$\hat{N}$	FN	TN	$\rightarrow$ Negative predictive value TN/(FN + TN)
	Sensitivity (Recall) TP/(TP + FN)	Specificity TN/(FP + TN)	Accuracy $\frac{\text{TP+TN}}{\text{TP+TN+FP+FN}}$

For notation convenience, we shall indicate a superscript $+$ or $-$ on the variable according to whether it belongs to class-1 (positive class) or class-0 (negative class), respectively. Among the m learning samples, suppose the samples x_i^+, $i = 1, 2, \ldots, m^+$, belong to the positive class while the samples x_j^-, $j = 1, 2, \ldots, m^-$, belong to the negative class. In other words, m^+ is the number of training samples x_i with label $y_i = 1$ and m^- is the number of training samples x_i with label $y_i = 0$. Similarly, for test data set, n^+ is the number of test samples x_j with label $y_j = 1$ and n^- is the number of test samples x_j with label $y_j = 0$ for $j = 1, 2, \ldots, n$.

Suppose a predictor $\hat{g}$ is chosen (trained), we can choose a threshold τ to classify the test samples into positive class (P) or negative class (N). Based on the given set of test samples, we calculate the following numbers for the chosen threshold τ in (4.7):

$$
\begin{aligned}
\text{TP} &= \text{`number of } x_j \text{ with } y_j = 1 \text{ and } cls(\hat{g}(x_j)) = 1\text{' (truepositive)} \\
\text{FN} &= \text{`number of } x_j \text{ with } y_j = 1 \text{ but } cls(\hat{g}(x_j)) = 0\text{' (falsenegative)} \\
\text{TN} &= \text{`number of } x_j \text{ with } y_j = 0 \text{ and } cls(\hat{g}(x_j)) = 0\text{' (truenegative)} \\
\text{FP} &= \text{`number of } x_j \text{ with } y_j = 0 \text{ but } cls(\hat{g}(x_j)) = 1\text{' (falsepositive)}
\end{aligned}
$$

and form a contingency table called *confusion* matrix (Table 4.1).

The FP and FN are also known as *type-I error* and *type-II error*, respectively. Mathematically, these numbers can be written as

$$
\text{TP} = \sum_{j=1}^{n^+} 1_{g(x_j^+) \geq \tau},
\tag{4.8}
$$

$$
\text{FN} = \sum_{j=1}^{n^+} 1_{g(x_j^+) < \tau},
\tag{4.9}
$$

Table 4.2 Confusion matrix for multi-class problems

	Prediction			
Ground truth	$P_{\hat{1}}$	$P_{\hat{2}}$	$\cdots$	$P_{\hat{C}}$
P_1	$P_{1,\hat{1}}$	$P_{1,\hat{2}}$	$\cdots$	$P_{1,\hat{C}}$
P_2	$P_{2,\hat{1}}$	$P_{2,\hat{2}}$	$\cdots$	$P_{2,\hat{C}}$
$\vdots$	$\vdots$	$\vdots$	$\ddots$	$\vdots$
P_C	$P_{C,\hat{1}}$	$P_{C,\hat{2}}$	$\cdots$	$P_{C,\hat{C}}$

$$\text{TN} = \sum_{j=1}^{n^-} 1_{g(x_j^-)\leqslant \tau}, \tag{4.10}$$

$$\text{FP} = \sum_{j=1}^{n^-} 1_{g(x_j^-)> \tau}, \tag{4.11}$$

where $1_{g(x_j)\geqslant \tau}$ gives a "1" whenever $g(x_j) \geqslant \tau$ and zero otherwise. These numbers can be normalized according to the population size of each class giving the *true positive rate* (TPR $= \text{TP}/n^+$), *true negative rate* (TNR $= \text{TN}/n^-$), *false positive rate* (FPR $= \text{FP}/n^-$), and *false negative rate* (FNR $= \text{FN}/n^+$). Since $n^+ = \text{TP} + \text{FN}$, $n^- = \text{TN} + \text{FP}$ and

$$\text{TPR} + \text{FNR} = \frac{\text{TP}}{n^+} + \frac{\text{FN}}{n^+} = \frac{(\text{TP} + \text{FN})}{n^+} = 1, \tag{4.12}$$

$$\text{TNR} + \text{FPR} = \frac{\text{TN}}{n^-} + \frac{\text{FP}}{n^-} = \frac{(\text{TN} + \text{FP})}{n^-} = 1, \tag{4.13}$$

it is sufficient to describe a classifier's performance using two out of the four rates, one from each of the above two equations. For instance, given TPR and TNR, one can derive FNR and FPR, respectively, based on $(1 - \text{TPR})$ using (4.12) and based on $(1 - \text{TNR})$ using (4.13).

The confusion matrix can be extended to multi-category problems. Suppose the data contains C classes and the number of samples in each class is, respectively, labeled as $P_1, P_2, \ldots, P_C$. Correspondingly, the predictions for these classes are labeled as $P_{\hat{1}}, P_{\hat{2}}, \ldots, P_{\hat{C}}$. Then the corresponding truth-prediction pair can be written as $P_{i,\hat{j}}$ where $i, j \in \{1, 2, \ldots, C\}$. These truth-prediction pairs can be tabulated as a confusion matrix as shown in Table 4.2.

When the P's are used as the number of samples for the corresponding prediction and ground truth, we can read $P_{i,\hat{j}}$ as the number of samples actually belonging to class-i which are predicted as class-j. The numbers in the diagonal entries thus correspond to those correctly predicted samples and they can be expressed as the complements of sum of off-diagonal entries in each row:

$$P_{i,\hat{i}} = P_i - \sum_{\hat{j}=1,\hat{j}\neq i}^{C} P_{i,\hat{j}}. \tag{4.14}$$

Similarly, for summation within each column, the following holds:

$$P_{\hat{j},\hat{j}} = P_{\hat{j}} - \sum_{i=1,i\neq\hat{j}}^{C} P_{i,\hat{j}}. \tag{4.15}$$

In terms of recognition rates or probability $p_{i,\hat{j}}$ which sums to 1, $\forall i$, (4.14) can be rewritten as

$$p_{i,\hat{i}} = 1 - \sum_{\hat{j}=1,\hat{j}\neq i}^{C} p_{i,\hat{j}}. \tag{4.16}$$

4.2.2 *Accuracy, Precision, and F-Measure*

In binary classification, the above rates define several frequently used performance measures such as those given in Table 4.3, where s is a skewing factor which is frequently taken to be the ratio between the size of negative samples and the size of positive samples (i.e., $s = n^-/n^+$).

Generally, *accuracy* refers to the degree of closeness of a measured value to its true value. *Precision*, on another hand, refers to the reproducibility of the measurements. In other words, precision is about how close the measured values are to each other. In regression problems, accuracy measures the distance between the true value and the mean of the measured values (see Fig. 4.1). In a two-dimensional representation, Fig. 4.2 illustrates the difference between a system with high accuracy but low precision and another system with high precision but low accuracy. In binary classification, accuracy is commonly adopted with $s = n^-/n^+$ as

$$Accuracy = \frac{TP + TN}{TP + TN + FP + FN} = \frac{TP + TN}{n^+ + n^-}. \tag{4.17}$$

The true positive rate (TPR) is also termed *sensitivity* or *recall* and the true negative rate (TNR) is also called *specificity*. Recall can be treated as a measure of completeness and precision can be treated as a measure of fidelity. In medical diagnostics, test sensitivity refers to the ability of a test to correctly identify those with the disease, and test specificity refers to the ability of the test to correctly identify those without the disease. In documents retrieval, precision is the fraction of the documents retrieved that are relevant to the user's information need and recall is the fraction of the documents that are relevant to the query being successfully retrieved. The F-measure is the harmonic mean of precision and recall.

4.3   Error Metrics and Loss Functions

4.3.1   Misclassification Error

Apart from the accuracy-related measures defined above, the performance of a classification system can also be described using the error measure. A commonly adopted error score function for misclassification is a counting function given by

$$\text{MCE}(g, y) = 1_{(g \neq y)}. \tag{4.18}$$

This *misclassification error* (MCE) or total misclassification counts, when normalized by the number of samples, is a dual of accuracy in two-category problems when $s = n^-/n^+$ in $\frac{TPR + s \cdot TNR}{1+s}$, i.e.:

$$\frac{\text{MCE}}{n} = 1 - Accuracy|_{s=n^-/n^+} = \frac{\text{FP} + \text{FN}}{\text{TP} + \text{TN} + \text{FP} + \text{FN}}. \tag{4.19}$$

For multi-category problems, the misclassification error is the sum of all off-diagonal elements in Table 4.2. The misclassification error rate is the misclassification error normalized by the total sample size.

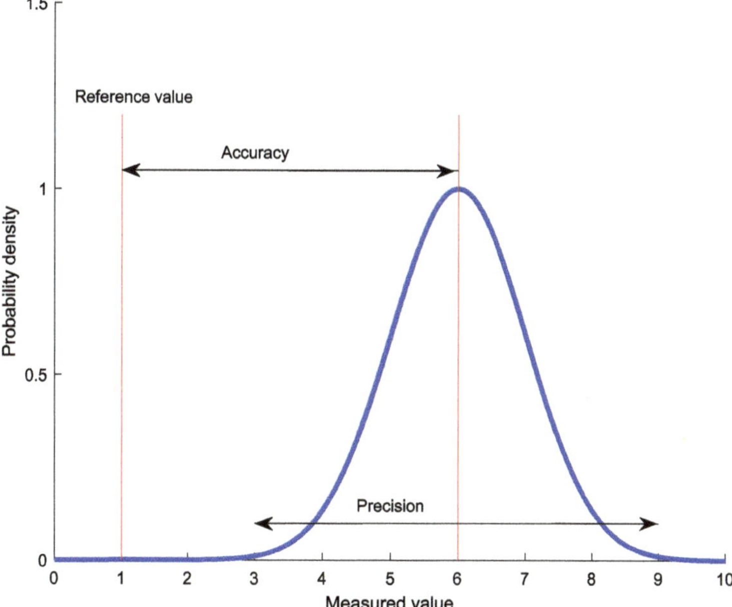

Fig. 4.1 Accuracy and precision in one-dimensional data. The horizontal axis represents the measured value and the vertical axis represents the probability density

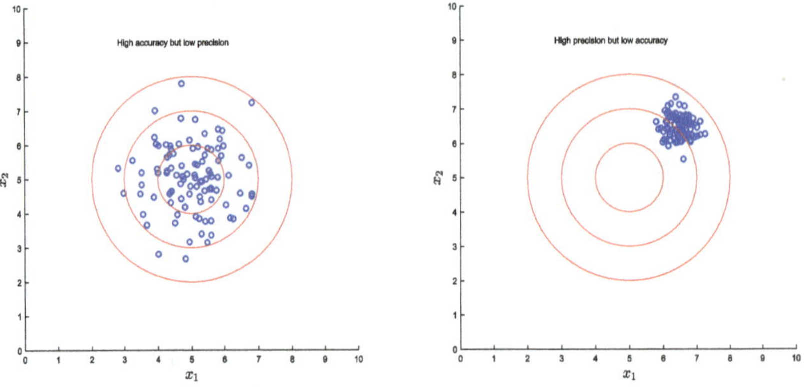

(a) High accuracy but low precision   (b) High precision but low accuracy

Fig. 4.2 Accuracy and precision in two-dimensional data

Table 4.3 Some performance measures for binary classifiers

Performance measures	Definition	$s = 1$	$s = n^-/n^+$
True positive rate	$\text{TPR} = \frac{\text{TP}}{n^+}$	–	–
True negative rate	$\text{TNR} = \frac{\text{TN}}{n^-}$	–	–
Accuracy	$\frac{\text{TPR}+s\cdot TNR}{1+s}$	$\frac{\text{TPR}+\text{TNR}}{2}$	$\frac{\text{TP}+\text{TN}}{n^++n^-}$
Precision	$\frac{\text{TPR}}{\text{TPR}+s\cdot\text{FPR}}$	$\frac{\text{TPR}}{\text{TPR}+\text{FPR}}$	$\frac{\text{TP}}{\text{TP}+\text{FP}}$
F-measure	$\frac{2\text{TPR}}{\text{TPR}+s\cdot\text{FPR}+1}$	$\frac{2\text{TPR}}{\text{TPR}+\text{FPR}+1}$	$\frac{2\text{TP}}{\text{TP}+\text{FP}+n^+}$

4.3.2 Total Error Rate

In binary classification, when the normalization is performed with respect to each
pattern class, an error rate known as the *total error rate* can be defined as follows:

$$\text{TER} = \text{FPR} + \text{FNR}. \tag{4.20}$$

Alternatively, the half total error rate $\text{HTER} = (\text{FPR} + \text{FNR})/2$ can be adopted when
a normalization between the two categories is considered.

As seen from Table 4.3, HTER is a *dual* of *accuracy* with a unit skewing factor
$s = 1$:

$$
\begin{aligned}
Accuracy &= \frac{\text{TPR} + \text{TNR}}{2} = \frac{1}{2}[(1 - \text{FNR}) + (1 - \text{FPR})] \\
&= \frac{1}{2}(2 - \text{TER}) = 1 - \text{HTER}.
\end{aligned}
\tag{4.21}
$$

When $s = n^-/n^+$,

$$Accuracy = \frac{TP + TN}{n^+ + n^-} = \frac{n^+ \left(\frac{TP}{n^+}\right) + n^- \left(\frac{TN}{n^-}\right)}{n^+ + n^-}$$

$$= \frac{n^+}{n}(1 - FNR) + \frac{n^-}{n}(1 - FPR). \tag{4.22}$$

This shows that accuracy at $s = n^-/n^+$ is related to weighted error rates.

4.3.3 Equal Error Rate

So far we have been relying on a predetermined threshold value to compute the error- and accuracy-related measures. Here, we introduce an error measure defined by a threshold derived from the data. Without loss of generality, suppose the output of binary predictor $\hat{g}$ is normalized within [0, 1]. Then, we can vary the threshold τ within [0, 1] to generate a series of error rates. Figure 4.3 illustrates a normalized plot showing those corresponding areas to the rates of FP, FN, TP, and TN obtained at a threshold value τ for two overlapping distributions of prediction output. By varying $\tau \in [0, 1]$, these rates (shaded areas) can vary from 0 to 100%. Given two overlapping distributions of prediction output with the mean of positive class higher than the mean of negative class as illustrated in Fig. 4.3, a typical FPR plot shows a decreasing trend while a FNR plot shows an increasing trend. Figure 4.4a shows a plot for the FPR and the FNR over variation of threshold τ. In this plot, the intersection point of FPR and FRR curves is called the *equal error rate* (EER).

In case of continuous distributions (also called density functions), both FPR and FNR can be expressed as a smooth function of τ, i.e., FPR(τ) and FNR(τ). Here there exists a threshold τ^* such that FPR(τ^*) = FNR(τ^*) where EER =FPR(τ^*) or EER = FNR(τ^*). For discrete distributions, FPR(τ^*) may not be equal to FNR(τ^*) and hence an approximation is often needed for EER value computation. The approximation can be obtained by averaging the four points around the tacit intersection point (see Fig. 4.4b).

Example 4.2 Given the target values $\mathbf{y} = [1, 0, 0, 1, 1, 1, 0, 0, 0, 0]^T$ and the predicted target $\hat{\mathbf{y}} = [0.1, 0.3, 0.1, 0.3, 0.6, 0.9, 0.2, 0.3, 0.8, 0.9]^T$ by a classifier. The EER can be computed in MATLAB as follows:

```
y = [1,0,0,1,1,1,0,0,0,0]; % 1 represents genuine-user 0 impostor

y_est = [0.1,0.3,0.1,0.3,0.6,0.9,0.2,0.3,0.8,0.9];

Equal_Error_Rate = EER(y,y_est)


function Err = EER(y,y_est)

    % Positive/Genuine and Negative/Impostor samples

    yp = y_est(y==1); yn = y_est(y==0);

    Np = length(yp); Nn = length(yn);

    % Use the smaller as threshold
```

```
if (Np < Nn) th = sort(yp);
else th = sort(yn); end
% Compute FAR and FRR
for k = 1:length(th)
    FA = (yn - th(k));
    FR = (th(k) - yp);
    FAR(k) = 100*length(find(FA>0))/Nn;
    FRR(k) = 100*length(find(FR>=0))/Np;
    Diff(k) = abs(FAR(k) - FRR(k));
end
[value, Indx] = min(Diff);
Threshold = th(Indx);
FAR_soln = FAR(Indx);
FRR_soln = FRR(Indx);
Err = (FAR_soln+FRR_soln)/2;
return
```

4.3.4 Loss Functions

Apart from the distance error loss given by $[g(\boldsymbol{x}_i, \boldsymbol{w}) - y_i]^2$ in regression and the binary counting loss given by (4.7) which is also known as "0–1" loss, several other types of loss function are also available for classification. We shall observe the relationship among these loss functions in this subsection.

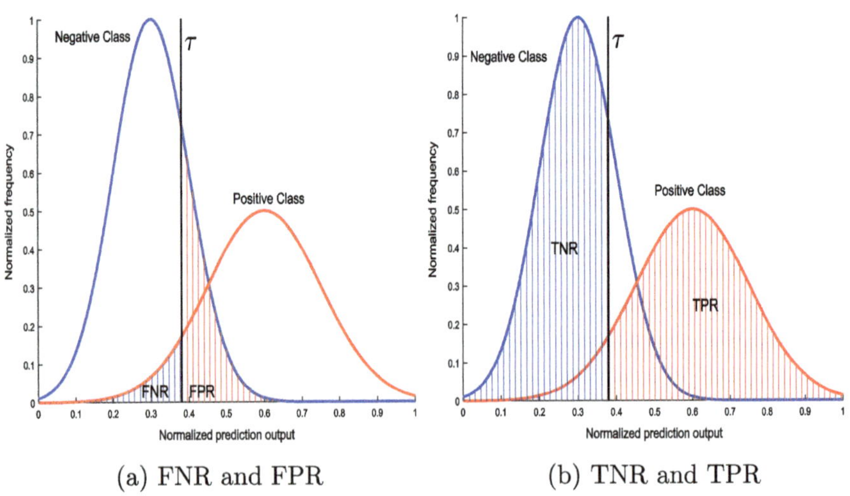

(a) FNR and FPR   (b) TNR and TPR

Fig. 4.3 Normalized distributions of binary prediction outputs

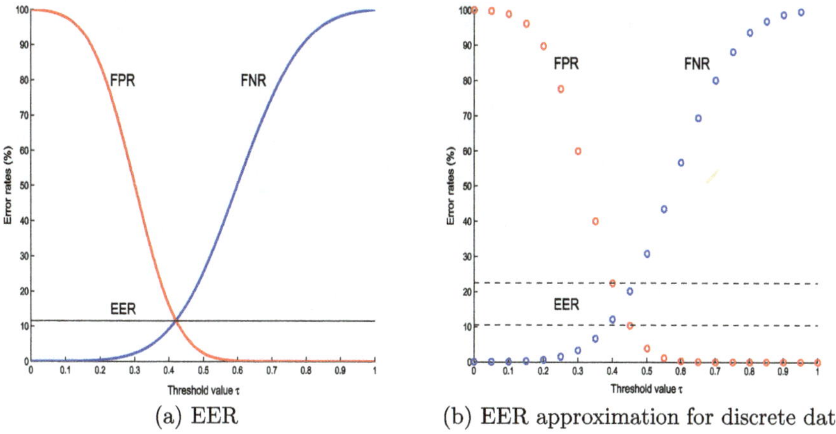

(a) EER   (b) EER approximation for discrete data

Fig. 4.4 Error rates versus threshold

Suppose we have a set of binary classification learning samples $\{x_i, y_i\}_{i=1}^m$ with $y_i \in \{-1, +1\}$. Let $g(x_i) \in \mathbb{R}$ be the prediction output of the classifier that we built. Then, based on the *margin* given by $y_i g(x_i)$, a learning *loss* term indicating the success or failure of classifying each sample can be defined as

$$L_{margin}(y_i, x_i) = y_i g(x_i), \quad i = 1, \ldots, m, \tag{4.23}$$

where a positive margin with y_i and $g(x_i)$ having the same sign indicates a correct classification and a negative margin indicates an incorrect classification. Here, we note that an implicit zero threshold has been assumed between the positive and negative values.

The loss term can be applied to both the training set and the testing set depending on whether to use it for training or to use it for test error evaluation. While omitting the sample indexing for simplicity, the "0–1" loss function can be written in terms of the margin as

$$L_{01}(y, g(x)) = \begin{cases} 0 \text{ if } yg(x) \geqslant 0 \\ 1 \text{ if } yg(x) < 0. \end{cases} \tag{4.24}$$

In linear regression, a *squared* loss term such as $L_2(x) = (g(x) - y)^2 = (y - g(x))^2$ is often adopted. For classification with $y \in \{-1, 1\}$, the squared loss can be expressed in terms of margin as

$$\begin{aligned} L_2(y, g(x)) &= (g(x) - y)^2 \\ &= (g(x))^2 - 2yg(x) + y^2 \\ &= 1 + (g(x))^2 - 2yg(x). \end{aligned} \tag{4.25}$$

Frequently, the squared loss function is used in the sum of squared errors to form the minimization objective function (4.26):

$$J(\boldsymbol{w}) = \sum_{i=1}^{m} (g(\boldsymbol{x}_i) - y_i)^2. \tag{4.26}$$

To deal with the case of different costs associated with different misclassifications, a weighting factor ω_i (Lowe & Webb, 1990; Bishop, 1995) can be incorporated for each training sample giving a weighted sum of squares:

$$J(\boldsymbol{w}) = \sum_{i=1}^{m} \sum_{k=1}^{c} \omega_i (g_k(\boldsymbol{x}_i) - y_{k,i})^2. \tag{4.27}$$

Apart from the square loss, another popular choice of loss term for classification is the *hinge* loss adopted in the Support Vector Machines (SVM):

$$L_{hinge}(yg(\boldsymbol{x})) = \max(0, 1 - yg(\boldsymbol{x})). \tag{4.28}$$

This hinge loss, sometimes expressed as $L_{hinge}(m) = |1 - yg(\boldsymbol{x})|_+$, can be considered a partial version of the *absolute value* loss given by $L_{abs}(\boldsymbol{x}) = |g(\boldsymbol{x}) - y|$ where the negative output values are clipped at zero.

Expressed in terms of the margin $\mu = yg(\boldsymbol{x})$ (we use a simpler notation for the margin here), the *logistic loss* can be written as

$$L_{logistic}(\mu) = \log(1 + e^{-\mu}) = \log(1 + e^{-yg(\boldsymbol{x})}), \tag{4.29}$$

or

$$L_{logistic}(\mu) = (\log 2)^{-1} \log(1 + e^{-\mu}), \tag{4.30}$$

where the constant term $(\log 2)^{-1}$ ensures that all losses for classification equal 1 for $g(\boldsymbol{x}) = 0$. The logistic loss is used in the LogitBoost algorithm. A variant of the above, the *exponential loss* which is used in the AdaBoost algorithm, is given by

$$L_{exp}(\mu) = \exp(-\mu) = \exp(-yg(\boldsymbol{x})). \tag{4.31}$$

In *linear logistic regression*, a linear parametric model is used for fitting the *log-odds* function. For instance, fitting the linear model $g(\boldsymbol{x}) = \alpha_0 + \boldsymbol{w}^T \boldsymbol{x}$ to the *log-odds* gives rise to

$$\begin{aligned} \log(p/(1-p)) &= g(\boldsymbol{x}) \\ \Rightarrow \frac{p}{1-p} &= e^{g(\boldsymbol{x})}, \end{aligned} \tag{4.32}$$

where the probability distribution of the model output response is given by

$$p = \sigma(g(\boldsymbol{x})) = \frac{1}{1 + e^{-g(\boldsymbol{x})}} = \frac{e^{g(\boldsymbol{x})}}{1 + e^{g(\boldsymbol{x})}}. \tag{4.33}$$

The term $\sigma(\cdot) = \frac{1}{1+e^{-\cdot}}$ is known as the *logistic function* or *sigmoid function*. The *binary log-likelihood* for maximization is defined as

$$L_{log-likelihood}(p) = y \log p + (1 - y) \log(1 - p) \tag{4.34}$$

with its dual form for minimization as *binary cross-entropy*:

$$L_{cross-entropy}(p) = -y \log p - (1 - y) \log(1 - p). \tag{4.35}$$

For multi-category problems with C classes, the multi-class cross-entropy is defined as

$$L_{cross-entropy}(p) = -\sum_{i=1}^{C} y_i \log(p_i), \tag{4.36}$$

where p_i is the prediction probability of the ith class.

Figure 4.5 shows a plot for several loss functions which can be expressed in terms of the margin.

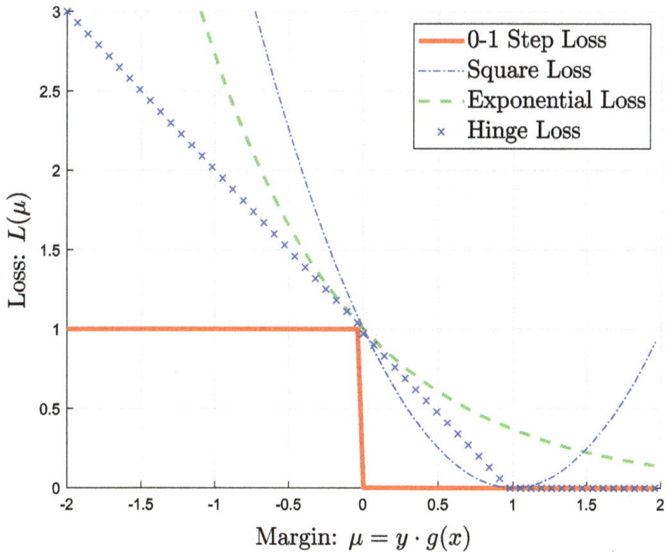

Fig. 4.5 Loss functions: margin-loss plot

4.3.5 Regularization Term

The regularization term in a regression or classification cost or objective function can be considered a loss term which accounts for penalizing the size of the parameters. In general, such a cost function can be written as composing of two loss components, namely, $L(\cdot)$ and $R(\cdot)$ as follows:

$$J(w) = \sum_{i=1}^{m} L(\mu_i(w)) + \lambda R(w), \tag{4.37}$$

where λ is a scalar regularization weighting factor, w denotes the adjustable weight parameter vector of the classifier g, and μ_i is the margin-loss term for each sample given by $y_i g(x_i, w)$. While a focused discussion of penalized learning shall be provided in Chap. 6, a brief coverage of the regularization loss is presented here.

The regularization term in (4.37) plays an important role in obtaining a stable classification solution in the sense that it controls the "magnitude" of the solution. In other words, a minimization of (4.37) is a joint minimization of both the loss term and the regularization term which is related to the "magnitude" of parametric solution.

The following lists some popular regularization terms:

$$R_2 = \frac{1}{2}\|w\|_2^2, \qquad \text{ridge regression,} \tag{4.38}$$

$$R_1 = \sum_i |w_i|, \qquad \text{lasso,} \tag{4.39}$$

$$R_0 = |\{i : w_i \neq 0\}|, \qquad \text{subset selection.} \tag{4.40}$$

In general, the above regularization terms can be interpreted as members of the generic family:

$$R_q = \left(\sum_i |w_i|^q\right)^{\frac{1}{q}}. \tag{4.41}$$

According to (4.41), a square root of R_2 is obtained for $q = 2$. However, for computational convenience, it is often easier to work with the norm squared as shown in (4.38). Next, R_0 is obtained when q approaches 0. Figure 4.6 shows a single-dimensional plot for R_1 and R_2. Figure 4.7 provides several examples of regularization functions for a two-dimensional parameter space.

As seen from Fig. 4.7, the impact of various regularization terms ($R_{0.1}$, $R_{0.5}$, R_1, R_2, R_3, R_{10}) on classifier learning can be understood from the different distortions of parameter space during optimization. A low-order regularization (e.g., $R_{0.1}$, $R_{0.5}$,

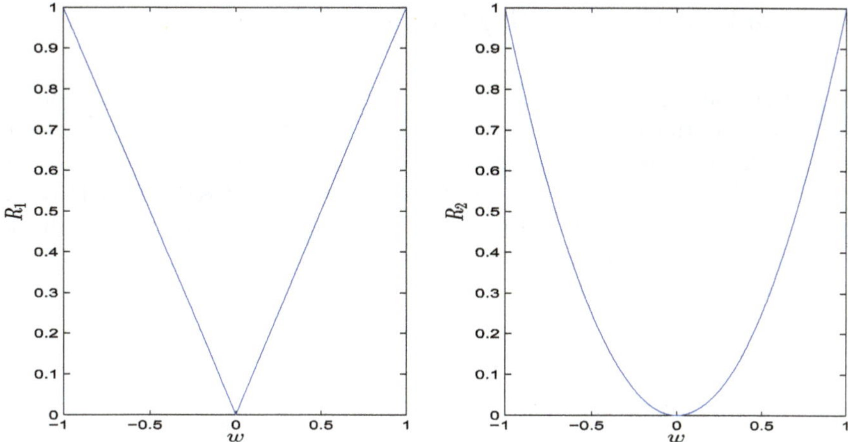

Fig. 4.6 Regularization functions (R_1 and R_2) for one-dimensional w

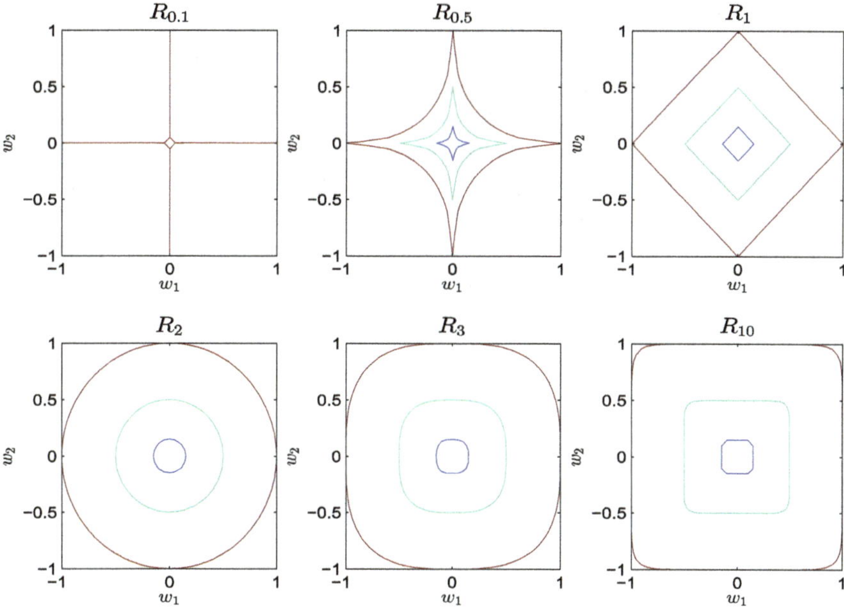

Fig. 4.7 Regularization functions ($R_{0.1}$, $R_{0.5}$, R_1, R_2, R_3, R_{10}) for two-dimensional $\boldsymbol{w}$. The outermost contour indicates one unit distance from the origin

and R_1) tends to ignore dimension interaction while a high-order regularization (e.g., R_2, R_3, and R_{10}) tends to capitalize on dimension interaction.

4.4 Ranking and Operating Characteristics

Apart from the above types of cost functions which measure either the magnitude or distance of error for regression or the count of correctly or incorrectly classified samples, there is another type of performance criterion which is based on the ranking order of comparison. The *receiver operating characteristic* curve provides such a measure of ranking order for every operating threshold of the classifier. In the following, several classification cost functions are presented with the relationship among them.

4.4.1 Receiver Operating Characteristics

Instead of plotting two separate curves, the $FNR(\tau)$ mentioned in Sect. 4.3.3 can be plotted over the $FPR(\tau)$ for the entire operating range of threshold τ values to obtain a single curve (Fig. 4.8a) in binary classification. This plot is called a *receiver operating characteristic* (ROC) curve. Here, the more bending is the curve is toward the origin, the better is the performance of a classifier. Alternatively, $TPR(\tau) = 1 - FNR(\tau)$ can be plotted over $FPR(\tau)$ (Fig. 4.8b) where the more bow is the curve toward the upper left corner, the better is the performance of a classifier.

Some graphs use a nonlinearly scaled (such as logarithmic scaled) x- and y-axes on Fig. 4.8a to highlight differences of importance in the critical operating region. This yields trade-off curves that appear more "linear" than ROC curves and they are called *detection error trade-off* (DET) curves. When the decision threshold and related decision parameters are not well defined, the ROC or DET offers a relevant tool for assessment of the overall performance of a classifier.

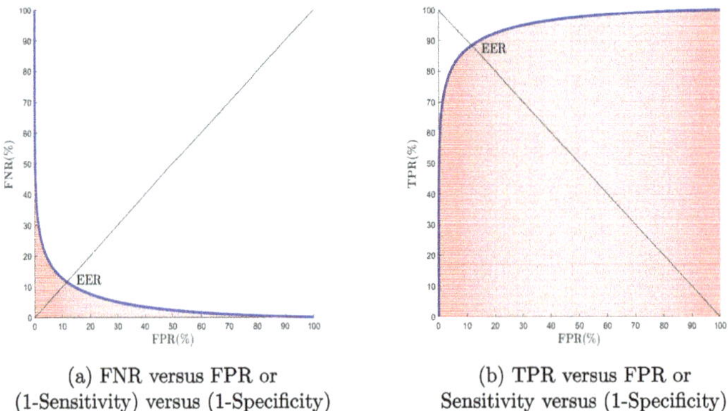

(a) FNR versus FPR or
(1-Sensitivity) versus (1-Specificity)   (b) TPR versus FPR or
Sensitivity versus (1-Specificity)

Fig. 4.8 ROC curves

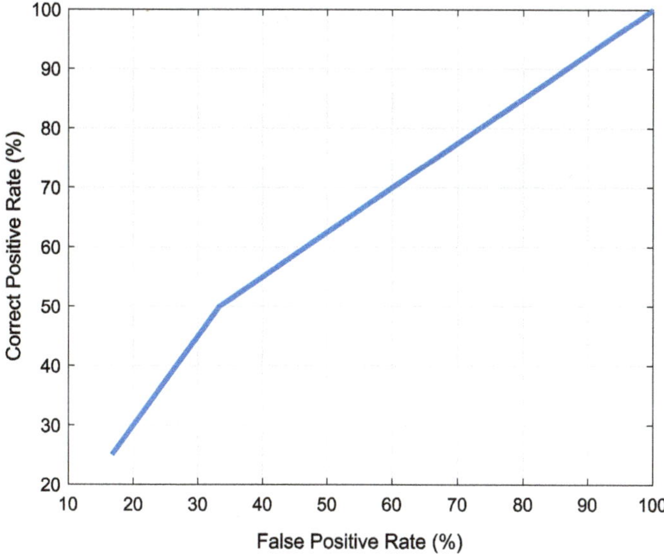

Fig. 4.9 The ROC curve for a small number of samples

Example 4.3 Consider a classification task with an imbalance data set. For example, in biometrics, the given sample size for the genuine-user category (positive class) is frequently found to be much smaller than the size of the impostor category (negative class). Since the ROC is a plot of FNR over FPR, the bottleneck of the plot's resolution is the resolution of FNR. In other words, the resolution of an ROC plot is determined by the smaller size data set among the two categories, that is, the genuine-user data set. The scores from the genuine-user data set can then be sorted in either ascending or descending order for threshold setting. Suppose a high score indicates a genuine match and a low score indicates an impostor match. Then the FNR obtained at a certain threshold value based on a certain genuine-user score accounts for those genuine-user scores falling below this threshold (the adopted genuine-user value). On the other hand, the FPR accounts for those impostor scores falling above the threshold. The codes below show an example of MATLAB codes for generating such a plot. Here we note that when the samples for both the genuine users and impostors are large (e.g., more than thousands), then a semilog plot would be useful to show a higher resolution on the low FPR side (Fig. 4.9).

```
% Plot ROC
y = [1,0,0,1,1,1,0,0,0,0]; % 1 represents genuine-user 0 impostor
y_est = [0.1,0.3,0.1,0.3,0.6,0.9,0.2,0.3,0.8,0.9];
genuine = y_est(y==1);
imposter = y_est(y==0);
length_gen = length(genuine); length_imp = length(imposter);
FP=zeros(1,length_gen); FN=zeros(1,length_gen);
genuinesorted=sort(genuine); % use smaller genuines as thresholds
for i=1:length_gen
    for j=1:length_imp
```

```
        if imposter(j)>=genuinesorted(i) FP(i)=FP(i)+1; end
    end
    for k=1:length_gen
        if genuine(k)<genuinesorted(i) FN(i)=FN(i)+1; end
    end
end
FPR=100*FP/length_imp; FNR=100*FN/length_gen;
figure, plot(FPR,100-FNR)
%figure, semilogx(FPR,100-FNR)
xlabel('False Positive Rate (%)')
ylabel('Correct Positive Rate (%)')
% End plot
```

$\square$

4.4.2 Area Under the ROC

Since the ROC provides a range of performance values instead of a single value, the Area Under the ROC Curve (AUC) is a natural choice as the performance measure for overall classifier comparison. The AUC can be viewed as the probability that the score of the positive class observation is greater than that of the negative class observation. In the discrete case, this probability (and hence AUC) can be computed via the fraction of positive-negative pairs that are ranked correctly.

Consider a set of binary training data indexed by $i = 1, 2, \ldots, m^+$ for positive class and $j = 1, 2, \ldots, m^-$ for negative class. Let us denote $\xi_{ij} = g(x_i^+) - g(x_j^-)$ which is the difference in the value of the classifier between a positive sample x_i^+ and a negative sample x_j^-. Thus, $\xi_{ij} > 0$ indicates that the positive sample x_i^+ and the negative sample x_j^- are correctly ranked by the classifier g. Consider also a Heaviside step function given by

$$u(\xi) = \begin{cases} 1, & \text{if } \xi > 0, \\ 0.5 & \text{if } \xi = 0, \\ 0, & \text{if } \xi < 0. \end{cases} \tag{4.42}$$

The AUC for m training examples can then be expressed as

$$\text{AUC} = \frac{1}{m^+ m^-} \sum_{i=1}^{m^+} \sum_{j=1}^{m^-} u(\xi_{ij}) \, , \tag{4.43}$$

which is the fraction of positive-negative pairs that are ranked correctly by classifier g. In other words, what matters for AUC is the score order but not the score value itself. The AUC is equal to one when all the samples are ranked correctly. Conversely, AUC$= 0$ when all the samples are ranked incorrectly. The AUC value in (4.43) is also known as the Wilcoxon-Mann-Whitney statistic.

Here, we note that some approaches approximated the computation of the step function as follows:

$$u'(\xi) = \begin{cases} 1, & \text{if } \xi > 0, \\ 0, & \text{otherwise.} \end{cases} \tag{4.44}$$

The shaded region in Fig. 4.8a shows the *area above the ROC* (AAC) curve and the shaded region in Fig. 4.8b shows the *area under the ROC* (AUC) curve. The AAC is the *dual* of the AUC since AAC$= 1-$AUC.

Example 4.4 Suppose we are given the target values $\mathbf{y} = [1, 1, 0, 1, 1, 1, 0, 0]^T$ and their estimated values $\hat{\mathbf{y}} = [1.1, 0.8, 0.1, 0.3, 0.6, 0.9, 0.2, 0.3]^T$ from a certain classifier. The AUC of this learned result can be computed in MATLAB as follows:

```matlab
y = [1,1,0,1,1,1,0,0];
y_est = [1.1,0.8,0.1,0.25,0.6,0.9,0.2,0.3];
Area = AUC(y,y_est)


function Area = AUC(y, y_est)
% y_est: raw prediction output values (not categorized)
    if (min(y)==0) % y\in{0,1}
        y_est(y_est>=0.5) = 1; % categorized into +ve class
        y_est(y_est<0.5) = -1; % categorized into -ve class
        yp = y_est(y==1); % +ve class samples
        yn = y_est(y==0); % -ve class samples
    else % y\in{-1,+1}
        y_est(y_est>=0) = 1; % categorized into +ve class
        y_est(y_est<0) = -1; % categorized into -ve class
        yp = y_est(y==1); % +ve class samples
        yn = y_est(y==-1); % -ve class samples
    end
    Np = length(yp); Nn = length(yn);
    A = 0;
    for i = 1:Np
        for j = 1:Nn
            A = A + max(0,sign(yp(i)-yn(j)));
        end
    end
    Area = A/(Np*Nn)
return
```

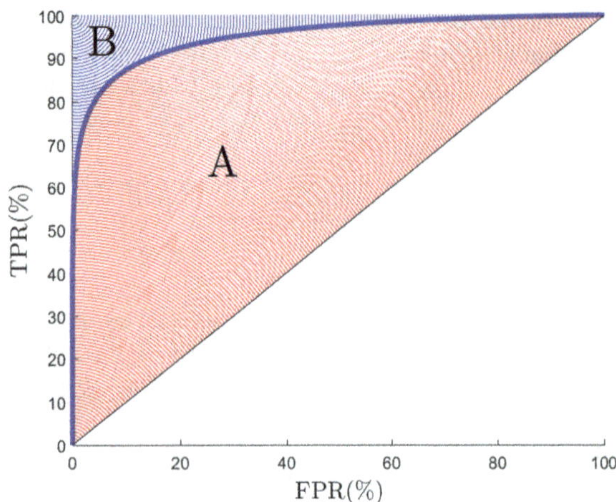

Fig. 4.10 Sub-areas on the ROC plot

4.4.3 *Lorenz Curve and Gini Coefficient*

In economics, the Lorenz curve is a graphical plot to represent the distribution of wealth. Derived from the Lorenz curve, the Gini coefficient, which is also called the Gini index, is a linear re-scaling of AUC given by $2 \times \text{AUC} - 1$. As shown in Fig. 4.10, the Gini coefficient is the ratio of areas given by G=A/(A+B). Since (A+B)= 0.5, then half of this area ratio (i.e., A itself and not divided by (A+B)) is equal to AUC−0.5. In economic and econometric terms, a Gini coefficient of zero (AUC= 0.5) means that income is evenly distributed. This can be interpreted as having a non-discriminative classifier when AUC= 0.5.

4.4.4 *Relationships Among MSE, MCE, TER, and AUC*

The MSE, MCE, TER, and AUC cost functions for training are listed together below for comparison:

$$\text{MSE} : J(\boldsymbol{w}) = \frac{1}{m} \sum_{i=1}^{m} (g(\boldsymbol{x}_i, \boldsymbol{w}) - y_i)^2, \quad \text{(regression)}, \tag{4.45}$$

$$\text{MCE} : J(\boldsymbol{w}) = \frac{1}{m} \sum_{i=1}^{m} (u(y_i g(\boldsymbol{x}_i, \boldsymbol{w})) - 1)^2, \quad \text{(classification)}, \tag{4.46}$$

$$\text{TER} : J(\boldsymbol{w}) = \frac{1}{m^-} \sum_{i=1}^{m^-} u\big(g(\boldsymbol{x}_i^-, \boldsymbol{w}) - \tau\big) + \frac{1}{m^+} \sum_{i=1}^{m^+} u\big(\tau - g(\boldsymbol{x}_i^+, \boldsymbol{w})\big),$$

$$\text{(weighted classification)}, \qquad (4.47)$$

$$\text{AUC} : J(\boldsymbol{w}) = \frac{1}{m^+ m^-} \sum_{i=1}^{m^+} \sum_{j=1}^{m^-} u\big(g(\boldsymbol{x}_i^+, \boldsymbol{w}) - g(\boldsymbol{x}_j^-, \boldsymbol{w})\big),$$

$$\text{(classification ranking order)}, \qquad (4.48)$$

where τ is a decision threshold and

$$u(\xi) = \begin{cases} 1, & \text{if } \xi > 0, \\ 0.5 & \text{if } \xi = 0, \\ 0, & \text{if } \xi < 0. \end{cases} \qquad (4.49)$$

Noting that the squared error term $(g(\boldsymbol{x}_i, \boldsymbol{w}) - y_i)^2$ can be written in terms of the margin $\mu = y_i g(\boldsymbol{x}_i, \boldsymbol{w})$ as $(y_i g(\boldsymbol{x}_i, \boldsymbol{w}) - 1)^2$ when $y_i \in \{-1, +1\}$, we see that the MCE (misclassification error) is related to MSE (mean squared error) in terms of the utilization of a binary quantized margin term $u(y_i g(\boldsymbol{x}_i, \boldsymbol{w}))$. The TER (total error rate) is a class-specific version with $s = 1$ of the Accuracy (Acc) (recall the skewing factor s in Sect. 4.2.2). The TER is a special case of the AUC when a fixed reference is adopted for each class (e.g., by alternatingly treating one of the terms $g(\boldsymbol{x}_i^+, \boldsymbol{w})$ and $g(\boldsymbol{x}_j^-, \boldsymbol{w})$ as a fixed reference threshold) in the double summation of the AUC (Area Under the ROC) counting.

4.5 Bibliographic Notes

In 1905, Max O. Lorenz proposed a visual method, now called the Lorenz curve, for measuring the concentration of income. In the original version, the cumulated percentage of the population number (ranking from poorest to richest) was plotted over the percentage of total income (Lorenz, 1905). When the income is distributed unequally, the curve will be bent in the middle. There were several versions of the Lorenz curve (Derobert & Thieriot, 2003; Kleiber, 2008) where Gini (1914) showed that these various representations were merely different ways of expressing the same phenomenon (Derobert & Thieriot, 2003).

The area between the Lorenz curve and the diagonal line is called Gini coefficient, Gini index, or Gini concentration ratio (Flach, 2003; Santos & Javier Busto Guerrero, 2010; Giorgi, 1990; Giorgi, 2005). The Gini coefficient is a measure of inequality in a population. A Gini coefficient of zero means perfect equality whereas a Gini coefficient of one means maximal inequality. The relation between AUC and Gini index is given by $G = 2 \times \text{AUC} - 1$ (Flach, 2010).

Based on the aggregated percentage populations obtained from thresholding two distributions, the Receiver Operating Characteristic (ROC) curve shows certain resemblance to the Lorenz curve which is based on a single distribution but using two attributes, namely, percentage of population and percentage of wealth or total income. The Lorenz curve is a popular tool for statistical distribution analysis in economics and econometrics (Kleiber, 2008) while the ROC curve is commonly used in analysis related to signal detection which can be dated back to World War II where it was used for analysis of radar signals (Flach, 2010).

The ROC soon found its use in psychology, medicine, biomedical informatics, radiology, biometrics, and other areas related to machine learning and data mining research after World War II. In psychology, the ROC was used to account for perceptual detection of stimuli (Swets, 1996). In clinical medicine and biomedical informatics, the ROC curve was used for evaluating classification and prediction models for decision support, diagnosis, and prognosis (Zweig & Campbell, 1993; Lasko et al., 2005; Zou et al., 2007) of a laboratory test. In radiology, the ROC curve was used as a measure of the accuracy of a diagnostic test (Obuchowski, 2003). In biometrics, the system performance at all operating threshold values was assessed using the ROC curve (Jain et al., 2002; 2004; 2011).

In continuous observations, the AUC can be expressed as the probability that the score of the positive class observation is greater than that of the negative class observation (Bamber, 1975). The AUC can also be considered a normalized form of the Wilcoxon-Mann-Whitney sum of rank test (see Chap. 9) (Mason & Graham, 2002). In multibiometrics, the area under the ROC curve (AUC) was adopted as an optimizing objective for fusion classifier design (Toh et al., 2008b).

4.6 Exercises

4.1 What is the difference between a regression loss and a classification loss?
4.2 What is the difference between a loss function and an error function?
4.3 What is the difference between a loss function and a cost function?
4.4 What are the common loss functions for regression and that for classification?
4.5 The plot below shows the decision boundary of Classifier A. Write down the confusion matrix for Classifier A below.

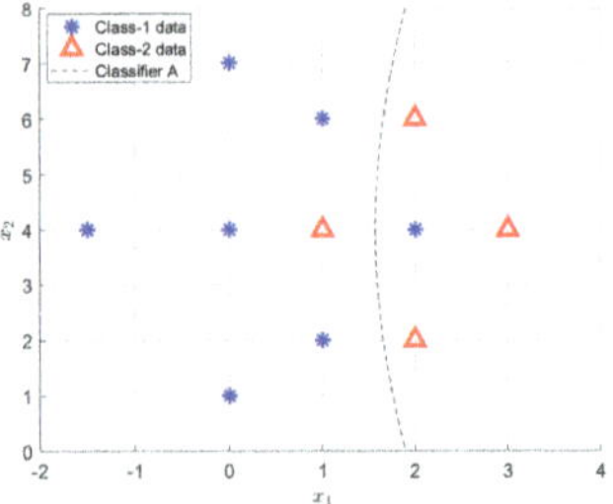

4.6 Consider the experimental observations of the training error rate (Tr) and the validation error rate (Va) for a machine learning algorithm for each setting of the parameter θ. Choose a parameter (P) based on Tr and Va.

θ	Tr	Va
0.1	0.08	0.23
0.2	0.33	0.38
0.3	0.21	0.18
0.4	0.13	0.26
0.5	0.16	0.18

Which value of parameter θ will you choose based on the above observation?

4.7 Consider the binary classification problem, which you are dealing with, has highly imbalanced classes. The majority class has 99 hundred samples and the minority class has 1 hundred samples. Which of the following metric(s) would you choose for assessing the classification performance? (select all relevant metric(s))

(a) Accuracy
(b) Cost-sensitive accuracy
(c) Precision and recall
(d) Type-I and type-II errors
(d) None of these

4.8 According to the plots of Figs. 4.8b and 4.10 (reproduced below), what is the relationship between the Gini coefficient and the Area Under the ROC (AUC)?

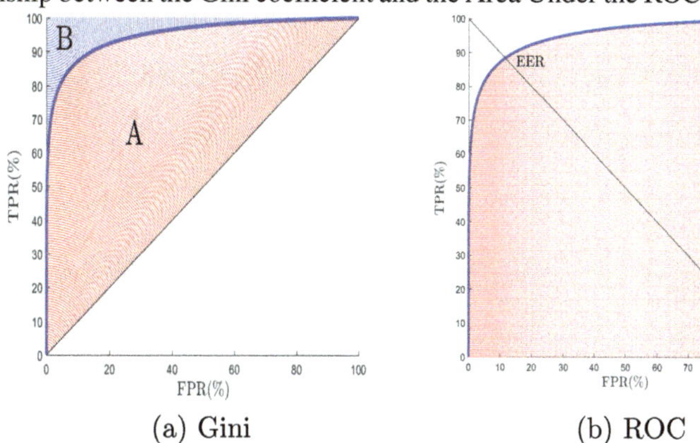

(a) Gini (b) ROC

4.9 Given the following confusion matrix for a binary classifier:

	Class($+1$)	Class(-1)
Predicted Class($+1$)	990	20
Predicted Class(-1)	10	380

Let n^+ and n^- denote the numbers of Class($+1$) and Class(-1) samples, respectively. Calculate the

(a) True positive rate
(b) True negative rate
(c) Accuracy at $s = n^-/n^+$ (this is the most widely adopted accuracy)
(d) Accuracy at $s = 1$
(e) Precision at $s = n^-/n^+$

4.10 The Misclassification Error (MCE) is the count of samples classified incorrectly. Suppose the binary targets are labeled as $\{-1, +1\}$ and $g(x)$ denotes the predictor output. Express the MCE in terms of the margin $y_i g(x_i)$.

Chapter 5
Analytic Learning

The main goal of learning is to make use of the available training samples, to form an inference engine for prediction. The underlying assumption is thus hinged upon the fidelity of unseen data's distribution with respect to the training data's distribution. In general, those learning score functions discussed in Chap. 4 can be adopted to form the cost or criterion functions for regression predictor or classifier learning. We shall focus on several foundation analytic learning and estimation methods in this chapter.

5.1 Regression Error-Based Learning Methods

By utilizing the square of the difference between the learner and the target as the loss function, the *error magnitude-based learning* method can be applied to both regression and classification problems. Under this approach, it is the learning target y that differentiates between regression and classification applications. When y has continuous real values such as $y \in \mathbb{R}$, the learning is for regression. When y has discrete values such as $y \in \{0, 1\}$ or $y \in \{-1, +1\}$, the learning is for classification. Apart from whether the learning target is continuous or discrete, the classification, in contrast to regression, has an additional step of target labeling or encoding for prediction. A typical labeling or encoding process is to apply a decision threshold to convert an estimated $\hat{y}$ value from the continuous form into a discrete form so that $\hat{y} \in \{0, 1\}$ or $\hat{y} \in \{-1, +1\}$. For multiple outputs, the one-hot encoding can be adopted. In the sequel, while working on discrete target values for classification, we note that the learning process applies equally well to target with continuous values for regression.

© The Author(s), under exclusive license to Springer Nature Singapore Pte Ltd. 2025   89
K.-A. Toh et al., *Analytic Learning Methods for Pattern Recognition*,
https://doi.org/10.1007/978-981-96-2151-4_5

5.1.1 SSE Minimization for Over-Determined Systems

For both regression and classification tasks, formulations based on the single output and the multiple outputs can be utilized for learning. Learning of a single output target with discrete binary labels such as $y \in \{0, 1\}$ or $y \in \{-1, +1\}$ corresponds to binary classifier learning. Learning of multiple outputs with one-hot encoded labels corresponds to multi-category learning.

Binary Learning: Consider a binary or two-category classification problem. Given a training data set which consists of m examples (x_i, y_i), $i = 1, \ldots, m$, where $x_i \in \mathbb{R}^d$ denotes the ith sample of the feature vector, and $y_i \in \{0, 1\}$ (or $y_i \in \{-1, +1\}$) denotes the corresponding *class indicator* or *target label*. In other words, the value y_i can be viewed as the class associated with x_i. Utilizing this training set, the goal is to determine a predictor $g(x, \hat{w})$ at a parameter or weight setting $\hat{w}^1$ and a threshold value $\hat{\tau}$ such that a correct class prediction can be obtained for novel feature samples x_j not found in the training set. An ideal classifier is such that $cls(g(x_j, \hat{w}) \geqslant \tau) = y_j{}^2$ for all $j = 1, 2, \ldots, n$, which relates each sample of unseen data x_j to its true target label y_j.

The well-known *Least Squares* (LS) method for predictor learning formulates the problem as to minimize a *Sum of Squared Errors* (SSE) term by adjusting the weighting parameters (packed as vector w) which correspond to the importance of the feature terms (packed as vector x). For the given set of training data containing m samples and the predictor $g(x_i, w)$ which operates on data x_i with an adjustable w vector, the SSE for this training set is defined as

$$J_{\text{SSE}}(w) = \sum_{i=1}^{m} (y_i - g(x_i, w))^2 . \tag{5.1}$$

Here, we pay particular attention to prediction models which can be written in the generalized linear parametric form

$$g(x_i, w) = w^T p(x_i), \quad i = 1, 2, \ldots, m, \tag{5.2}$$

where $p(x_i) = [1, p_1(x_i), \ldots, p_D(x_i)]^T$ is any linear or nonlinear mapping of x_i and $w \in \mathbb{R}^{D+1}$ is the corresponding weight coefficients vector in this case. In a compact matrix-vector notation, Eq. (5.1), which adopts the linear parametric model (5.2), can be written as

$$J_{\text{SSE}}(w) = \|\mathbf{y} - \mathbf{P}w\|_2^2 = (\mathbf{y} - \mathbf{P}w)^T (\mathbf{y} - \mathbf{P}w), \tag{5.3}$$

where

---

[1] As mentioned in Chap. 3, $\hat{w}$ indicates a solution point in the space of parameter vector $w \in \mathbb{R}^d$.

[2] Suppose $y_i \in \{0, 1\}$, then $cls(g(x_j, \hat{w}) \geqslant \tau) = \begin{cases} 1 \text{ if } g(x_j, \hat{w}) \geqslant \tau \\ 0 \text{ else} \end{cases}$, where $\tau = 0.5$ is a common choice.

$$\mathbf{y} = \begin{bmatrix} y_1 \\ y_2 \\ \vdots \\ y_m \end{bmatrix}, \mathbf{P} = \begin{bmatrix} 1 & p_1(\mathbf{x}_1) & \cdots & p_D(\mathbf{x}_1) \\ 1 & p_1(\mathbf{x}_2) & \cdots & p_D(\mathbf{x}_2) \\ \vdots & \vdots & \ddots & \vdots \\ 1 & p_1(\mathbf{x}_m) & \cdots & p_D(\mathbf{x}_m) \end{bmatrix}, \mathbf{w} = \begin{bmatrix} w_0 \\ w_1 \\ \vdots \\ w_D \end{bmatrix}. \tag{5.4}$$

Here, each row of $\mathbf{P}$ constitutes an augmented feature sample and this sample corresponds to a target element in $\mathbf{y}$. The augmentation by means of a bias or intercept term corresponds to the 1's added to facilitate translation of the learned output toward an affine transformation (i.e., being able to move away from the origin).

A necessary condition for minimization of $J_{\mathrm{SSE}}(\mathbf{w})$ in (5.3) requires that its first-order derivative with respect to the adjustable parameter vector be put to zero, i.e., $\nabla_{\mathbf{w}} J_{\mathrm{SSE}}(\mathbf{w}) = \mathbf{0}$, leading to the normal equation given by

$$-2\mathbf{P}^T(\mathbf{y} - \mathbf{P}\mathbf{w}) = \mathbf{0}. \tag{5.5}$$

For an over-determined system (i.e., $m > D + 1$) with independent observations such that $(\mathbf{P}^T\mathbf{P})$ is non-singular, (5.5) admits a unique solution given by

$$\hat{\mathbf{w}} = (\mathbf{P}^T\mathbf{P})^{-1}\mathbf{P}^T\mathbf{y}. \tag{5.6}$$

As mentioned, the notation $\hat{\mathbf{w}}$ in (5.6) indicates that $\hat{\mathbf{w}}$ is a fixed point determined by values in matrix $\mathbf{P}$ and vector $\mathbf{y}$. Since the quadratic objective function (5.1) or (5.3) is convex, (5.6) is an optimal solution point.

The training regressor vectors $\mathbf{p}(\mathbf{x}_i)$, $i = 1, 2, \ldots, m$ when projected onto the solution vector $\hat{\mathbf{w}}$, yields the trained outputs

$$g(\mathbf{x}_i, \hat{\mathbf{w}}) = \hat{\mathbf{w}}^T \mathbf{p}(\mathbf{x}_i), \ i = 1, 2, \ldots, m. \tag{5.7}$$

The entire set of training input features can be denoted by a stacked matrix as $\mathbf{X} = [\mathbf{x}_1, \mathbf{x}_2, \ldots \mathbf{x}_m]^T$ and its transformed feature matrix corresponds to $\mathbf{P}$ in (5.4) where the trained outputs can be obtained in stacked vector form as

$$\mathbf{g}(\mathbf{X}, \hat{\mathbf{w}}) = \mathbf{P}\hat{\mathbf{w}}. \tag{5.8}$$

For testing or prediction of unseen data given by $\mathbf{x}_j$, $j = 1, 2, \ldots, n$, the prediction output can be obtained by projecting $\mathbf{p}(\mathbf{x}_j)$ onto the learned $\hat{\mathbf{w}}$ giving

$$g(\mathbf{x}_j, \hat{\mathbf{w}}) = \hat{\mathbf{w}}^T \mathbf{p}(\mathbf{x}_j), \ j = 1, 2, \ldots, n. \tag{5.9}$$

The stacked form of the above formulation is given by

$$\mathbf{g}(\mathbf{X}_t, \hat{\mathbf{w}}) = \mathbf{P}_t\hat{\mathbf{w}}. \tag{5.10}$$

where the unseen test data set x_j, $j = 1, 2, \ldots, n$ has been transformed and stacked in $\mathbf{P}_t$ as the test feature matrix. For classification, a threshold process (such as utilizing the $cls(\cdot)$ function) is followed.

Multi-Category Learning: Consider a problem with C-categories. Then, the above learning can be extended to train a concatenation of C number of two-category predictors given by

$$\mathbf{G} = \mathbf{PW}, \tag{5.11}$$

where

$$
\begin{aligned}
\mathbf{G} &= [g(x, w_1), \; g(x, w_2), \ldots, \; g(x, w_C)] \\
&= \begin{bmatrix}
g(x_1, w_1) & g(x_1, w_2) & \cdots & g(x_1, w_C) \\
g(x_2, w_1) & g(x_2, w_2) & \cdots & g(x_2, w_C) \\
\vdots & & \ddots & \vdots \\
g(x_m, w_1) & g(x_m, w_2) & \cdots & g(x_m, w_C)
\end{bmatrix},
\end{aligned}
\tag{5.12}
$$

with each of its column corresponds to learning a category label, and

$$
\begin{aligned}
\mathbf{W} &= [w_1, \; w_2, \ldots, \; w_C] \\
&= \begin{bmatrix}
w_{0,1} & w_{0,2} & \cdots & w_{0,C} \\
w_{1,1} & w_{1,2} & \cdots & w_{1,C} \\
\vdots & & \ddots & \vdots \\
w_{D,1} & w_{D,2} & \cdots & w_{D,C}
\end{bmatrix},
\end{aligned}
\tag{5.13}
$$

is a concatenation of each category's weight parameter vector. The feature matrix $\mathbf{P}$ is similar to that in (5.4). Here, the SSE learning objective is generalized as

$$
\begin{aligned}
J_{\mathrm{SSE}}(\mathbf{W}) &= \mathrm{trace}\{(\mathbf{Y} - \mathbf{G})^T (\mathbf{Y} - \mathbf{G})\} \\
&= \mathrm{trace}\{(\mathbf{Y} - \mathbf{PW})^T (\mathbf{Y} - \mathbf{PW})\},
\end{aligned}
\tag{5.14}
$$

where the trace of a square matrix is defined to be the sum of elements on the main diagonal, and $\mathbf{Y}$ is a target indicator matrix given by

$$
\mathbf{Y} = \begin{bmatrix} \mathbf{Y}_1 \\ \mathbf{Y}_2 \\ \vdots \\ \mathbf{Y}_C \end{bmatrix} = \left. \begin{matrix} \overbrace{\begin{bmatrix} 1\,0\,\cdots\,0 \\ 1\,0\,\cdots\,0 \\ \vdots\,\vdots\,\cdots\,\vdots \\ 0\,1\,\cdots\,0 \\ 0\,1\,\cdots\,0 \\ \vdots\qquad\vdots \\ 0\,0\,\ddots\,1 \\ 0\,0\,\cdots\,1 \end{bmatrix}}^{C} \begin{matrix} \left.\vphantom{\begin{matrix}1\\1\end{matrix}}\right\}\mathbf{Y}_1 \\ \left.\vphantom{\begin{matrix}1\\1\end{matrix}}\right\}\mathbf{Y}_2 \\ \vdots \\ \left.\vphantom{\begin{matrix}1\\1\end{matrix}}\right\}\mathbf{Y}_C \end{matrix} \end{matrix}\right\} m. \tag{5.15}
$$

In this illustration, $\mathbf{Y}$ consists of stacks of block matrices $\mathbf{Y}_k$, $k = 1, 2, \ldots, C$ corresponding to each class and all of the elements of $\mathbf{Y}_k$ are zero except for those in the kth column, which are unity. In general, each sample row of $\mathbf{P}$ drawn from a particular category just need to be labeled by the corresponding sample row of $\mathbf{Y}$. In other words, only row correspondence is needed between $\mathbf{P}$ and $\mathbf{Y}$ and no sorting of samples to form stacked blocks is required.

Let $\mathbf{E} = \mathbf{Y} - \mathbf{PW}$, then

$$
\begin{aligned}
\mathrm{trace}\{(\mathbf{Y} - \mathbf{PW})^T (\mathbf{Y} - \mathbf{PW})\} &= \mathrm{trace}\{\mathbf{E}^T \mathbf{E}\} \\
&= \mathrm{trace}\left(\begin{bmatrix} \mathbf{e}_1^T \\ \vdots \\ \mathbf{e}_C^T \end{bmatrix} [\mathbf{e}_1, \ldots, \mathbf{e}_C] \right) \\
&= \sum_{k=1}^{C} \mathbf{e}_k^T \mathbf{e}_k,
\end{aligned}
\tag{5.16}
$$

where $\mathbf{e}_k = (\mathbf{y} - \mathbf{P}\boldsymbol{w}) \in \mathbb{R}^{m \times 1}$. Since the solution for each $\mathbf{e}_k^T \mathbf{e}_k$ is $\hat{\boldsymbol{w}}_k = (\mathbf{P}^T \mathbf{P})^{-1} \mathbf{P}^T \mathbf{y}_k$ for $k = 1, \ldots, C$ when $\mathbf{P}^T \mathbf{P}$ is non-singular, they can be stacked together as

$$
[\hat{\boldsymbol{w}}_1, \hat{\boldsymbol{w}}_2, \ldots, \hat{\boldsymbol{w}}_C] = (\mathbf{P}^T \mathbf{P})^{-1} \mathbf{P}^T [\mathbf{y}_1, \mathbf{y}_2, \ldots, \mathbf{y}_C]. \tag{5.17}
$$

In other words, the solution for minimizing (5.14) can be packed as

$$
\hat{\mathbf{W}} = (\mathbf{P}^T \mathbf{P})^{-1} \mathbf{P}^T \mathbf{Y}, \tag{5.18}
$$

where $\hat{\mathbf{W}} = [\hat{\boldsymbol{w}}_1, \hat{\boldsymbol{w}}_2, \ldots, \hat{\boldsymbol{w}}_C]$. The training and test outputs of the predictor are respectively obtained as

$$
\hat{\mathbf{G}} = \mathbf{P}\hat{\mathbf{W}} \tag{5.19}
$$

and

$$
\hat{\mathbf{G}}_t = \mathbf{P}_t \hat{\mathbf{W}}. \tag{5.20}
$$

For classification, a decision-making process (see the multi-category case in Sect. 5.1.7) is followed.

Example 5.1(a)

Given the following samples of data features (x_1, x_2) and their corresponding target outputs (y_1, y_2, y_3) for training.

x_1:	−5.5	−4.5	−3.3	−2.5	−1.3	−0.5	0.2	1.5	2.8	3.3
x_2:	6.8	5.2	4.6	3.2	2.8	2.1	1.9	1.6	1.3	0.8
y_1:	1	1	1	1	0	0	0	0	0	0
y_2:	0	0	0	0	1	1	1	1	0	0
y_3:	0	0	0	0	0	0	0	0	1	1

We want to predict the outputs of $(x_1, x_2) = (-1, 3.3)$ and $(x_1, x_2) = (2.5, 0.5)$ using a full polynomial model of second order.

The first step is to pack the data in matrix form:

$$\mathbf{X} = \begin{bmatrix} -5.5 & 6.8 \\ -4.5 & 5.2 \\ -3.3 & 4.6 \\ -2.5 & 3.2 \\ -1.3 & 2.8 \\ -0.5 & 2.1 \\ 0.2 & 1.9 \\ 1.5 & 1.6 \\ 2.8 & 1.3 \\ 3.3 & 0.8 \end{bmatrix}, \quad \mathbf{Y} = \begin{bmatrix} 1 & 0 & 0 \\ 1 & 0 & 0 \\ 1 & 0 & 0 \\ 1 & 0 & 0 \\ 0 & 1 & 0 \\ 0 & 1 & 0 \\ 0 & 1 & 0 \\ 0 & 1 & 0 \\ 0 & 0 & 1 \\ 0 & 0 & 1 \end{bmatrix}.$$

With the second-order polynomial expansion given by $\boldsymbol{p} = [1, x_1, x_2, x_1^2, x_1 x_2, x_2^2]$, the data matrix $\mathbf{X}$ is next transformed into the polynomial matrix given by

$$\mathbf{P} = \begin{bmatrix} 1.0000 & -5.5000 & 6.8000 & 30.2500 & -37.4000 & 46.2400 \\ 1.0000 & -4.5000 & 5.2000 & 20.2500 & -23.4000 & 27.0400 \\ 1.0000 & -3.3000 & 4.6000 & 10.8900 & -15.1800 & 21.1600 \\ 1.0000 & -2.5000 & 3.2000 & 6.2500 & -8.0000 & 10.2400 \\ 1.0000 & -1.3000 & 2.8000 & 1.6900 & -3.6400 & 7.8400 \\ 1.0000 & -0.5000 & 2.1000 & 0.2500 & -1.0500 & 4.4100 \\ 1.0000 & 0.2000 & 1.9000 & 0.0400 & 0.3800 & 3.6100 \\ 1.0000 & 1.5000 & 1.6000 & 2.2500 & 2.4000 & 2.5600 \\ 1.0000 & 2.8000 & 1.3000 & 7.8400 & 3.6400 & 1.6900 \\ 1.0000 & 3.3000 & 0.8000 & 10.8900 & 2.6400 & 0.6400 \end{bmatrix}.$$

The system is over-determined since there are more samples than features. The training is performed using (5.18) to give

$$\hat{\mathbf{W}} = (\mathbf{P}^T\mathbf{P})^{-1}\mathbf{P}^T\mathbf{Y} = \begin{bmatrix} 0.2999 & -0.3268 & 1.0268 \\ -0.8781 & 1.8294 & -0.9512 \\ -0.4796 & 1.6369 & -1.1572 \\ 0.1717 & -0.3934 & 0.2217 \\ 0.3572 & -0.8690 & 0.5118 \\ 0.1579 & -0.4619 & 0.3040 \end{bmatrix}.$$

The prediction of the two test samples, with transformed feature matrix $\mathbf{P}_t$, is thus

$$\hat{\mathbf{G}} = \mathbf{P}_t\hat{\mathbf{W}} = \begin{bmatrix} 0.3080 & 0.6892 & 0.0028 \\ -0.5763 & 1.4049 & 0.1714 \end{bmatrix}.$$

MATLAB codes:

```
% Training data
X = [-5.5, 6.8; -4.5, 5.2; -3.3, 4.6; -2.5, 3.2; -1.3, 2.8; -0.5, 2.1;
     0.2, 1.9; 1.5, 1.6; 2.8, 1.3; 3.3, 0.8];
Y = [1 0 0; 1 0 0; 1 0 0; 1 0 0; 0 1 0; 0 1 0; 0 1 0; 0 1 0; 0 0 1; 0 0 1];
% Test data
Xt =  [-1.0, 3.3; 2.5, 0.5];
Yt = [0 1 0; 0 0 1];
% Generate polynomial matrices using X and Xt
P = [ones(10,1), X(:,1), X(:,2), X(:,1).^2, X(:,1).*X(:,2), X(:,2).^2];
Pt = [ones(2,1), Xt(:,1), Xt(:,2), Xt(:,1).^2, Xt(:,1).*Xt(:,2), Xt(:,2).^2];
% Training
W = inv(P'*P)*P'*Y;
% Prediction
Yts_est = Pt*W
```

$\square$

5.1.2 SSE Minimization for Under-Determined Systems

Binary Learning: The above LS solution (5.6) often works well for systems with more data samples than the number of parameters, i.e., it is useful for over-determined systems with $m > D + 1$. For the case of under-determined systems with $m < D + 1$, the covariance matrix $\mathbf{P}^T\mathbf{P} \in \mathbb{R}^{(D+1)\times(D+1)}$ is singular since $\mathbf{P}$ has at most m independent rows. In other words, the rank of $\mathbf{P}$ is at most m and this leads to difficulty in estimating w using (5.6) or (5.18).

A practical way to handle the situation is by restricting the feasible solution set to limit the effective size of parameters to less than or equal to the data sample size m. This can be achieved by mapping the original parameters from the larger $\mathbb{R}^{D+1}$ space onto the smaller $\mathbb{R}^m$ space using

$$w = \mathbf{P}^T \alpha, \tag{5.21}$$

where $\alpha \in \mathbb{R}^m$. This mapping of w can be substituted into (5.3) where the minimization problem

$$\min_{w} J_{\text{SSE}}(w) = \min_{w}(\mathbf{y} - \mathbf{P}w)^T(\mathbf{y} - \mathbf{P}w) \tag{5.22}$$

is modified to adjust α instead of w as

$$\min_{\alpha} J_{\text{SSE}}(\alpha) = \min_{\alpha}(\mathbf{y} - \mathbf{P}\mathbf{P}^T \alpha)^T(\mathbf{y} - \mathbf{P}\mathbf{P}^T \alpha). \tag{5.23}$$

By taking the first partial derivative of $J_{\text{SSE}}(\alpha)$ (5.23) with respect to α and put it to zero gives

$$- \mathbf{P}\mathbf{P}^T(\mathbf{y} - \mathbf{P}\mathbf{P}^T \alpha) = \mathbf{0}$$
$$\Rightarrow \quad \hat{\alpha} = (\mathbf{P}\mathbf{P}^T)^{-1}\mathbf{y}, \tag{5.24}$$

when $\mathbf{P}\mathbf{P}^T$ has full rank. By substituting (5.24) back into the subspace mapping (5.21) gives

$$\hat{w} = \mathbf{P}^T(\mathbf{P}\mathbf{P}^T)^{-1}\mathbf{y}. \tag{5.25}$$

This solution is similar to the least norm solution which solves

$$\min_{w} \|w\|_2^2 \quad \text{subject to } \mathbf{y} - \mathbf{P}w = \mathbf{0}, \tag{5.26}$$

or

$$\min_{w} \|w\|_2^2 + (\mathbf{y} - \mathbf{P}w)^T \alpha, \tag{5.27}$$

where taking the first derivative with respect to w implies

$$2w - \mathbf{P}^T \alpha = \mathbf{0}$$
$$w = \frac{1}{2}\mathbf{P}^T \alpha, \tag{5.28}$$

that can be substituted into $\mathbf{y} = \mathbf{P}w$ to give

$$\mathbf{y} = \mathbf{P}\left(\frac{1}{2}\mathbf{P}^T \alpha\right)$$
$$\Rightarrow \quad \hat{\alpha} = 2(\mathbf{P}\mathbf{P}^T)^{-1}\mathbf{y}, \tag{5.29}$$

and finally putting (5.29) back into (5.28) gives (5.25).

Multi-Category Learning: For multi-category problems, the learning objective can be stacked similar to that in (5.14) where the solution for each of the two-category problems (i.e., class-k and non-class-k, $k = 1, \ldots, C$) can be stacked as

$$\hat{\mathbf{W}} = \mathbf{P}^T (\mathbf{P}\mathbf{P}^T)^{-1}\mathbf{Y}, \tag{5.30}$$

where $\mathbf{Y}$ is the target indicator matrix defined in (5.15) and $\mathbf{P}$ is the regressor matrix defined in (5.4) with full rank. Here, each column of $\hat{\mathbf{W}}$ as defined in (5.13) corresponds to the learning weights for each two-category problem with elements belonging to either class-k or non-class-k for $k = 1, \ldots, C$.

Example 5.2(a)

Given the following samples of data features (x_1, x_2) and their corresponding target outputs (y_1, y_2, y_3) for training.

x_1:	-3.3	-2.5	-1.3	0.2	2.8	3.3
x_2:	4.6	3.2	2.8	1.9	1.3	0.8
y_1:	1	1	0	0	0	0
y_2:	0	0	1	1	0	0
y_3:	0	0	0	0	1	1

We want to predict the outputs of $(x_1, x_2) = (-1, 3.3)$ and $(x_1, x_2) = (2.5, 0.5)$ using a full polynomial model of third order.

Similar to the previous example, the first step is to pack the data in matrix form:

$$\mathbf{X} = \begin{bmatrix} -3.3 & 4.6 \\ -2.5 & 3.2 \\ -1.3 & 2.8 \\ 0.2 & 1.9 \\ 2.8 & 1.3 \\ 3.3 & 0.8 \end{bmatrix}, \quad \mathbf{Y} = \begin{bmatrix} 1 & 0 & 0 \\ 1 & 0 & 0 \\ 0 & 1 & 0 \\ 0 & 1 & 0 \\ 0 & 0 & 1 \\ 0 & 0 & 1 \end{bmatrix}.$$

With the third-order polynomial expansion given by $\boldsymbol{p} = [1, x_1, x_2, x_1^2, x_1 x_2, x_2^2, x_1^3, x_1^2 x_2, x_1 x_2^2, x_2^3]$, the data matrix $\mathbf{X}$ is next transformed into the polynomial matrix given by

$$\mathbf{P} = \begin{bmatrix} 1 & -3.3 & 4.60e{+}00 & 1.089e{+}01 & -1.518e{+}01 & 2.116e{+}01 & -3.5937e{+}01 & 5.0094e{+}01 & -6.9828e{+}01 & 9.7336e{+}01 \\ 1 & -2.5 & 3.20e{+}00 & 6.250e{+}00 & -8.000e{+}00 & 1.024e{+}01 & -1.5625e{+}01 & 2.0000e{+}01 & -2.5600e{+}01 & 3.2768e{+}01 \\ 1 & -1.3 & 2.80e{+}00 & 1.690e{+}00 & -3.640e{+}00 & 7.840e{+}00 & -2.1970e{+}00 & 4.7320e{+}00 & -1.0192e{+}01 & 2.1952e{+}01 \\ 1 & 0.2 & 1.90e{+}00 & 4.000e{-}02 & 3.800e{-}01 & 3.610e{+}00 & 8.0000e{-}03 & 7.6000e{-}02 & 7.2200e{-}01 & 6.8590e{+}00 \\ 1 & 2.8 & 1.30e{+}00 & 7.840e{+}00 & 3.640e{+}00 & 1.690e{+}00 & 2.1952e{+}01 & 1.0192e{+}01 & 4.7320e{+}00 & 2.1970e{+}00 \\ 1 & 3.3 & 8.00e{-}01 & 1.089e{+}01 & 2.640e{+}00 & 6.400e{-}01 & 3.5937e{+}01 & 8.7120e{+}00 & 2.1120e{+}00 & 5.1200e{-}01 \end{bmatrix}.$$

The system is under-determined since there are less samples than features. The training is performed using (5.30) to give

$$\hat{\mathbf{W}} = \mathbf{P}^T (\mathbf{P}\mathbf{P}^T)^{-1}\mathbf{Y} = \begin{bmatrix} 0.09713828 & 0.20075237 & -0.06491357 \\ 0.07577593 & 0.19742008 & -0.06843 \\ 0.09630898 & 0.18836854 & -0.05122601 \\ 0.09042469 & 0.08425601 & -0.02269605 \\ 0.03381779 & 0.16158149 & -0.04495802 \\ 0.05589557 & 0.09667595 & -0.00549766 \\ -0.03486167 & -0.0274169 & 0.02569465 \\ -0.0112531 & -0.14726259 & 0.0703693 \\ -0.07041111 & -0.06834405 & 0.0724417 \\ -0.06729266 & 0.00712169 & 0.02273085 \end{bmatrix}.$$

The prediction of the two test samples with transformed feature matrix $\mathbf{P}_t$ is thus

$$\hat{\mathbf{G}} = \mathbf{P}_t \hat{\mathbf{W}} = \begin{bmatrix} -0.72648499 & 1.77165404 & 0.13432329 \\ 0.32307727 & 0.60922216 & 0.20907129 \end{bmatrix}.$$

Python codes:

```python
import numpy as np
from numpy.linalg import inv
from sklearn.preprocessing import PolynomialFeatures
X = np.array([[-3.3,4.6],[-2.5,3.2],[-1.3,2.8],[0.2,1.9],[2.8,1.3],[3.3,0.8]])
Y = np.array([[1,0,0],[1,0,0],[0,1,0],[0,1,0],[0,0,1],[0,0,1]])
Xt =  np.array([[-1.0,3.3],[2.5,0.5]])
## Generate polynomial features
poly = PolynomialFeatures(3)
P = poly.fit_transform(X)
Pt = poly.fit_transform(Xt)
## Training
W1 = P.T @ inv(P @ P.T) @ Y
## Prediction
Yt_predict1 = Pt @ W1
print(Yt_predict1)
```

$\square$

5.1.3   Primal Ridge Regression

Binary Learning: With the goal of shrinking the parametric weight values in $\boldsymbol{w}$ during estimation to avoid over amplification of the output $(g(\boldsymbol{w}, \boldsymbol{x}_i) = \boldsymbol{w}^T \boldsymbol{p}(\boldsymbol{x}_i))$ where instability may occur, the error objective function in (5.3) can be modified to include a weight-decay term $\|\boldsymbol{w}\|_2^2$ giving

$$\begin{aligned} J_{\text{SSEr}}(\boldsymbol{w}) &= \|\mathbf{y} - \mathbf{P}\boldsymbol{w}\|_2^2 + \lambda\|\boldsymbol{w}\|_2^2 \\ &= (\mathbf{y} - \mathbf{P}\boldsymbol{w})^T (\mathbf{y} - \mathbf{P}\boldsymbol{w}) + \lambda\boldsymbol{w}^T \boldsymbol{w}, \end{aligned} \tag{5.31}$$

where a scalar regularization constant λ is used to control the importance of $\|\boldsymbol{w}\|_2^2$ during minimization. The solution to minimizing this regularized cost function via putting its first-order derivative to zero ($\nabla_{\boldsymbol{w}} J_{\text{SSEr}}(\boldsymbol{w}) = \mathbf{0}$) is given by

$$\hat{\boldsymbol{w}} = (\mathbf{P}^T \mathbf{P} + \lambda \mathbf{I})^{-1} \mathbf{P}^T \mathbf{y}. \tag{5.32}$$

We observe here that the weight-decay term appears in the solution as a scaled identity matrix within the matrix inverse term and this stabilizes the inversion even when the covariance $\mathbf{P}^T \mathbf{P}$ is singular in theory. This formulation is known as *ridge regression* or *weight-decay regularization*.

In practice, the ridge regression applies well for over-determined systems when the size of training data samples is larger than the size of projection dimension, i.e., when $m > D + 1$ (recall that $\mathbf{P} \in \mathbb{R}^{m \times (D+1)}$ in (5.4)). However, the situation may become numerically ill-conditioned for under-determined systems when $m < D + 1$ even with weight-decay regularization. Particularly, the prediction may become very inaccurate.

Multi-Category Learning: The learning objective of primal ridge regression for multi-category case is similar to that of SSE except for the inclusion of weight-decay term:

$$J_{\text{SSEr}}(\mathbf{W}) = \text{trace}\{(\mathbf{Y} - \mathbf{PW})^T (\mathbf{Y} - \mathbf{PW}) + \lambda \mathbf{W}^T \mathbf{W}\}, \tag{5.33}$$

and its solution is

$$\hat{\mathbf{W}} = (\mathbf{P}^T \mathbf{P} + \lambda \mathbf{I})^{-1} \mathbf{P}^T \mathbf{Y}. \tag{5.34}$$

The training and test outputs computation is similar to that in (5.19) and (5.20) respectively.

Example 5.1(b)

The over-determined system of Example 5.1(a) is next trained with primal ridge regression using (5.34) to give

$$\hat{\mathbf{W}} = (\mathbf{P}^T \mathbf{P} + 0.001\mathbf{I})^{-1} \mathbf{P}^T \mathbf{Y} = \begin{bmatrix} 0.2371 & -0.1819 & 0.8951 \\ -0.8396 & 1.7396 & -0.8742 \\ -0.4231 & 1.5058 & -1.0415 \\ 0.1665 & -0.3811 & 0.2114 \\ 0.3408 & -0.8306 & 0.4795 \\ 0.1457 & -0.4333 & 0.2795 \end{bmatrix}.$$

The test output is

$$\hat{\mathbf{G}} = \mathbf{P}_t \hat{\mathbf{W}} = \begin{bmatrix} 0.3085 & 0.6886 & 0.0053 \\ -0.5709 & 1.3915 & 0.1792 \end{bmatrix}.$$

MATLAB codes:

```
% Training ridge regression
W = inv(P'*P+0.001*eye(6))*P'*Y;
% Prediction
Yts_est = Pt*W
```

5.1.4 Dual Ridge Regression

Binary Learning: An alternative solution to the above ridge regression is to perform the inverse in the smallest space of the two possibilities, i.e., either in the space of the feature dimension $(D + 1)$ or in the space of data sample size dimension (m). This alternative approach is called the dual ridge regression and is given by the following manipulation:

$$
\begin{aligned}
(\mathbf{P}^T \mathbf{P} + \lambda \mathbf{I})w &= \mathbf{P}^T \mathbf{y} \\
\Rightarrow \quad w &= \lambda^{-1}(\mathbf{P}^T \mathbf{y} - \mathbf{P}^T \mathbf{P} w) \\
\Rightarrow \quad w &= \lambda^{-1}\mathbf{P}^T(\mathbf{y} - \mathbf{P}w) \\
\Rightarrow \quad w &= \mathbf{P}^T \alpha
\end{aligned}
$$
(5.35)

where

$$
\begin{aligned}
\alpha &= \lambda^{-1}(\mathbf{y} - \mathbf{P}w) \\
\Rightarrow \quad \lambda \alpha &= (\mathbf{y} - \mathbf{P}w) \\
\Rightarrow \quad \lambda \alpha &= (\mathbf{y} - \mathbf{P}\mathbf{P}^T \alpha) \\
\Rightarrow \quad \mathbf{P}\mathbf{P}^T \alpha + \lambda \alpha &= \mathbf{y} \\
\Rightarrow \quad (\mathbf{P}\mathbf{P}^T + \lambda \mathbf{I})\alpha &= \mathbf{y} \\
\Rightarrow \quad \hat{\alpha} &= (\mathbf{P}\mathbf{P}^T + \lambda \mathbf{I})^{-1}\mathbf{y} \\
\Rightarrow \quad \hat{\alpha} &= (\mathbf{K} + \lambda \mathbf{I})^{-1}\mathbf{y}
\end{aligned}
$$
(5.36)

The term $\mathbf{K} = \mathbf{P}\mathbf{P}^T$ is called the Gram matrix. Notice that $\hat{w} = \mathbf{P}^T \hat{\alpha}$ and each trained output can be obtained as

$$g(\boldsymbol{x}_i, \hat{\boldsymbol{w}}) = \hat{\boldsymbol{w}}^T \boldsymbol{p}(\boldsymbol{x}_i)$$
$$= \left(\mathbf{P}^T \hat{\boldsymbol{\alpha}}\right)^T \boldsymbol{p}(\boldsymbol{x}_i)$$
$$= \hat{\boldsymbol{\alpha}}^T \mathbf{P} \boldsymbol{p}(\boldsymbol{x}_i), \quad i = 1, 2, \ldots, m. \tag{5.37}$$

Recall that $\mathbf{P}$ is a function of training inputs $\mathbf{X} = [\boldsymbol{x}_1, \boldsymbol{x}_2, \ldots \boldsymbol{x}_m]^T$ (see (5.4)), the entire set of trained outputs can also be expressed in stacked matrix form as follows:

$$\boldsymbol{g}(\mathbf{X}, \hat{\boldsymbol{w}}) = \mathbf{P}\hat{\boldsymbol{w}}$$
$$= \mathbf{P}\mathbf{P}^T \hat{\boldsymbol{\alpha}}. \tag{5.38}$$

For each unseen test data sample $\boldsymbol{x}_j$ of $j = 1, 2, \ldots, n$, the prediction is given by

$$g(\boldsymbol{x}_j, \hat{\boldsymbol{w}}) = \hat{\boldsymbol{w}}^T \boldsymbol{p}(\boldsymbol{x}_j)$$
$$= \left(\mathbf{P}^T \hat{\boldsymbol{\alpha}}\right)^T \boldsymbol{p}(\boldsymbol{x}_j)$$
$$= \hat{\boldsymbol{\alpha}}^T \mathbf{P} \boldsymbol{p}(\boldsymbol{x}_j), \quad j = 1, 2, \ldots, n. \tag{5.39}$$

In a similar way, the test prediction output can be stacked like those in training case (5.38) as follows:

$$\boldsymbol{g}(\mathbf{X}_t, \hat{\boldsymbol{w}}) = \mathbf{P}_t\hat{\boldsymbol{w}}$$
$$= \mathbf{P}_t\mathbf{P}^T \hat{\boldsymbol{\alpha}}. \tag{5.40}$$

The main difference between (5.37) and (5.39) (or between (5.38) and (5.40)) is the pure use of training data in (5.37) or (5.38) and the mixed used of training and test data in (5.39) or (5.40).

In summary, the estimation of $\boldsymbol{w}$ (with $(D + 1)$-dimension) has been transformed to estimation of $\boldsymbol{\alpha}$ (with m-dimension, or the sample dimension). This dual ridge regression provides a better conditioned formulation than that of ridge regression when $m < D + 1$.

Multi-Category Learning: Essentially, the above dual ridge regression performs estimation first at the lower dimensional space ($\hat{\boldsymbol{\alpha}} = (\mathbf{K} + \lambda\mathbf{I})^{-1}\mathbf{y}$) and then convert it to the higher dimensional space via the mapping $\hat{\boldsymbol{w}} = \mathbf{P}^T \hat{\boldsymbol{\alpha}}$. By stacking $\hat{\mathbf{W}} = [\hat{\boldsymbol{w}}_1, \hat{\boldsymbol{w}}_2, \ldots, \hat{\boldsymbol{w}}_C]$ and $\hat{\mathbf{A}} = [\hat{\boldsymbol{\alpha}}_1, \hat{\boldsymbol{\alpha}}_2, \ldots, \hat{\boldsymbol{\alpha}}_C]$ from each category, the multi-category parameters can be estimated using

$$\hat{\mathbf{W}} = \mathbf{P}^T \hat{\mathbf{A}}, \tag{5.41}$$

where $\hat{\mathbf{A}}$ is estimated via

$$\hat{\mathbf{A}} = (\mathbf{K} + \lambda\mathbf{I})^{-1}\mathbf{Y}. \tag{5.42}$$

The resulted training and test estimation outputs are respectively

$$\hat{\mathbf{G}} = \mathbf{P}\hat{\mathbf{W}} = \mathbf{P}\mathbf{P}^T\hat{\mathbf{A}} \tag{5.43}$$

and

$$\hat{\mathbf{G}}_t = \mathbf{P}_t\hat{\mathbf{W}} = \mathbf{P}_t\mathbf{P}^T\hat{\mathbf{A}}. \tag{5.44}$$

For classification, a decision-making process (see the multi-category case of Sect. 5.1.7) is followed.

Example 5.2(b)

The under-determined system of Example 5.2(a) is next trained with dual ridge regression using (5.41) and (5.42) with $\lambda = 0.001$ to give

$$\hat{\mathbf{W}} = \mathbf{P}^T\hat{\mathbf{A}} = \mathbf{P}^T(\mathbf{P}\mathbf{P}^T + 0.001\mathbf{I})^{-1}\mathbf{Y}$$

$$= \begin{bmatrix} 0.096921 & 0.20029239 & -0.06474344 \\ 0.07552804 & 0.19689304 & -0.06823502 \\ 0.09613258 & 0.1879961 & -0.05108843 \\ 0.09031235 & 0.08402066 & -0.02260897 \\ 0.03361644 & 0.16115132 & -0.04479921 \\ 0.05586837 & 0.09662081 & -0.00547768 \\ -0.0348168 & -0.02732227 & 0.02566001 \\ -0.01114985 & -0.14704293 & 0.07028742 \\ -0.07025671 & -0.06801809 & 0.07231998 \\ -0.06722926 & 0.00725292 & 0.022682 \end{bmatrix}.$$

By utilizing the transformed text matrix $\mathbf{P}_t$, the test output is computed as

$$\hat{\mathbf{G}} = \mathbf{P}_t\hat{\mathbf{W}} = \begin{bmatrix} -0.7265 & 1.7717 & 0.1343 \\ 0.3231 & 0.6092 & 0.2091 \end{bmatrix}.$$

Python codes (see Example 5.2(a) for the formation of matrices $\mathbf{P}$, $\mathbf{P}_t$ and $\mathbf{Y}$):

```
Lambda = 0.001
W2 = P.T @ inv(P @ P.T + Lambda*np.eye(P.shape[0])) @ Y
% Prediction
Yt_predict2 = Pt @ W2
print(Yt_predict2)
```

$\square$

5.1.5 Kernel Regression

Binary Learning: As mentioned in Sect. 3.6.1 that according to the representer theorem, the minimizer of any arbitrary cost function plus a strict monotonic increasing transformation of the minimizer under a normed RKHS admits a representation of the form as a linear combination of kernelized training data points. In other words, the output g of a projected data point $\boldsymbol{p} := \boldsymbol{p}(\boldsymbol{x})$ can be expressed in terms of the given set of projected training data $\boldsymbol{p}_i := \boldsymbol{p}(\boldsymbol{x}_i), i = 1, \ldots, m$ in inner product form to fit the target y with an arbitrary error ϵ. Mathematically,

$$
\begin{aligned}
y &= g(\boldsymbol{p}, \boldsymbol{w}) + \epsilon \\
&= \boldsymbol{p}^T \boldsymbol{w} + \epsilon \\
&= \sum_{i=1}^{m} \alpha_i k(\boldsymbol{p}, \boldsymbol{p}_i) + \epsilon,
\end{aligned}
\tag{5.45}
$$

with a kernel function defined by $k(\boldsymbol{p}, \boldsymbol{p}_i) = \langle \boldsymbol{p}, \boldsymbol{p}_i \rangle$ instead of a parametric form governed by $\boldsymbol{w} \in \mathbb{R}^{(D+1)}$.

Collectively, for m number of training points stacked as a matrix $\mathbf{P} = [\boldsymbol{p}_1, \ldots, \boldsymbol{p}_m]^T \in \mathbb{R}^{m \times (D+1)}$, and a target vector $\mathbf{y} = [y_1, \ldots, y_m]^T \in \mathbb{R}^m$ with errors $\boldsymbol{\epsilon} = [\epsilon_1, \ldots, \epsilon_m]^T \in \mathbb{R}^m$, we can relate the training target vector and the training matrix by

$$
\begin{aligned}
\mathbf{y} &= \mathbf{P}\boldsymbol{w} + \boldsymbol{\epsilon} \\
&= \mathbf{P}\mathbf{P}^T \boldsymbol{\alpha} + \boldsymbol{\epsilon} \\
&= \mathbf{K}\boldsymbol{\alpha} + \boldsymbol{\epsilon},
\end{aligned}
\tag{5.46}
$$

where

$$
\mathbf{K} = \mathbf{P}\mathbf{P}^T, \quad \boldsymbol{w} = \mathbf{P}^T \boldsymbol{\alpha},
\tag{5.47}
$$

with $\boldsymbol{\alpha} \in \mathbb{R}^m$ and $\mathbf{K} \in \mathbb{R}^{m \times m}$. Similar to the estimation for under-determined systems, the representation by $\boldsymbol{w} = \mathbf{P}^T \boldsymbol{\alpha}$ can be considered a reparameterization from $\boldsymbol{w} \in \mathbb{R}^{(D+1)}$ in feature dimension to $\boldsymbol{\alpha} \in \mathbb{R}^m$ in sample dimension.

Consequently, the learning problem can be formulated in kernel form to solve for the parameter $\boldsymbol{\alpha}$ in the dual space ($\mathbb{R}^m$) with respect to the given training data. Suppose $\mathbf{K}$ is non-singular and consider the rewritten error function (in terms of $\mathbf{K}$ and $\boldsymbol{\alpha}$) to be minimized:

$$
J(\boldsymbol{\alpha}) = \frac{1}{2}\boldsymbol{\epsilon}^T \boldsymbol{\epsilon} = \frac{1}{2}(\mathbf{y} - \mathbf{K}\boldsymbol{\alpha})^T (\mathbf{y} - \mathbf{K}\boldsymbol{\alpha}).
\tag{5.48}
$$

By setting the first partial derivative of $J(\boldsymbol{\alpha})$ (5.48) with respect to $\boldsymbol{\alpha}$ to zero, we have

$$\mathbf{K}^T(\mathbf{y} - \mathbf{K}\alpha) = \mathbf{0}$$
$$\Rightarrow \quad \mathbf{K}^T\mathbf{y} - \mathbf{K}^T\mathbf{K}\alpha = \mathbf{0}$$
$$\Rightarrow \quad \hat{\alpha} = (\mathbf{K}^T\mathbf{K})^{-1}\mathbf{K}^T\mathbf{y}$$
$$\Rightarrow \quad \hat{\alpha} = \mathbf{K}^{-1}\mathbf{y}, \tag{5.49}$$

since $\mathbf{K}$ is non-singular and $(\mathbf{K}^T\mathbf{K})^{-1} = \mathbf{K}^{-1}(\mathbf{K}^T)^{-1}$. Typically in practice, the inverse term in (5.49) exists for data with $m < D + 1$ such that $\mathbf{K}$ has full rank.

Having a learned $\hat{\alpha}$, the model can be applied to predict a new set of unseen data $\mathbf{P}_t = [\boldsymbol{p}_{t1}, \ldots, \boldsymbol{p}_{tn}]^T$ using

$$g(\hat{\alpha}, \boldsymbol{p}_t) = \mathbf{P}_t\hat{\boldsymbol{w}}$$
$$= \mathbf{P}_t\mathbf{P}^T\hat{\alpha}$$
$$= k(\mathbf{P}_t, \mathbf{P})\mathbf{K}^{-1}\mathbf{y}. \tag{5.50}$$

Multi-Category Learning: The learning for multi-category problems is similar to all the above cases where $\mathbf{y}$ is replaced by $\mathbf{Y}$ as follows:

$$\hat{\mathbf{A}} = \mathbf{K}^{-1}\mathbf{Y}, \tag{5.51}$$

where $\hat{\mathbf{A}} = [\hat{\alpha}_1, \ldots, \hat{\alpha}_C]$ packs the estimation for all categories. The parameter matrix is then given by

$$\hat{\mathbf{W}} = \mathbf{P}^T\hat{\mathbf{A}}. \tag{5.52}$$

Essentially, the kernel regression corresponds to SSE minimization for underdetermined systems but from the kernel perspective. This means that we can work directly in terms of kernels (we might interpret it as a certain distance between the data of interest to the set of training data) and avoid explicit manipulation of the feature vector $\boldsymbol{p}(\boldsymbol{x})$. This allows an implicit utilization of very high or even infinite feature dimensionality.

5.1.6 Kernel Ridge Regression

Binary Learning: From the kernel perspective, the parameter vector $\boldsymbol{w} \in \mathbb{R}^{D+1}$ can be projected onto a subspace $\alpha \in \mathbb{R}^m$ via

$$\boldsymbol{w} = \mathbf{P}^T\alpha, \tag{5.53}$$

when $m < D + 1$ where we can write

$$\|\boldsymbol{w}\|_2^2 = \boldsymbol{w}^T\boldsymbol{w} = \alpha^T\mathbf{P}\mathbf{P}^T\alpha. \tag{5.54}$$

By defining $\mathbf{K} = \mathbf{P}\mathbf{P}^T = \mathbf{K}^T$ (Gram matrix), the regularized objective function can be written as

$$J(\alpha) = \frac{\lambda}{2}\alpha^T \mathbf{K}^T \alpha + \frac{1}{2}(\mathbf{y} - \mathbf{K}\alpha)^T(\mathbf{y} - \mathbf{K}\alpha), \tag{5.55}$$

where $\alpha \in \mathbb{R}^m$ corresponds to the parameter vector to be adjusted for minimizing J. Assuming $\mathbf{K}$ is non-singular and by setting the first partial derivative of $J(\alpha)$ with respect to α to zero, we have

$$\lambda \mathbf{K}^T \alpha - \mathbf{K}^T(\mathbf{y} - \mathbf{K}\alpha) = \mathbf{0}$$
$$\Rightarrow \quad \lambda \mathbf{K}^T \alpha - \mathbf{K}^T \mathbf{y} + \mathbf{K}^T \mathbf{K}\alpha = \mathbf{0}$$
$$\Rightarrow \quad \lambda \mathbf{K}^T \alpha + \mathbf{K}^T \mathbf{K}\alpha = \mathbf{K}^T \mathbf{y}$$
$$\Rightarrow \quad \mathbf{K}^T(\mathbf{K} + \lambda \mathbf{I})\alpha = \mathbf{K}^T \mathbf{y}$$
$$\Rightarrow \quad \hat{\alpha} = (\mathbf{K} + \lambda \mathbf{I})^{-1}(\mathbf{K}^T)^{-1}\mathbf{K}^T \mathbf{y}$$
$$\Rightarrow \quad \hat{\alpha} = (\mathbf{K} + \lambda \mathbf{I})^{-1}\mathbf{y}. \tag{5.56}$$

For prediction of unseen test data points $\mathbf{X}_t$ packed within $\mathbf{P}_t$, we have

$$\begin{aligned} g(\mathbf{X}_t, \hat{w}) &= \mathbf{P}_t w \\ &= \mathbf{P}_t \mathbf{P}^T \alpha \\ &= k(\mathbf{P}_t, \mathbf{P})(\mathbf{K} + \lambda \mathbf{I})^{-1}\mathbf{y}. \end{aligned} \tag{5.57}$$

Multi-Category Learning: Similar to the dual ridge regression, the multi-category prediction needs a mere replacement of single output vector $\mathbf{y}$ by a packed matrix $\mathbf{Y}$ which contains multiple output vectors:

$$\hat{\mathbf{G}}_t = k(\mathbf{P}_t, \mathbf{P})(\mathbf{K} + \lambda \mathbf{I})^{-1}\mathbf{Y}. \tag{5.58}$$

A decision process by the one-versus-all technique (see Sect. 5.1.7) can be adopted to generate the final output class label.

For the case of asymmetric kernel model given by (3.130) which is extended below for the multi-category (C-classes) case:

$$\mathbf{G}(w, x) = \phi(\mathbf{X})\mathbf{X}^T \mathbf{A}, \tag{5.59}$$

with $\mathbf{A} \in \mathbb{R}^{m \times C}$ being the kernel parameter matrix, the estimation can be made using

$$\hat{\mathbf{A}} = (k(\phi(\mathbf{X}), \mathbf{X}) + \lambda \mathbf{I})^{-1}\mathbf{Y} \tag{5.60}$$

and the corresponding prediction is

$$\hat{\mathbf{G}}_t = \phi(\mathbf{X}_t)\mathbf{X}^T \hat{\mathbf{A}} = k(\phi(\mathbf{X}_t), \mathbf{X})[k(\phi(\mathbf{X}), \mathbf{X}) + \lambda \mathbf{I}]^{-1}\mathbf{Y}. \tag{5.61}$$

5.1.7 Threshold and Classification Decision

The above least squares error-based learning gives rise to a predictor which processes each sample of input features and outputs a prediction value related to its distance or magnitude of each pattern class label. In other words, the output of a least squares error-based predictor can be interpreted as being related to the probability of belonging to each pattern class in terms of the magnitude of closeness to the class label. A decision is thus needed to convert such a distance or probabilistic value to the desired discrete pattern category in classification applications. We shall discuss this classification decision based on two-category problems and then extend it to multi-category problems.

Binary Classification: For two-category or two-class problems, a single-dimensional output $g(x, w) \in \mathbb{R}$ can be used to represent the desired target values of $y \in \{0, 1\} \in \mathbb{N}$ (or $y \in \{-1, +1\} \in \mathbb{N}$) for the two classes. A decision *threshold* τ can then be applied to $g(x, w)$ to decide whether an input x belongs to "class-0" (or class "-1") or "class-1" (or class "$+1$"):

$$cls(g(x, w)) = \begin{cases} 1 \text{ if } g(x, w) \geqslant \tau \\ 0 \text{ if } g(x, w) < \tau \end{cases}. \tag{5.62}$$

For normalized distribution of $g(x, w) \in [0, 1]$ or $g(x, w) \in [-1, +1]$, this threshold is often chosen as $\tau = 0.5$ for $y \in \{0, 1\}$ or $\tau = 0$ for $y \in \{-1, +1\}$ respectively.

Multi-Category Classification: For n test samples, the prediction output can be packed as

$$\hat{\mathbf{G}}_t = \begin{bmatrix} \mathbf{g}^T(x_1, \mathbf{W}) \\ \mathbf{g}^T(x_2, \mathbf{W}) \\ \vdots \\ \mathbf{g}^T(x_n, \mathbf{W}) \end{bmatrix} = \begin{bmatrix} g(x_1, w_1) & g(x_1, w_2) & \cdots & g(x_1, w_C) \\ g(x_2, w_1) & g(x_2, w_2) & \cdots & g(x_2, w_C) \\ \vdots & \cdots & \ddots & \vdots \\ g(x_n, w_1) & g(x_n, w_2) & \cdots & g(x_n, w_C) \end{bmatrix}. \tag{5.63}$$

The category decision for each row of data sample indexed by $j = 1, 2, \ldots, n$ is determined using

$$cls(\mathbf{g}(x_j, \mathbf{W})) = \arg\max_k g(x_j, w_k), \quad \forall k = 1, 2, \ldots, C. \tag{5.64}$$

This technique for category labeling is known as *one-versus-all*, *winner-takes-all*, or *one-hot encoding* technique.

Example 5.1(c)

The over-determined system of Example 5.1(b) has been trained with primal ridge regression. For category prediction, (5.64) is utilized to give the classification decision.

MATLAB codes:

```
Yts_est_class = zeros(2,3);
for k=1:2
    [M,I] = max(Yts_est_b(k,:));
    Yts_est_class(k,I) = 1;
end
Yts_est_class =
     0     1     0
     0     0     1
```

$\square$

Example 5.2(c)

The under-determined system of Example 5.2(b) has been trained with dual ridge regression. For category prediction, (5.64) is utilized to give the classification decision.

Python codes (see Example 2(a) and (b) for declarations and Yt_predict2):

```
## Class prediction
Yt_class_predict = np.zeros((2,3))
for k in range(2):
    idx = np.argmax(Yt_predict2[k,])
    Yt_class_predict[k,idx] = 1
print(Yt_class_predict)
  [[0. 1. 0.]
   [0. 0. 1.]]
```

$\square$

5.2 Classification Error Based Learning Methods

Apart from the above error magnitude-based methods, a more matching approach for classification is to directly formulate the classification metric as the learning objective. This approach has been a classical subject of study known as *misclassification error* (MCE) based methods. However, due to the counting nature of the classification metric (i.e., by accumulating the number of successful prediction counts of matching between the discrete target and the category decision using (5.62) and (5.64)), attempts to the problem had inevitably resulted in solving a nonlinear formulation. In this section, we introduce two classifier learning solutions which can be obtained in closed form and yet possess nonlinear mapping capability. The first method directly uses the *total error rate* (TER, see (4.47)) as the learning objective function while the second method uses the *area under the ROC curve* (AUC, see (4.48)) as the learning objective function.

5.2.1 Total Error Rate-Based Learning

Binary Learning: Consider the two-category classification problem. Given a training set which consists of m examples $(\boldsymbol{x}_i, y_i)$, $i = 1, \ldots, m$, where $\boldsymbol{x}_i \in \mathbb{R}^d$ denotes the ith input feature sample, and $y_i \in \{0, 1\}$ (or $y_i \in \{-1, +1\}$) denotes the corresponding *target label* or *class indicator*. Given a predictor which can be written in linear parametric form $g(\boldsymbol{x}_i, \boldsymbol{w}) = \boldsymbol{w}^T \boldsymbol{p}(\boldsymbol{x}_i)$ where $\boldsymbol{p}(\boldsymbol{x}_i) = [1, p_1(\boldsymbol{x}_i), \ldots, p_D(\boldsymbol{x}_i)]^T$ is a linear or nonlinear mapping of $\boldsymbol{x}_i$ and $\boldsymbol{w} \in \mathbb{R}^{D+1}$ is the corresponding weight coefficient vector.

Our goal here is to train the predictor directly according to the classification objective by adjusting the parameter vector $\boldsymbol{w}$. As discussed in the previous chapter, the total error rate (TER $=$ FPR $+$ FNR, (4.20)) would be a suitable candidate for objective function formulation since it is directly accumulating the classification error count in normalized form. Without loss of generality, let the target labels be $y \in \{-1, +1\}$. Then, we can indicate the variables ($\boldsymbol{x}$ and m) of samples labeled as positive (with "$+1$" label) and that labeled as negative (with "-1" label) using superscripts $+$ and $-$ respectively. By expressing in terms of classification error counting, the TER can be written as

$$\text{TER} = \text{FPR} + \text{FNR}$$

$$= \frac{1}{m^-} \sum_{j=1}^{m^-} u(g(\boldsymbol{x}_j^-, \boldsymbol{w}) \geq \tau) + \frac{1}{m^+} \sum_{i=1}^{m^+} u(g(\boldsymbol{x}_i^+, \boldsymbol{w}) < \tau), \quad (5.65)$$

where $u(\cdot)$ denotes a zero-one step counting function giving "1" whenever $(\cdot)$ holds true and "0" otherwise.[3] In other words, $u(g(x_j^-, w) \geqslant \tau) = 1$ whenever $g(x_j^-, w) \geqslant \tau$ holds true and $u(g(x_j^-, w) \geqslant \tau) = 0$ otherwise. Similarly, $u(g(x_i^+, w) < \tau) = 1$ whenever $g(x_i^+, w) < \tau$ holds true and $u(g(x_i^+, w) < \tau) = 0$ otherwise. Here we note that since we take $y \in \{-1, +1\}$, the threshold for category decision for $g(x_i, w)$ is $\tau = 0$ by default.

By observing the symmetry between $u(g(x_j^-, w) \geqslant \tau)$ and $u(g(x_i^+, w) < \tau)$, a simple change of variables can unify these two counting functions $u(g^- \geqslant \tau)$ and $u(g^+ < \tau)$ into one loss form. Define $\epsilon_j = g(x_j^-, w) - \tau$ for $j = 1, 2, \ldots, m^-$ and $\varepsilon_i = \tau - g(x_i^+, w) - \Delta$ for $i = 1, 2, \ldots, m^+$. Here, $\Delta \to 0$ which is to account for the strict inequality in FNR of (5.65) can be ignored in practice. Then the TER loss function can be expressed in terms of ϵ_j and ε_i as follows

$$\text{TER} = \frac{1}{m^-} \sum_{j=1}^{m^-} u(\epsilon_j \geqslant 0) + \frac{1}{m^+} \sum_{i=1}^{m^+} u(\varepsilon_i \geqslant 0). \tag{5.66}$$

The goal to train a predictor as a classifier is then to find the minimizer of TER given by

$$\hat{w} = \arg\min_{w} \text{TER}$$

$$= \arg\min_{w} \left\{ \frac{1}{m^-} \sum_{j=1}^{m^-} u(\epsilon_j \geqslant 0) + \frac{1}{m^+} \sum_{i=1}^{m^+} u(\varepsilon_i \geqslant 0) \right\}. \tag{5.67}$$

The formulation in (5.67) is not a differentiable function where deterministic optimization can apply. A natural way to solve this problem is to approximate the zero-one step loss function by a smooth *sigmoid* function given by

$$\sigma(\theta) = \frac{1}{1 + e^{-\gamma\theta}}, \quad \gamma > 0 \tag{5.68}$$

where $\theta \in \mathbb{R}$ is an input variable (such as ϵ_j and ε_i in (5.67)) and $\gamma \in \mathbb{R}$ controls the slope of the activation (see Fig. 5.1). The minimization problem then becomes a smooth and differentiable one given by

$$\min_{w} \text{TER} \approx \min_{w} \left\{ \frac{1}{m^-} \sum_{j=1}^{m^-} \sigma(\epsilon_j \geqslant 0) + \frac{1}{m^+} \sum_{i=1}^{m^+} \sigma(\varepsilon_i \geqslant 0) \right\}. \tag{5.69}$$

However, such a sigmoidal approximation is often associated with problems related to solution search. The first problem is the resulted nonlinear formulation with respect to the learning parameter vector w (i.e., $\sum \sigma(w)$), recall that ϵ in $\sigma(\epsilon \geqslant 0)$ is

---

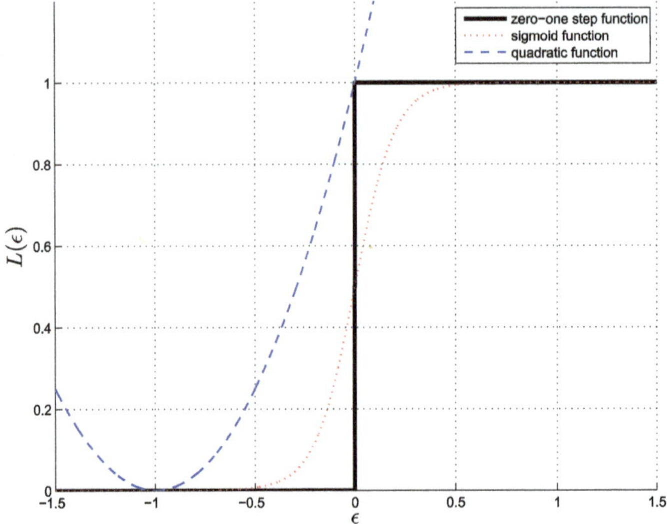

Fig. 5.1 Sigmoidal and quadratic approximations to the zero-one step loss function. The sigmoidal function is adopted as $1/(1 + \exp(-10\epsilon))$

a function of $\boldsymbol{w}$). This in turn gives rise to local solutions where an iterative search can be employed but different initializations may end up with different local solutions. Laborious trial and error efforts might hence be incurred in order to reach an acceptable setting. The second problem comes from the local plateaus resulting from summing the flat regions of the sigmoid. A lot of search efforts may be spent upon making little progress in these locally flat regions.

Several other techniques (see the bibliography section toward the end of this chapter) are also available for the counting step function approximation. However, due to their nonlinear nature of formulation, both the local minima issue and the search convergence issue are inevitable.

Fortunately, the optimization problem can be approached from matching the *link-loss functional pair* perspective. In statistics, a link function in the *generalized linear model* (GLM[4]) refers to a function which connects the response (output) variable and the effects (inputs). Mathematically, the TER formulation in (5.65) or (5.67) constitutes a *functional*[5] since it is a composite function, i.e., $u(g(\cdot))$, where u is a loss function and g is a link function. By matching the *linear link* ($g(\boldsymbol{x}, \boldsymbol{w}) = \boldsymbol{w}^T \boldsymbol{x}$) with a *quadratic loss* ($L(\cdot) = (\cdot)^2$), a convex formulation can be obtained. Particularly, in this linear-quadratic pair, a closed-form solution can be permissible. However, we recall that the quadratic function is by no means closed to the loss property of having a heavier penalty for an incorrect output than a correct output where the zero-one

---

[4] GLM is a flexible generalization of ordinary linear regression which allows for response variables having other than a normal distribution.

[5] In mathematics, a *functional* is often referred to as a *function* of a *function*.

function is possessed. In other words, a monotonic loss function is necessary for the error counting loss formulation.

Nevertheless, from data perspective, we can seek to manipulate the data such that only one quadratic arm is utilized during estimation. To achieve this goal, a uniform offset can be introduced to the entire bulk of input data so that they are projected onto a single quadratic arm. The dashed curve in Fig. 5.1 illustrates a shifted (from the origin) quadratic function plotted along with the zero-one step loss function. Translating this idea into our error rate counting loss function, the following quadratic approximation to (5.67) can be adopted in our minimization formulation:

$$\min_{w} \mathrm{TER}(w) \approx \min_{w} \left\{ \frac{1}{2\,m^-} \sum_{j=1}^{m^-} (\epsilon_j + \eta)^2 + \frac{1}{2\,m^+} \sum_{i=1}^{m^+} (\varepsilon_i + \eta)^2 \right\}. \quad (5.70)$$

The normalizing factor of 2 in both the false negative and false positive terms accounts for the differentiation factor resulted from the two square terms.

Recall that $\epsilon_j := g(x_j^-, w) - \tau$, $j = 1, 2, \ldots, m^-$ and $\varepsilon_i := \tau - g(x_i^+, w) - \Delta$, $i = 1, 2, \ldots, m^+$. Given a linear predictor $g(x_i, w) = w^T p(x_i)$ and considering a weight decay regularization by adding $\frac{b}{2} \|w\|_2^2$, we have

$$\min_{w} \mathrm{TER}(w) \approx \min_{w} \left\{ \frac{b}{2} \|w\|_2^2 + \frac{1}{2m^-} \sum_{j=1}^{m^-} [(w^T p(x_j^-) - \tau) + \eta]^2 \right.$$

$$\left. + \frac{1}{2m^+} \sum_{i=1}^{m^+} [(\tau - w^T p(x_i^+)) + \eta]^2 \right\},$$

$$(5.71)$$

where value of b controls the amount of regularization. The parameter vector $\hat{w}$ which minimizes this approximated TER can be obtained by putting its first partial derivative terms to zeros:

$$\frac{\partial \mathrm{TER}(x^+, x^-, w)}{\partial w} = 0 \quad (5.72)$$

where it implies that

$$bw + \frac{1}{m^-} \sum_{j=1}^{m^-} p(x_j^-) \left[p^T(x_j^-)w - \tau + \eta \right]$$

$$+ \frac{1}{m^+} \sum_{i=1}^{m^+} -p(x_i^+) \left[\tau - p^T(x_i^+)w + \eta \right] = 0$$

$$\Rightarrow \quad \left[b\mathbf{I} + \frac{1}{m^-} \sum_{j=1}^{m^-} p(x_j^-) p^T(x_j^-) + \frac{1}{m^+} \sum_{i=1}^{m^+} p(x_i^+) p^T(x_i^+) \right] w$$

$$+ \frac{(\eta - \tau)}{m^-} \sum_{j=1}^{m^-} p(x_j^-) - \frac{(\eta + \tau)}{m^+} \sum_{i=1}^{m^+} p(x_i^+) = \mathbf{0}. \qquad (5.73)$$

Abbreviating the column basis vectors $p_j = p(x_j^-) \in \mathbb{R}^{D+1}$ and $p_i = p(x_i^+) \in \mathbb{R}^{D+1}$, the solution for w which minimizes (5.71) can be written as

$$\hat{w} = \left[b\mathbf{I} + \frac{1}{m^-} \sum_{j=1}^{m^-} p_j p_j^T + \frac{1}{m^+} \sum_{i=1}^{m^+} p_i p_i^T \right]^{-1} \left[\frac{(\tau - \eta)}{m^-} \sum_{j=1}^{m^-} p_j + \frac{(\tau + \eta)}{m^+} \sum_{i=1}^{m^+} p_i \right]$$

$$(5.74)$$

where $\mathbf{I}$ is an identity matrix of $(D + 1) \times (D + 1)$ size (recall that $D + 1$ is the total number of transformed feature terms). In a more compact matrix notation, (5.74) can be written as

$$\hat{w} = \left(b\mathbf{I} + \frac{1}{m^-} \mathbf{P}_-^T \mathbf{P}_- + \frac{1}{m^+} \mathbf{P}_+^T \mathbf{P}_+ \right)^{-1} \left(\frac{(\tau - \eta)}{m^-} \mathbf{P}_-^T \mathbf{1}_- + \frac{(\tau + \eta)}{m^+} \mathbf{P}_+^T \mathbf{1}_+ \right) \qquad (5.75)$$

where

$$\mathbf{P}_+ = \begin{bmatrix} p^T(x_1^+) \\ p^T(x_2^+) \\ \vdots \\ p^T(x_{m^+}^+) \end{bmatrix}, \qquad \mathbf{P}_- = \begin{bmatrix} p^T(x_1^-) \\ p^T(x_2^-) \\ \vdots \\ p^T(x_{m^-}^-) \end{bmatrix}, \qquad (5.76)$$

and $\mathbf{1}_+ = [1, \ldots, 1]^T \in \mathbb{N}^{m^+}$, $\mathbf{1}_- = [1, \ldots, 1]^T \in \mathbb{N}^{m^-}$.

It is noted that the decision threshold τ has been included in the optimization process when determining w. This is in contrast to many conventional classifiers such as support vector machines and neural networks which do not include the decision threshold explicitly during classifier design. Moreover, the solution here for minimizing TER is deterministic as it does not require initialization and an iterative search.

Multi-category Learning: Consider the multi-class problem with C output categories. An indicator matrix can be formed for the above TER formulation similar to that of (5.15) in LSE formulation. Assume a common setting for threshold (τ) and bias (η) throughout all C outputs. Then, we have for each kth-class ($k = 1, 2, \ldots, C$), a positive output $\mathbf{y}_k^+ = (\tau + \eta)\mathbf{1}_+ \in \mathbb{R}^{m^+}$ which corresponds to a kth-class indicator, and a negative output $\mathbf{y}_k^- = (\tau - \eta)\mathbf{1}_- \in \mathbb{R}^{m^-}$ which corresponds to a non-kth-class indicator. By packing the solutions with corresponding $\mathbf{P}_{k+}$ and $\mathbf{P}_{k-}$ matrices, Eq. (5.75) can be extended to multi-category as

$$\hat{\mathbf{W}} = [\hat{\boldsymbol{w}}_1, \ldots, \hat{\boldsymbol{w}}_C], \tag{5.77}$$

where

$$\hat{\boldsymbol{w}}_k = \left(b\mathbf{I} + \frac{1}{m_k^-}\mathbf{P}_{k-}^T\mathbf{P}_{k-} + \frac{1}{m_k^+}\mathbf{P}_{k+}^T\mathbf{P}_{k+}\right)^{-1} \times \left(\frac{1}{m_k^-}\mathbf{P}_{k-}^T\boldsymbol{y}_k^- + \frac{1}{m_k^+}\mathbf{P}_{k+}^T\boldsymbol{y}_k^+\right),$$
$$k = 1, \ldots, C.$$

Here we note that $\hat{\mathbf{W}}$ in (5.77) is analogous to that of (5.34) which can be directly applied to unseen test data regressor $\mathbf{P}_t$ to generate the test output. A classification decision can then be made using the *one-versus-all* technique in (5.64).

Example 5.3

Consider the following training data points for two categories,

$$\begin{bmatrix} x_1 \\ x_2 \end{bmatrix} : \begin{bmatrix} 1 \\ 2 \end{bmatrix}, \begin{bmatrix} 2 \\ 0 \end{bmatrix}, \begin{bmatrix} 1 \\ 1 \end{bmatrix}, \begin{bmatrix} 3 \\ 1 \end{bmatrix}, \begin{bmatrix} 2 \\ 3 \end{bmatrix}, \begin{bmatrix} 3 \\ 3 \end{bmatrix}$$

which correspond to labels $y : \{1, 1, 1, 0, 0, 0\}$ respectively. Using a linear model and the TER method, our goal is to predict the class labels of $\mathbf{x}_{t1} = \begin{bmatrix} 4 \\ 1 \end{bmatrix}$ (red "triangle" in Fig. 5.2) and $\mathbf{x}_{t2} = \begin{bmatrix} 1 \\ 0 \end{bmatrix}$ (black "triangle" in Fig. 5.2) which have true labels $y_{t1} = 0$, $y_{t2} = 1$ respectively.

For illustration purpose, we shall treat the two-category problem as multiple two-class problems and use the one-hot indicator (5.15) where the data are split into two sets of +ve-class and $-$ve-class matrices:

$$\mathbf{X}_+ = \begin{bmatrix} 1, 1, 2 \\ 1, 2, 0 \\ 1, 1, 1 \end{bmatrix}, \quad \mathbf{y}_+ = \begin{bmatrix} 1 \\ 1 \\ 1 \end{bmatrix} \quad \text{and} \quad \mathbf{X}_- = \begin{bmatrix} 1, 3, 1 \\ 1, 2, 3 \\ 1, 3, 3 \end{bmatrix}, \quad \mathbf{y}_- = \begin{bmatrix} 0 \\ 0 \\ 0 \end{bmatrix} \quad \text{for} \quad k = 1.$$

$$\mathbf{X}_- = \begin{bmatrix} 1, 1, 2 \\ 1, 2, 0 \\ 1, 1, 1 \end{bmatrix}, \quad \mathbf{y}_- = \begin{bmatrix} 0 \\ 0 \\ 0 \end{bmatrix} \quad \text{and} \quad \mathbf{X}_+ = \begin{bmatrix} 1, 3, 1 \\ 1, 2, 3 \\ 1, 3, 3 \end{bmatrix}, \quad \mathbf{y}_+ = \begin{bmatrix} 1 \\ 1 \\ 1 \end{bmatrix} \quad \text{for} \quad k = 2.$$

- TER learning (need to iterate for each k since the matrices/vectors $\mathbf{X}_+, \mathbf{X}_-$, $\mathbf{y}_+, \mathbf{y}_-$ corresponding to each k are differently defined): $\hat{\boldsymbol{w}}_k = (\frac{1}{m^+}\mathbf{X}_+^T\mathbf{X}_+ + \frac{1}{m^-}\mathbf{X}_-^T\mathbf{X}_-)^{-1}(\frac{1}{m^+}\mathbf{X}_+^T\mathbf{y}_+ + \frac{1}{m^-}\mathbf{X}_-^T\mathbf{y}_-)$, $k = 1, 2$ for over-determined systems. Pack parameters as $\hat{\mathbf{W}} = [\hat{\boldsymbol{w}}_1, \hat{\boldsymbol{w}}_2]$.
- TER Prediction:

(i) Single sample: $\hat{\mathbf{y}}^T = \mathbf{x}_t^T\hat{\mathbf{W}}$ for unseen data sample $\mathbf{x}_t \Rightarrow$ The column number of the row vector $\hat{\mathbf{y}}^T$ with the largest value is the class number. The predictions for the two samples are:
$\hat{\mathbf{y}}_1^T = \mathbf{x}_{t1}^T\hat{\mathbf{W}} = [-0.2529, \mathbf{1.2529}]$, $\hat{\mathbf{y}}_2^T = \mathbf{x}_{t2}^T\hat{\mathbf{W}} = [\mathbf{1.3000} \; -0.3000]$.
The category predictions are therefore class-0 and class-1 respectively.

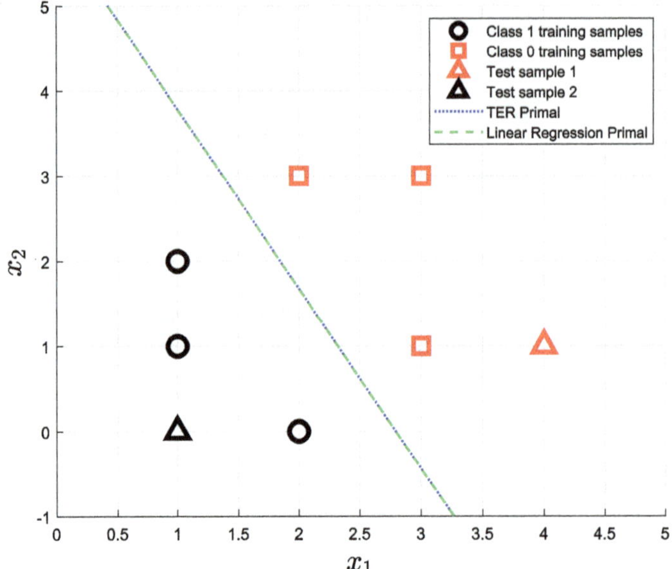

Fig. 5.2 Learning decision contour at level [0]

(ii) Multiple samples: $\hat{\mathbf{Y}} = \mathbf{X}_t \hat{\mathbf{W}}$ for unseen data matrix $\mathbf{X}_t$ packed in similar manner as that in regression.

- Linear Regression Prediction:

(i) The prediction results for LSE or Linear Regression (LR) are also included:
$\hat{\mathbf{y}}_1^T = \mathbf{x}_{t1}^T \hat{\mathbf{W}} = [-0.2529, \; \mathbf{1.2529}], \hat{\mathbf{y}}_2^T = \mathbf{x}_{t2}^T \hat{\mathbf{W}} = [\mathbf{1.3000} \; -0.3000]$.

(ii) This shows that for data with balanced categories, the TER and LR have the same prediction. This is because the class-specific weighting of TER becomes indistinguishable between the two categories.

MATLAB Codes:

```
# Disclaimer:
# The user assumes responsibility for determining appropriate use of the code,
# for consequences of its use, and for checking results against other reliable
# sources. No warrantee is given.
X = [1,2; 2,0; 1,1; 3,1; 2,3; 3,3];
Y = [1,0; 1,0; 1,0; 0,1; 0,1; 0,1];
Xt = [4,1; 1,0];
Yt = [0,1; 1,0];
Xb = [ones(length(X(:,1)),1),X]; % X with bias/intercept
Xtb = [ones(length(Xt(:,1)),1),Xt]; % Xt with bias/intercept
% TER primal
W_primal = TERtrainPrimalbinary(Xb,Y,0); % Threshold explicitly set at 0.5
[Yt_output_predict, Yt_class_predict, TestAccuracy] = TERtest(Xtb, Yt, W_primal)
% Linear regression
W_regression = inv(Xb'*Xb)*Xb'*Y;
[Yt_est, Yt_class_est, TestAccuracy] = TERtest(Xtb, Yt, W_regression)
```

```
function W = TERtrainPrimalbinary(P,Y,b,Sp,Sm)
% Traing by TER method
if (nargin==2) % 'b' and 'Sp, Sm' scaling not given
    Sp=1; Sm=1; b = 1e-4;
elseif (nargin==3) % 'Sp, Sm' scaling not given
    Sp=1; Sm=1;
end
[m,k] = size(Y);
tau = 0.5; eta = 0.5;
for jj = 1:k
    y = Y(:,jj);
    Pplus = P(y==1,:);
    Pminus = P(y==0,:);
    [Mp, Pdim] = size(Pplus);
    [Mm, Pdim] = size(Pminus);
    Mp = Sp*Mp; Mm = Sm*Mm;
    II = eye(Pdim,Pdim);
    W(:,jj) = inv(Pplus'*Pplus/Mp + Pminus'*Pminus/Mm + b*II)*...
    ( ((tau+eta)/Mp)*(sum(Pplus,1))' + ((tau-eta)/Mm)*(sum(Pminus,1))' );
end
return


function [Yt_output_predict, Yt_class_predict, TestAccuracy] = TERtest(Pt, Yt, W)
%--- prediction ---%
Yt_output_predict = Pt*W;
[nrow, ncol] = size(Yt);
if (ncol==1)
    %--- Binary ---%
    Yt_class_predict = zeros(nrow,1);
    Yt_class_predict(Yt_output_predict >= (max(Yt) + min(Yt))/2) = max(Yt);
    Yt_class_predict(Yt_output_predict < (max(Yt) + min(Yt))/2) = min(Yt);
    TestAccuracy = 100*length(find(Yt==Yt_class_predict))/nrow;
else
    %--- one-versus-all ---%
    Yt_class_predict = zeros(nrow,ncol);
    for jj=1:nrow
        [jval,jindx]=max(Yt_output_predict(jj,:));
        Yt_class_predict(jj,jindx) = 1;
    end
    [M1, I1] = max(Yt');
    [M2, I2] = max(Yt_class_predict');
    TestAccuracy = (mean(I1 == I2))*100;
end
return
```

□

Total Error Rate-Based Learning in the Dual Space: By defining two class-specific diagonal weighting matrices

$$
\mathbf{M}_- =
\begin{bmatrix}
\frac{1}{m^-} & & & & 0 \\
& \ddots & & & \\
& & \frac{1}{m^-} & & \\
& & & 0 & \\
& & & & \ddots & \\
0 & & & & & 0
\end{bmatrix}
\left.\begin{matrix} \\ \\ \end{matrix}\right\} m^-
\left.\begin{matrix} \\ \\ \end{matrix}\right\} m^+
\tag{5.78}
$$

$$
\mathbf{M}_+ =
\begin{bmatrix}
0 & & & & 0 \\
& \ddots & & & \\
& & 0 & & \\
& & & \frac{1}{m^+} & \\
& & & & \ddots & \\
0 & & & & & \frac{1}{m^+}
\end{bmatrix}
\left.\begin{matrix} \\ \\ \end{matrix}\right\} m^-
\left.\begin{matrix} \\ \\ \end{matrix}\right\} m^+
\;,
\tag{5.79}
$$

and having the target vector ordered as

$$
\mathbf{y} = \begin{bmatrix} \mathbf{y}^- \\ \mathbf{y}^+ \end{bmatrix} = \left[\frac{(\tau - \eta)}{m^-}, \ldots, \frac{(\tau - \eta)}{m^-}, \frac{(\tau + \eta)}{m^+}, \ldots, \frac{(\tau + \eta)}{m^+} \right]^T,
\tag{5.80}
$$

the Total Error Rate based binary learning can be rewritten as

$$
\begin{aligned}
\hat{w} &= \left[b\mathbf{I} + \mathbf{P}^T(\mathbf{M}_- + \mathbf{M}_+)\mathbf{P} \right]^{-1} \left[\mathbf{P}^T(\mathbf{M}_- + \mathbf{M}_+)\mathbf{y} \right] \\
&= \left(b\mathbf{I} + \mathbf{P}^T\mathbf{M}\mathbf{P} \right)^{-1} \mathbf{P}^T\mathbf{M}\mathbf{y},
\end{aligned}
\tag{5.81}
$$

where $\mathbf{M} = \mathbf{M}_- + \mathbf{M}_+$ and the row samples of $\mathbf{P}$ being ordered according to the two classes. By putting $w = \mathbf{P}^T\mathbf{M}\alpha$ and using the derivation steps of (5.36), we have

$$
\hat{\alpha} = (b\mathbf{I} + \mathbf{P}\mathbf{P}^T\mathbf{M})^{-1}\mathbf{y}.
\tag{5.82}
$$

This $\hat{\alpha}$ can be substituted back into $w = \mathbf{P}^T\mathbf{M}\alpha$ to give

$$
\hat{w} = \mathbf{P}^T\mathbf{M}(b\mathbf{I} + \mathbf{P}\mathbf{P}^T\mathbf{M})^{-1}\mathbf{y}.
\tag{5.83}
$$

For multi-category case, the solution for parameter estimation can be packed as

$$
\hat{\mathbf{W}} = [\hat{w}_1, \ldots, \hat{w}_C]
\tag{5.84}
$$

where

$$
\hat{w}_k = \mathbf{P}_k^T\mathbf{M}_k(b\mathbf{I} + \mathbf{P}_k\mathbf{P}_k^T\mathbf{M}_k)^{-1}\mathbf{y}_k, \quad k = 1, 2, \ldots, C.
\tag{5.85}
$$

Here, it is noted that for each $\hat{\mathbf{w}}_k$, $k = 1, \ldots, C$ the sample order of matrices $\mathbf{P}_k$, $\mathbf{M}_k$ needs to be congruence to the sample order of $\mathbf{y}_k$ in terms of their distinct set of indicating target labels for each $k \in \{1, 2, \ldots, C\}$.

Assuming $\mathbf{P}$ has full rank. For under-determined case where the number of samples (m, number of rows of $\mathbf{P}$) is smaller than the feature dimension ($D + 1$, number of columns of $\mathbf{P}$), i.e., $m < D + 1$, then the dual ridge solution provides a better conditioned solution than that of the primal ridge solution since the inverse is performed at the lower dimension m. Moreover, as mentioned, it is more computationally cost effective to compute the estimation at the lower dimension of the two possibilities.

Example 5.4

Consider the following training data points for two categories,

$$\begin{bmatrix} x_1 \\ x_2 \end{bmatrix} : \begin{bmatrix} 2 \\ 2 \end{bmatrix}, \begin{bmatrix} 1 \\ 2 \end{bmatrix}, \begin{bmatrix} 1 \\ 1 \end{bmatrix}, \begin{bmatrix} 3 \\ 1 \end{bmatrix}, \begin{bmatrix} 2 \\ 3 \end{bmatrix}$$

which correspond to labels $y : \{1, 0, 1, 0, 0\}$ respectively. Using a third-order polynomial model and the TER method, our goal is to predict the class labels of $\mathbf{x}_{t1} = \begin{bmatrix} 2.5 \\ 4 \end{bmatrix}$ (red "triangle" in Fig. 5.3) and $\mathbf{x}_{t2} = \begin{bmatrix} 1 \\ 0 \end{bmatrix}$ (black "triangle" in Fig. 5.3) which have true labels $y_{t1} = 0$, $y_{t2} = 1$ respectively.

Figure 5.3 shows the decision boundary at threshold 0 according to the one-hot implementation of the binary target. This threshold is different from that in Example 3 which has an explicit threshold setting at 0.5. The Python codes for TER learning which can operate in either the Primal or the Dual modes are shown below.

Fig. 5.3 Learning decision contour at level [0]

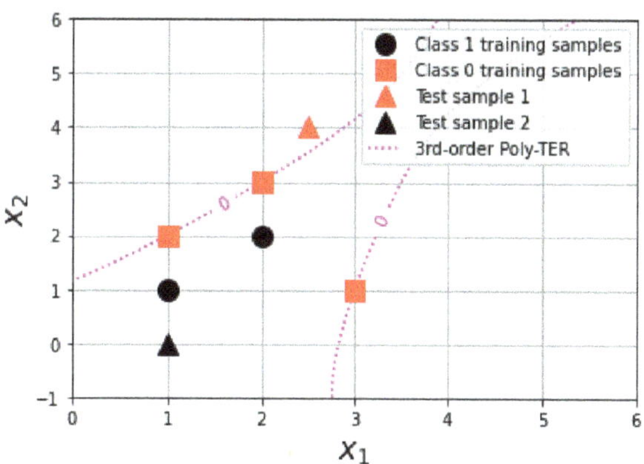

```python
# Disclaimer:
# The user assumes responsibility for determining appropriate use of the code,
# for consequences of its use, and for checking results against other reliable
# sources. No warrantee is given.
import numpy as np
from numpy.linalg import inv

def TERtrain(P,Y,b=None,Sp=None,Sm=None):
    # Training by TER method Primal/Dual
    if b is None and Sp is None and Sm is None:
        # 'b' and 'Sp, Sm' scaling not given
        Sp=1; Sm=1; b = 1e-4 # Use default settings
    elif Sp is None and Sm is None:
        # only 'Sp, Sm' scaling not given
        Sp=1; Sm=1
    m,K = Y.shape
    W = np.zeros((P.shape[1], K))
    for jj in range(K):
        max_value = np.max(Y[:, jj])
        min_value = np.min(Y[:, jj])
        Mp, D = P[Y[:,jj]==max_value,:].shape
        Mm, D = P[Y[:,jj]==min_value,:].shape
        Mvector = np.ones(m)
        Mvector[Y[:,jj]==min_value] = 1/(Sm*Mm)
        Mvector[Y[:,jj]==max_value] = 1/(Sp*Mp)
        M = np.diag(Mvector)
        if m>=D: # Primal
            II = np.eye(D,D)
            W[:,jj] = inv(P.T @ M @ P + b*II) @ P.T @ M @ Y[:,jj]
        else: # Dual
            II = np.eye(m,m)
            W[:,jj] = P.T @ inv(M @ P @ P.T + b*II) @ M @ Y[:,jj]
    return W

def TERtest(Pt, Yt, W):
    #--- prediction ---%
    Yt_out_pred = Pt @ W
    nrow, ncol = Yt.shape
    if ncol == 1:
        #--- Binary ---%
        Yt_class_pred = np.zeros((nrow, 1))
        Yt_class_pred[Yt_out_pred>=(np.max(Yt)+np.min(Yt))/2] = np.max(Yt)
        Yt_class_pred[Yt_out_pred<(np.max(Yt)+np.min(Yt))/2] = np.min(Yt)
        TestAccuracy = 100 * len(np.where(Yt == Yt_class_pred)[0])/nrow
    else:
        #--- one-versus-all ---%
        Yt_class_pred = np.zeros((nrow,ncol))
        for jj in range(nrow):
            jindx = np.argmax(Yt_out_pred[jj,:]),np.argmax(Yt_out_pred[jj,:])
            Yt_class_pred[jj,jindx] = 1
        M1, I1 = np.max(Yt, axis=1), np.argmax(Yt, axis=1)
        M2, I2 = np.max(Yt_class_pred, axis=1), np.argmax(Yt_class_pred, axis=1)
        TestAccuracy = (np.mean(I1 == I2))*100
    return Yt_out_pred, Yt_class_pred, TestAccuracy

from sklearn.preprocessing import PolynomialFeatures
X = np.array([[2,2],[1,2],[1,1],[3,1],[2,3]])
Y = np.array([[1,0],[0,1],[1,0],[0,1],[0,1]])
Xt =  np.array([[2.5,4],[1,0]])
Yt =  np.array([[0,1],[1,0]])
poly = PolynomialFeatures(3)
P = poly.fit_transform(X)
Pt = poly.fit_transform(Xt)
```

```
W_dual = TERtrain(P,Y,0)
Yt_predict, Yt_class_predict = TERtest(Pt, Yt, W_dual)
M1, I1 = np.max(Yt, axis=1), np.argmax(Yt, axis=1)
M2, I2 = np.max(Yt_class_predict, axis=1), np.argmax(Yt_class_predict, axis=1)
TestAccuracy = (np.mean(I1 == I2))*100
print(Yt_predict)
print(Yt_class_predict)
print(TestAccuracy)
   [[-0.67427395  1.51920227]
    [ 1.17037166 -0.18958447]]
   [[0. 1.]
    [1. 0.]]
   100.0
```

5.2.2 AUC Based Learning

Binary Learning: Apart from the error rate-based cost function measured at single operating threshold, the receiver operating characteristic (ROC) curve (see Sect. 4.4), which provides an overview of error costs over all operating thresholds, can also be utilized for learning. However, the ROC gives a range of operating points instead of a scalar value, so it is not convenient for optimization. The Area Under the ROC Curve (AUC) is thus a natural alternative to construct the cost function for classifier learning concerning all operating points.

For binary data samples indexed by $i = 1, 2, \ldots, m^+$ and $j = 1, 2, \ldots, m^-$ in respective categories, let us denote the difference between the predictor values of a positive sample x_i^+ and that of a negative sample x_j^- by $\xi_{ij} = g(x_i^+) - g(, x_j^-)$. Here, $\xi_{ij} > 0$ indicates that the positive sample x_i^+ and the negative sample x_j^- are correctly ranked by the predictor g. Consider the Heaviside step function given by

$$u(\xi) = \begin{cases} 1, & \text{if } \xi > 0, \\ 0.5 & \text{if } \xi = 0, \\ 0, & \text{if } \xi < 0. \end{cases} \tag{5.86}$$

Then, the AUC for $m = m^+ + m^-$ training examples can be expressed as

$$\text{AUC}(w, x) = \frac{1}{m^+ m^-} \sum_{i=1}^{m^+} \sum_{j=1}^{m^-} u(\xi_{ij})$$

$$= \frac{1}{m^+ m^-} \sum_{i=1}^{m^+} \sum_{j=1}^{m^-} u(g(x_i^+) - g(, x_j^-)), \tag{5.87}$$

which is the fraction of positive-negative pairs that are ranked correctly by predictor $g(\cdot)$. The value in (5.87) is also known as the Wilcoxon-Mann-Whitney statistic in literature.

The AUC can be maximized with respect to $\boldsymbol{w}$ of the parametric predictor $g(\boldsymbol{x}, \boldsymbol{w}) = \boldsymbol{w}^T \boldsymbol{p}(\boldsymbol{x})$ in (5.2) for effective classification considering the entire operating threshold range. Alternatively, the following equivalent minimization of the *Area Above the ROC Curve* (AAC) can be used:

$$\min_{\boldsymbol{w}} \mathrm{AAC}(\boldsymbol{w}, \boldsymbol{x}) = \min_{\boldsymbol{w}} \frac{1}{m^+ m^-} \sum_{i=1}^{m^+} \sum_{j=1}^{m^-} u(-\xi_{ij}) \ . \tag{5.88}$$

Similar to the TER approximation in Sect. 5.2.1, when $g(\cdot)$ is in linear parametric form such as (5.7), the minimization problem given by (5.88) can be solved by a quadratic approximation to the non-differentiable step function $u(\cdot)$ with a weight-decay regularization:

$$\min_{\boldsymbol{w}} \mathrm{AAC}(\boldsymbol{w}, \boldsymbol{x})$$

$$= \min_{\boldsymbol{w}} \left\{ \frac{b}{2} \|\boldsymbol{w}\|_2^2 + \frac{1}{2 m^+ m^-} \sum_{i=1}^{m^+} \sum_{j=1}^{m^-} \left[\left(\boldsymbol{p}(\boldsymbol{x}_j^-) - \boldsymbol{p}(\boldsymbol{x}_i^+) \right)^T \boldsymbol{w} + \eta \right]^2 \right\} . \tag{5.89}$$

When the predictor is normalized so that $g \in [0, 1]$, then the ranking difference value is $\xi_{ij} \in [-1, +1]$. An effective approximation to (5.88) requires a monotonic function for the ranking process, i.e., a cost function should give a "high" output whenever $-\xi_{ij} = g(\boldsymbol{w}, \boldsymbol{x}_i^-) - g(\boldsymbol{w}, \boldsymbol{x}_j^+) > 0$ (wrongly ranked) and a "low" output whenever $-\xi_{ij} = g(\boldsymbol{w}, \boldsymbol{x}_i^-) - g(\boldsymbol{w}, \boldsymbol{x}_j^+) < 0$ (correctly ranked). However, taking a quadratic on $-\xi_{ij}$ alone will always result in a "high" output regardless of its sign. The situation can be managed, however, by including an offset $\eta > 0$ to the quadratic function, such that $(-\xi_{ij} + \eta)^2 > (-\xi_{kl} + \eta)^2$ whenever $\xi_{ij} < 0$ and $\xi_{kl} > 0$, for all ξ_{ij} and ξ_{kl} in $[-1, +1]$.

Figure 5.4a shows an illustration of the quadratic output values of those wrongly ranked data ($\xi_{ij} < 0$, which is indicated by solid line) being higher than those generated from correctly ranked data ($\xi_{kl} > 0$, which is indicated by a dashed line), i.e., $(-\xi_{ij} + \eta)^2 > (-\xi_{kl} + \eta)^2$ whenever $\xi_{ij} < 0$ and $\xi_{kl} > 0$ when $\eta = +1$. Conversely, when $\eta = -1$ in Fig. 5.4b, those wrongly ranked data ($\xi_{ij} < 0$, indicated by solid line) give lower quadratic output values $(-\xi_{ij} + \eta)^2$ than those given by $(-\xi_{kl} + \eta)^2$, which are generated by the correctly ranked data ($\xi_{kl} > 0$, indicated by dashed line). This means that only one arm of the quadratic function has been utilized for monotonic ranking approximation.

Under this setting, the minimization in (5.89) can be solved by setting its gradient with respect to $\boldsymbol{w}$ equals to zero, giving

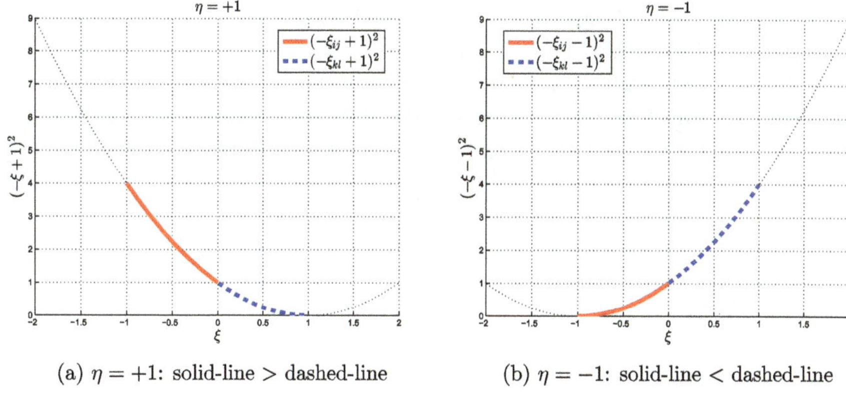

(a) $\eta = +1$: solid-line > dashed-line   (b) $\eta = -1$: solid-line < dashed-line

Fig. 5.4 **a** $\eta = +1$: $(-\xi_{ij} + \eta)^2 > (-\xi_{kl} + \eta)^2$ whenever $\xi_{ij} < 0$ and $\xi_{kl} > 0$; **b** $\eta = -1$: $(-\xi_{ij} + \eta)^2 < (-\xi_{kl} + \eta)^2$ whenever $\xi_{ij} < 0$ and $\xi_{kl} > 0$

$$
\hat{w} = \left[b\mathbf{I} + \frac{1}{m^+ m^-} \sum_{i=1}^{m^+} \sum_{j=1}^{m^-} \left(\boldsymbol{p}_j^- - \boldsymbol{p}_i^+ \right) \left(\boldsymbol{p}_j^- - \boldsymbol{p}_i^+ \right)^T \right]^{-1}
$$

$$
\times \left[\frac{-\eta}{m^+ m^-} \sum_{i=1}^{m^+} \sum_{j=1}^{m^-} \left(\boldsymbol{p}_j^- - \boldsymbol{p}_i^+ \right) \right], \qquad (5.90)
$$

with the projection column vectors being abbreviated by $\boldsymbol{p}_j^- := p(\boldsymbol{x}_j^-)$ and $\boldsymbol{p}_i^+ := p(\boldsymbol{x}_i^+)$.

The solution $\hat{w}$ in (5.90) for optimal AUC classification decision can be obtained in a single evaluation step which is also least squares optimal but in the AUC sense. This solution can be used to compute the outputs for an unseen test data set $\{\boldsymbol{x}_1, \ldots, \boldsymbol{x}_n\}$ based on the stacked projection matrix $\mathbf{P}_t$ as that in the least squares regression, i.e., $\hat{\boldsymbol{g}}(\{\boldsymbol{x}_1, \ldots, \boldsymbol{x}_n\}) = \mathbf{P}_t \hat{\boldsymbol{w}}$.

The above solution for AUC or AAC is optimal with respect to the full range of operating threshold decision points. Hence, there is no explicit threshold being defined in (5.90). However, in practice, a specific choice of threshold is often needed for decision-making. To do this, an optimal threshold with respect to the total error rate can be obtained by minimizing the approximated TER (5.70) with respect to τ and utilizing the above AAC solution (5.90), giving

$$
\tau = \frac{1}{2m^-} \sum_{j=1}^{m^-} \hat{\boldsymbol{w}}^T \boldsymbol{p}_j^- + \frac{1}{2m^+} \sum_{i=1}^{m^+} \hat{\boldsymbol{w}}^T \boldsymbol{p}_i^+. \qquad (5.91)
$$

In summary, the AAC minimizer (5.90) provides an optimal solution for all operating classification thresholds. This is useful when the threshold needs to be varied according to needs without compromising the optimality in terms of the overall operating performance. To cater for specific operating requirements, an optimal threshold with respect to minimizing the TER can nevertheless be chosen based on (5.91).

Example 5.5

Consider the following training data points for two categories,

$$\begin{bmatrix} x_1 \\ x_2 \end{bmatrix} : \begin{bmatrix} 2 \\ 2 \end{bmatrix}, \begin{bmatrix} 1 \\ 2 \end{bmatrix}, \begin{bmatrix} 1 \\ 1 \end{bmatrix}, \begin{bmatrix} 3 \\ 1 \end{bmatrix}, \begin{bmatrix} 2 \\ 3 \end{bmatrix}$$

which correspond to labels $y : \{1, 0, 1, 0, 0\}$ respectively. Our goal here is to use AUC learning to predict the class labels of $\mathbf{x}_{t1} = \begin{bmatrix} 2 \\ 4 \end{bmatrix}$ (red "triangle" in Fig. 5.3) and $\mathbf{x}_{t2} = \begin{bmatrix} 3 \\ 0.3 \end{bmatrix}$ (black "triangle" in Fig. 5.3) which have true labels $y_{t1} = 0$, $y_{t2} = 1$ respectively.

The learning based on AUC has been implemented in Python with a decision threshold based on the TER criterion. In other words, the AUC has utilized all decision thresholds of the ROC curve for learning, and hence a specific threshold is needed to find the accuracy as a specific point on the ROC curve. Apart from the AUC-based learning, the LSE or Linear Regression (LR) is also included for comparison.

Figure 5.5 shows the decision boundaries for AUC and LR learning. The result shows that LR has an incorrect prediction for the second test sample for $\mathbf{x}_{t2}$ while the AUC has both predictions correct. Due to the weighting according to each class size, the AUC which adopted a threshold based on the TER has enlarged its decision

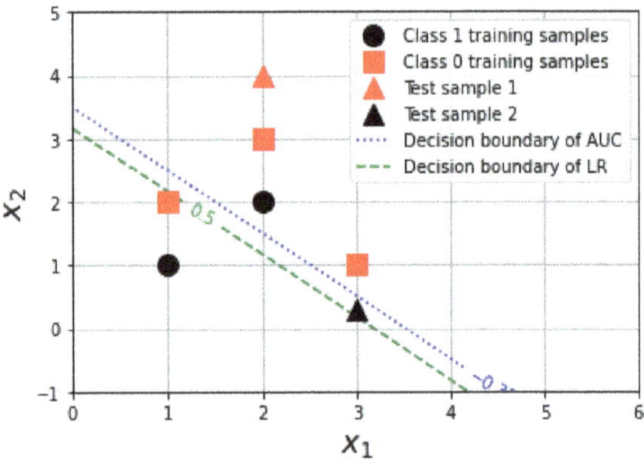

Fig. 5.5 Learning decision contours

region for the minority category, thereby pushing the decision boundary away from the centroid of the minority category to count $\mathbf{x}_{t2}$ as class-1 instead of class-0. Python codes:

```python
# Disclaimer:
# The user assumes responsibility for determining appropriate use of the code,
# for consequences of its use, and for checking results against other reliable
# sources. No warrantee is given.
import numpy as np

def AUCtrain(P,y,b=None):
    if b is None:
        b = 1e-4 # Use default setting
    eta = 1
    Pp = P[y[:,0]==1,:] # +ve class samples
    Pm = P[y[:,0]==0,:] # -ve class samples
    Mp, D = Pp.shape
    Mm, D = Pm.shape
    M = np.zeros((D,D))
    N = np.zeros((D,1))
    for i in range(Mp):
        for j in range(Mm):
            p_i = Pp[i,:]
            p_j = Pm[j,:]
            M = M + (p_j - p_i).T.dot(p_j - p_i)
            N = N + (p_j - p_i).reshape(-1,1)
    M = M/(Mp*Mm)
    N = -eta*N/(Mp*Mm)
    w = np.linalg.inv(M + b*np.eye(M.shape[0])).dot(N)
    threshold = sum(Pm.dot(w))/(2*Mm) + sum(Pp.dot(w))/(2*Mp)
    return w, threshold

def TERtest(Pt, Yt, W, th=None):
    (*** Pleasse see TERtest() in Example 4) ***)
    return Yt_out_pred, Yt_class_pred, TestAccuracy

# Example 5 data
X = np.array([[2,2], [1,2], [1,1], [3,1], [2,3]])
y = np.array([[1], [0], [1], [0], [0]])
Xt = np.array([[2,4], [3,0.3]])
yt = np.array([[0], [1]])
# AUC learning
w_AUC, th = AUCtrain(X,y,0.0001)
# AUC Prediction
yt_predict, yt_class_predict, TestAccuracy = TERtest(Xt, yt, w_AUC, th)
print(yt_predict, yt_class_predict)
print(TestAccuracy)
# Linear Regression
w_regression = np.linalg.inv(Xb.T.dot(Xb)).dot(Xb.T).dot(y)
# LR Prediction
Yt_est, Yt_class_est, TestAccuracy = TERtest(Xtb, yt, w_regression)
print(Yt_est, Yt_class_est)
print(TestAccuracy)

# Plotting
import matplotlib.pyplot as plt
X1, X2 = np.meshgrid(np.arange(0, 6.1, 0.1), np.arange(-1, 6.1, 0.1))
nrow, ncol = X1.shape
XX = np.hstack((X1.reshape(nrow*ncol, 1), X2.reshape(nrow*ncol, 1)))
XXb = np.hstack((np.ones((len(XX), 1)), XX))
y_est_AUC_col = XX.dot(w_AUC)
```

```python
y_est_AUC_contour = y_est_AUC_col[:,0].reshape(nrow, ncol)
y_est_linear_regression_col = XXb.dot(w_regression)
y_est_linear_regression_contour
              = y_est_linear_regression_col[:,0].reshape(nrow, ncol)
fig = plt.figure(10)
plt.plot(X[[0,2],0], X[[0,2],1], 'ko', markersize=12, linewidth=3,
              label='Class 1 training samples')
plt.plot(X[[1,3,4],0], X[[1,3,4],1], 'rs', markersize=12, linewidth=3,
              label='Class 0 training samples')
plt.plot(Xt[0,0], Xt[0,1], 'r^', markersize=12, linewidth=3,
              label='Test sample 1')
plt.plot(Xt[1,0], Xt[1,1], 'k^', markersize=12, linewidth=3,
              label='Test sample 2')
c1 = plt.contour(X1,X2,y_est_AUC_contour,th,linestyles=':',
              colors='b',linewidths=1.5)
c2 = plt.contour(X1,X2,y_est_linear_regression_contour,[0.5],linestyles='--',
              colors='g',linewidths=1.5)
plt.clabel(c1, inline=1, fontsize=10)
plt.clabel(c2, inline=1, fontsize=10)
label1 = ['Decision boundary of AUC']
label2 = ['Decision boundary of LR']
for i in range(len(label1)):
    c1.collections[i].set_label(label1[i])
for i in range(len(label2)):
    c2.collections[i].set_label(label2[i])
plt.legend(loc='upper right')
plt.axis([0, 6,- 1 ,5])
plt.grid()
  [[-0.64284337]
  [-0.35356385]]
  [[0.]
  [1.]]
  100.0
  [[-0.15384615]
  [ 0.46923077]]
  [[0.]
  [0.]]
  50.0}
```

□

Multi-category Learning: The generalization of ROC from binary to multi-category case is not straightforward. The "area" under the multi-category "curve" is termed "*volume under surface*" (VUS) since it involves higher dimension than two. This topic is beyond the scope of current book.

5.3 A Generalized Learning Framework

Based on the structural resemblance among the LSE, TER, and AUC solutions, it is observed that these classifiers can be linked together by simple data manipulation. Here, we introduce an additive form of linearly transformed data at the feature level for the AUC formulation. By mere variation of coefficient settings of the additive terms, we show that several linear classifiers can be interrelated. The knowledge of this relationship leads to a generalized classifier which allows variation of its coefficient settings beyond the above-known settings.

5.3.1 Data Transformation

Consider each sample of input feature vector given by $x \in \mathbb{R}^d$ with a corresponding target label given by $y \in \{0, 1\}$, a transformation on the data x can be performed as follows:

$$\tilde{x} = Tx + a, \tag{5.92}$$

where $a = [a_1, a_2, \ldots, a_d]^T \in \mathbb{R}^d$ is a column vector for translation and T is a $d \times d$ matrix.

Next consider m number of training samples given by (x_i, y_i), where $i = 1, 2, \ldots, m$. Suppose there are K distinct scaling matrices T_k and translating vectors $a_k, k = 1, 2, \ldots, K$. This gives rise to $m \times K$ training data samples:

$$\tilde{x}_{ik} = T_k x_i + a_k, i = 1, \ldots, m, \ k = 1, \ldots, K. \tag{5.93}$$

In a similar fashion, we could also work on the transformed projection vectors of the generalized linear models:

$$\tilde{p}(x) = T p(x) + \mathbf{a}, \tag{5.94}$$

where $p(x) \in \mathbb{R}^{D+1}$ projects x into a nonlinear hyperplane, $\mathbf{T}$ is $(D+1) \times (D+1)$ matrix and $\mathbf{a} \in \mathbb{R}^{D+1}$.

Now, consider a linear parametric model applied respectively to the transformed data in (5.93) and the transformed projection vector (5.94), we have

$$g(w, \tilde{x}) = \sum_{j=1}^{d} w_j \tilde{x}_j = \tilde{x}^T w \tag{5.95}$$

and

$$\tilde{g}(w, x) = \sum_{j=1}^{D+1} w_j \tilde{p}_j(x) = \tilde{p}(x)^T w \tag{5.96}$$

respectively.

In the following discussion, we shall use (5.96) for our development while noticing that a similar deployment can be carried out by replacing $\tilde{p}(x)$ by $\tilde{x}$. To simplify the notations, we shall use

$$\tilde{p}_{kj}^- = \tilde{p}_k(x_j^-) = \mathbf{T}_k p(x_j^-) + \mathbf{a}_k,$$

$$\tilde{\boldsymbol{p}}_{ki}^{+} = \tilde{\boldsymbol{p}}_k(\boldsymbol{x}_i^{+}) = \mathbf{T}_k \boldsymbol{p}(\boldsymbol{x}_i^{+}) + \mathbf{a}_k,$$

for $k = 1, 2, \ldots, K$, $j = 1, 2, \ldots, m^{-}$, and $i = 1, 2, \ldots, m^{+}$. The new AUC-based (AAC, to be specific) learning formulation based on Eq. (5.96) is thus

$$\min_{\boldsymbol{w}} \mathrm{A\tilde{A}C}(\boldsymbol{w}, \boldsymbol{x}^{+}, \boldsymbol{x}^{-}) =$$

$$\min_{\boldsymbol{w}} \left\{ \frac{b}{2} \|\boldsymbol{w}\|_2^2 + \frac{1}{2 \, m^{+} m^{-}} \sum_{i=1}^{m^{+}} \sum_{j=1}^{m^{-}} \sum_{k=1}^{K} \left[\left(\tilde{\boldsymbol{p}}_{kj}^{-} - \tilde{\boldsymbol{p}}_{ki}^{+}\right)^{T} \boldsymbol{w} + \eta \right]^2 \right\} . \tag{5.97}$$

Then, according to (5.90), the solution $\tilde{\boldsymbol{w}}_{\mathrm{TAUC}}$ which minimizes the Transformed AUC (TAUC) in (5.97) is given by

$$\tilde{\boldsymbol{w}}_{\mathrm{TAUC}} = \left[b\mathbf{I} + \frac{1}{m^{+} m^{-}} \sum_{i=1}^{m^{+}} \sum_{j=1}^{m^{-}} \sum_{k=1}^{K} \left(\tilde{\boldsymbol{p}}_{kj}^{-} - \tilde{\boldsymbol{p}}_{ki}^{+}\right) \left(\tilde{\boldsymbol{p}}_{kj}^{-} - \tilde{\boldsymbol{p}}_{ki}^{+}\right)^{T} \right]^{-1}$$

$$\times \left[\frac{-\eta}{m^{+} m^{-}} \sum_{i=1}^{m^{+}} \sum_{j=1}^{m^{-}} \sum_{k=1}^{K} \left(\tilde{\boldsymbol{p}}_{kj}^{-} - \tilde{\boldsymbol{p}}_{ki}^{+}\right) \right] .$$

$$\tag{5.98}$$

Based on this generalized solution, we shall observe the relationships among several existing binary classifiers and predictors in the following subsections.

5.3.2   Between TAUC and AUC Solutions

Consider a transformation with $\mathbf{T}_k = ((1 - y)\beta_k + y\gamma_k)\mathbf{I}_{D+1}$ (where $\mathbf{I}_{D+1} \in \mathbb{R}^{(D+1)\times(D+1)}$) which is dependent on the label $y \in \{0, 1\}$, where $k = 1, 2, \ldots, K$. In other words the transformation uses the scaling transformation β_k when $y = 0$ and it uses γ_k when $y = 1$. Writing this transformation in terms of negative and positive labels, we have

$$\tilde{\boldsymbol{p}}_{kj}^{-} = \tilde{\boldsymbol{p}}_k(\boldsymbol{x}_j^{-}) = \beta_k \boldsymbol{p}(\boldsymbol{x}_j^{-}) + \mathbf{a}_k, \tag{5.99}$$

$$\tilde{\boldsymbol{p}}_{ki}^{+} = \tilde{\boldsymbol{p}}_k(\boldsymbol{x}_i^{+}) = \gamma_k \boldsymbol{p}(\boldsymbol{x}_i^{+}) + \mathbf{a}_k.$$

This means that each β_k corresponds to a global scaling of the negative samples while each γ_k corresponds to a global scaling of the positive samples.

When $K = 2$ (i.e., with two scaling transformations) and $\mathbf{a}_k = \mathbf{0}$ (which means no translation), the AUC solution that incorporates only scaling transformation can be written as

$$
\tilde{w} = \left[b\mathbf{I} + \frac{1}{m^+ m^-} \sum_{i=1}^{m^+} \sum_{j=1}^{m^-} \left(\beta_1 \boldsymbol{p}_j^- - \gamma_1 \boldsymbol{p}_i^+ \right) \left(\beta_1 \boldsymbol{p}_j^- - \gamma_1 \boldsymbol{p}_i^+ \right)^T \right.
$$
$$
\left. + \frac{1}{m^+ m^-} \sum_{i=1}^{m^+} \sum_{j=1}^{m^-} \left(\beta_2 \boldsymbol{p}_j^- - \gamma_2 \boldsymbol{p}_i^+ \right) \left(\beta_2 \boldsymbol{p}_j^- - \gamma_2 \boldsymbol{p}_i^+ \right)^T \right]^{-1}
$$
$$
\times \left[\frac{-\eta}{m^+ m^-} \sum_{i=1}^{m^+} \sum_{j=1}^{m^-} \left(\beta_1 \boldsymbol{p}_j^- - \gamma_1 \boldsymbol{p}_i^+ \right) + \frac{-\eta}{m^+ m^-} \sum_{i=1}^{m^+} \sum_{j=1}^{m^-} \left(\beta_2 \boldsymbol{p}_j^- - \gamma_2 \boldsymbol{p}_i^+ \right) \right].
$$

$$(5.100)$$

Here we see that the AUC solution (5.90) becomes a special case of (5.100) when $\beta_1 = \beta_2 = \gamma_1 = \gamma_2 = 1$. In other words, the TAUC solution generalizes the AUC solution by considering scaled data in the learning.

5.3.3 Between TAUC and TER Solutions

Now, consider another specific case for $K = 2$ with $\gamma_1 = \beta_2 = 0$ and $\beta_1 = \gamma_2 = 1$, we have

$$
\tilde{w} = \left[b\mathbf{I} + \frac{1}{m^+ m^-} \sum_{i=1}^{m^+} \sum_{j=1}^{m^-} \left(\beta_1 \boldsymbol{p}_j^- \right) \left(\beta_1 \boldsymbol{p}_j^- \right)^T + \frac{1}{m^+ m^-} \sum_{i=1}^{m^+} \sum_{j=1}^{m^-} \left(-\gamma_2 \boldsymbol{p}_i^+ \right) \left(-\gamma_2 \boldsymbol{p}_i^+ \right)^T \right]^{-1}
$$
$$
\times \left[\frac{-\eta}{m^+ m^-} \sum_{i=1}^{m^+} \sum_{j=1}^{m^-} \left(\beta_1 \boldsymbol{p}_j^- \right) + \frac{-\eta}{m^+ m^-} \sum_{i=1}^{m^+} \sum_{j=1}^{m^-} \left(-\gamma_2 \boldsymbol{p}_i^+ \right) \right]
$$
$$
= \left[b\mathbf{I} + \frac{1}{m^-} \sum_{j=1}^{m^-} \boldsymbol{p}_j^- \boldsymbol{p}_j^{-T} + \frac{1}{m^+} \sum_{i=1}^{m^+} \boldsymbol{p}_i^+ \boldsymbol{p}_i^{+T} \right]^{-1} \left[\frac{0-\eta}{m^-} \sum_{j=1}^{m^-} \boldsymbol{p}_j^- + \frac{0+\eta}{m^+} \sum_{i=1}^{m^+} \boldsymbol{p}_i^+ \right].
$$

$$(5.101)$$

Except for a default zero threshold ($\tau = 0$), the above solution (5.101) is similar to that of TER in (5.74).

5.3.4 Between TAUC and FLD Solutions

Consider, the scaled version of (5.100) plus a global translation for each class, i.e.,

$$
\tilde{\boldsymbol{p}}_k(\boldsymbol{x}) = ((1-y)\beta_k + y\gamma_k)\mathbf{I}_{D+1}\boldsymbol{p}(\boldsymbol{x}) + (1-y)\boldsymbol{u}_k + y\boldsymbol{v}_k, \text{ where } k = 1, 2,
$$

for each sample pair $(\boldsymbol{x}, y)$), we get

$$
\tilde{w} = \left[b\mathbf{I} + \frac{1}{m^+ m^-} \sum_{i=1}^{m^+} \sum_{j=1}^{m^-} \left((\beta_1 p_j^- - u_1) - (\gamma_1 p_i^+ - v_1) \right) \left((\beta_1 p_j^- - u_1) - (\gamma_1 p_i^+ - v_1) \right)^T \right.
$$

$$
\left. + \frac{1}{m^+ m^-} \sum_{i=1}^{m^+} \sum_{j=1}^{m^-} \left((\beta_2 p_j^- - u_2) - (\gamma_2 p_i^+ - v_2) \right) \left((\beta_2 p_j^- - u_2) - (\gamma_2 p_i^+ - v_2) \right)^T \right]^{-1}
$$

$$
\times \left[\frac{-\eta}{m^+ m^-} \sum_{i=1}^{m^+} \sum_{j=1}^{m^-} \left((\beta_1 p_j^- - u_1) - (\gamma_1 p_i^+ - v_1) \right) + \frac{-\eta}{m^+ m^-} \sum_{i=1}^{m^+} \sum_{j=1}^{m^-} \left((\beta_2 p_j^- - u_2) - (\gamma_2 p_i^+ - v_2) \right) \right].
$$

$$(5.102)$$

When $b = \gamma_1 = \beta_2 = 0$, $\beta_1 = \gamma_2 = 1$ and $\eta = 1/2$, by setting $u_2 = v_1 = \mathbf{0}$, $u_1 = \frac{1}{m^-} \sum_{i=1}^{m^-} p_i^- = \mu_-$ (mean of vectors p_i^-) while $v_2 = \frac{1}{m^+} \sum_{j=1}^{m^+} p_j^+ = \mu_+$ (mean of vectors p_j^+) , we arrive at

$$
\tilde{w} = \left[\frac{1}{m^-} \sum_{j=1}^{m^-} (p_j^- - \mu_-)(p_j^- - \mu_-)^T + \frac{1}{m^+} \sum_{i=1}^{m^+} (p_i^+ - \mu_+)(p_i^+ - \mu_+)^T \right]^{-1}
$$

$$
\times (\mu_+ - \mu_-).
$$

$$(5.103)$$

In particular, when $m^- = m^+ = 1$, the above solution is analogous to Fisher Linear Discriminant analysis (FLD, see Duda et al. (2001), p.120), but with embedding of nonlinear features within basis expansion vector p (i.e., $p(x)$ instead of x). Hence the solution to the transformed AUC space can be generalized to contain the FLD solution. Moreover, the FLD solution can be considered a centered version of the TER solution.

5.3.5 *Between TER and LSE Solutions*

Utilizing the weight matrices in (5.78)–(5.79) and without considering the weight-decay regularization, the TER solution (5.101) can be written using matrix notation, i.e.,

$$
\tilde{w} = \left[\mathbf{P}^T (\mathbf{M}_- + \mathbf{M}_+) \mathbf{P} \right]^{-1} \left[\mathbf{P}^T (\mathbf{M}_- + \mathbf{M}_+) \mathbf{y} \right]
$$

$$
= (\mathbf{P}^T \mathbf{M} \mathbf{P})^{-1} \mathbf{P}^T \mathbf{M} \mathbf{y}, \tag{5.104}
$$

where $\mathbf{M} = \mathbf{M}_- + \mathbf{M}_+$ and elements of $\mathbf{P}$ being ordered according to the two classes. The column vector $\mathbf{y}$ is the target vector given by

$$
\left[\frac{(0 - \eta)}{m^-}, \ldots, \frac{(0 - \eta)}{m^-}, \frac{(0 + \eta)}{m^+}, \ldots, \frac{(0 + \eta)}{m^+} \right]^T. \tag{5.105}
$$

Hence, the TER solution can be further related to a Weighted Least Squares (Weighted-LS) solution. Together with the well-known relationship between LS and Weighted-LS, this observation shows that the AUC in the scaling space (Eq. (5.100)) provides a generalized framework for estimation based on TER, Weighted-LS, and LS.

5.3.6 Relationship with Bayesian Inference

In statistics, the regression model described by Eq. (5.2) is often analyzed in the context of Bayesian inference with the assumption of likelihood and prior probability distributions. Consider a data set of input-target pairs $(\boldsymbol{x}_i, y_i)$, $i = 1, 2, \ldots, m$. We assume that the targets are samples from the model with additive noise e_i:

$$y_i = g(\boldsymbol{x}_i, \boldsymbol{w}) + e_i = \boldsymbol{x}_i^T \boldsymbol{w} + e_i, \tag{5.106}$$

where $e_i = y_i - g(\boldsymbol{x}_i, \boldsymbol{w})$ represent independent samples drawn from a certain noise process. We assume this noise process to be Gaussian with zero mean and variance σ^2, denoted as $\mathcal{N}(0, \sigma^2)$. Also, we assume independence of samples y_i, the corresponding likelihood of the complete data set can be written as Rasmussen and Williams (2006):

$$p(\mathbf{y}|\mathbf{X}, \boldsymbol{w}) = \prod_{i=1}^{m} p(y_i|\boldsymbol{x}_i, \boldsymbol{w})$$

$$= (2\pi\sigma^2)^{-m/2} \exp\left\{ -\frac{1}{2\sigma^2}(\mathbf{y} - \mathbf{X}\boldsymbol{w})^T(\mathbf{y} - \mathbf{X}\boldsymbol{w}) \right\}, \tag{5.107}$$

where $\mathbf{y}$, $\mathbf{X}$ and $\boldsymbol{w}$ are defined similarly to those stacked symbols in (3.25) and (5.4). $\mathbf{X}$ represents the $m \times d$ input data matrix (recall that $\boldsymbol{x} \in \mathbb{R}^d$ where the bias term is not utilized here for data distribution analysis). This likelihood function for the data is next combined with a prior belief (frequently chosen again, to be Gaussian such as $\boldsymbol{w} \sim \mathcal{N}(\mathbf{0}, \boldsymbol{\Sigma}_p)$) about the parameters according to Bayes' theorem to yield the posterior belief about the parameter $\boldsymbol{w}$:

$$p(\boldsymbol{w}|\mathbf{X}, \mathbf{y}) \sim \mathcal{N}(\bar{\boldsymbol{w}}, \boldsymbol{\Sigma}), \tag{5.108}$$

where $\bar{\boldsymbol{w}} = \frac{1}{\sigma^2}(\frac{1}{\sigma^2}\mathbf{P}^T\mathbf{P} + \boldsymbol{\Sigma}_p^{-1})^{-1}\mathbf{P}^T\mathbf{y}$ and $\boldsymbol{\Sigma} = (\frac{1}{\sigma^2}\mathbf{P}^T\mathbf{P} + \boldsymbol{\Sigma}_p^{-1})^{-1}$. While noting the resemblance of $\bar{\boldsymbol{w}}$ with that of LSE learning in (5.6) and (5.32), the interested readers are referred to Rasmussen and Williams (2006) for a detailed account to Gaussian processes and Tipping (2000, 2001), Bishop (2006) for a set of sparse solutions to both regression and classification tasks.

Recall the TER formulation (4.20) with quadratic approximation, which can be written in residual form as

$$\text{TER} = \frac{1}{m^-} \sum_{j=1}^{m^-} L(\epsilon_j) + \frac{1}{m^+} \sum_{i=1}^{m^+} L(\varepsilon_i)$$

$$\approx \frac{1}{m^-} \sum_{j=1}^{m^-} (\epsilon_j + \eta)^2 + \frac{1}{m^+} \sum_{i=1}^{m^+} (\varepsilon_i + \eta)^2, \tag{5.109}$$

where $\epsilon_j = g(\boldsymbol{x}_j^-) - \tau$ and $\varepsilon_i = \tau - g(\boldsymbol{x}_i^+)$ can both be assumed Gaussian with independency among samples within and between the two classes. Consider a linear projection model for g and let $y_j^- = \tau - \eta$ and $y_i^+ = \tau + \eta$, the above equation can be rewritten as

$$\text{TER} \approx \frac{1}{m^-} \sum_{j=1}^{m^-} \left((\boldsymbol{x}_j^-)^T \boldsymbol{w} - y_j^- \right)^2 + \frac{1}{m^+} \sum_{i=1}^{m^+} \left(y_i^+ - (\boldsymbol{x}_i^+)^T \boldsymbol{w} \right)^2. \tag{5.110}$$

Different from common regression, the negative and positive targets namely, y_j^-, $j = 1, \ldots, m^-$ and y_i^+, $i = 1, \ldots, m^+$ can be thought of as two constant target values corresponding to the two classes. Optimization of TER utilizing quadratic approximation in (5.110) can thus be treated as a *dual-Gaussian* regression:

$$p(\mathbf{y}|\mathbf{X}, \boldsymbol{w}) = \left(\prod_{j=1}^{m^-} p(y_j^- | \boldsymbol{x}_j^-, \boldsymbol{w}) \right) \left(\prod_{i=1}^{m^+} p(y_i^+ | \boldsymbol{x}_i^+, \boldsymbol{w}) \right)$$

$$= (2\pi\sigma_-^2)^{-m^-/2} \exp \left\{ -\frac{1}{2\sigma_-^2} (\mathbf{y}^- - \mathbf{X}_- \boldsymbol{w})^T (\mathbf{y}^- - \mathbf{X}_- \boldsymbol{w}) \right\}$$

$$\times (2\pi\sigma_+^2)^{-m^+/2} \exp \left\{ -\frac{1}{2\sigma_+^2} (\mathbf{y}^+ - \mathbf{X}_+ \boldsymbol{w})^T (\mathbf{y}^+ - \mathbf{X}_+ \boldsymbol{w}) \right\}. \tag{5.111}$$

Based on the Bayes' rule, the posterior probability of parameter estimate can be written as

$$p(\boldsymbol{w}|\mathbf{y}, \mathbf{X}) = \frac{p(\mathbf{y}|\mathbf{X}, \boldsymbol{w}) p(\boldsymbol{w})}{p(\mathbf{y}|\mathbf{X})}$$

$$\propto \exp \left\{ -\frac{1}{2\sigma_-^2} (\mathbf{y}^- - \mathbf{X}_- \boldsymbol{w})^T (\mathbf{y}^- - \mathbf{X}_- \boldsymbol{w}) \right\}$$

$$\times \exp \left\{ -\frac{1}{2\sigma_+^2} (\mathbf{y}^+ - \mathbf{X}_+ \boldsymbol{w})^T (\mathbf{y}^+ - \mathbf{X}_+ \boldsymbol{w}) \right\} \exp \left(-\frac{1}{2} \boldsymbol{w}^T \Sigma_p^{-1} \boldsymbol{w} \right). \tag{5.112}$$

Since

$$
\begin{aligned}
(\mathbf{y} &- \mathbf{X}w)^T (\mathbf{y} - \mathbf{X}w) \\
&= ((\mathbf{y} - \mathbf{X}\bar{w}) + (\mathbf{X}\bar{w} - \mathbf{X}w))^T ((\mathbf{y} - \mathbf{X}\bar{w}) + (\mathbf{X}\bar{w} - \mathbf{X}w)) \\
&= (\mathbf{y} - \mathbf{X}\bar{w})^T (\mathbf{y} - \mathbf{X}\bar{w}) + (w - \bar{w})^T (\mathbf{X}^T \mathbf{X})(w - \bar{w}) \\
&\qquad\qquad + \underbrace{2(\mathbf{X}\bar{w} - \mathbf{X}w)^T (\mathbf{y} - \mathbf{X}\bar{w})}
\end{aligned}
\tag{5.113}
$$

with the underbraced term equals to zero, and by retaining only those terms dependent on w, we have

$$
p(w|\mathbf{y}, \mathbf{X}) \propto \exp\left(-\frac{1}{2}(w - \bar{w})^T \left(\frac{1}{\sigma_-^2}\mathbf{X}_-^T \mathbf{X}_- + \frac{1}{\sigma_+^2}\mathbf{X}_+^T \mathbf{X}_+ + \Sigma_p^{-1} \right)(w - \bar{w}) \right),
\tag{5.114}
$$

where

$$
\bar{w} = \left(\frac{1}{\sigma_-^2}\mathbf{X}_-^T \mathbf{X}_- + \frac{1}{\sigma_+^2}\mathbf{X}_+^T \mathbf{X}_+ + \Sigma_p^{-1} \right)^{-1} \left(\frac{1}{\sigma_-^2}\mathbf{X}_-^T \mathbf{y}^- + \frac{1}{\sigma_+^2}\mathbf{X}_+^T \mathbf{y}^+ \right),
\tag{5.115}
$$

and the covariance is

$$
\Sigma = (\frac{1}{\sigma_-^2}\mathbf{X}_-^T \mathbf{X}_- + \frac{1}{\sigma_+^2}\mathbf{X}_+^T \mathbf{X}_+ + \Sigma_p^{-1})^{-1}.
\tag{5.116}
$$

The mean value given by $\bar{w}$ resembles the TER solution in (5.75).

Consequently, the quadratic approximation to the AUC-based objective in Eq. (5.88) and the TAUC extension can each be treated as a Gaussian process. This process takes pairwise data $(x_j^- - x_i^+)$ as regression input, along with a corresponding scaled version.

5.3.7 Summary of Classifier Relationships

Based on the scaling and translation settings of (5.102), several of the above classifiers can be related in terms of data manipulation. Table 5.1 shows the details of these parametric settings of (5.102) with a Novel classifier at random setting for both the scaling and translation parameters.

5.4 Recursive Learning

Under the situation where a regular training update is needed for every newly arrived data, a recursive learning formulation would be useful. Particularly, when dealing

Table 5.1 Summary of classifiers and settings

	Classifier	Class-size, Reg, Offset				Scaling				Translation			
		m^+	m^-	b	η	β_1	β_2	γ_1	γ_2	u_1	u_2	v_1	v_2
1.	LSE	1	1	b	1/2	1	0	0	1	**0**	**0**	**0**	**0**
2.	FLD	1	1	b	1/2	1	0	0	1	μ_-	**0**	**0**	μ_+
3.	TER	m^+	m^-	b	1/2	1	0	0	1	**0**	**0**	**0**	**0**
4.	AUC	m^+	m^-	b	1/2	1	1	1	1	**0**	**0**	**0**	**0**
5.	Novel	m^+	m^-	b	1/2	r	r	r	r	r	r	r	r

m^+, m^- : sizes of positive class and negative class training samples respectively,

b : weight-decay regularization parameter which is chosen as 10^{-4} according to Toh & Eng (2008),

η, r : η is quadratic offset as seen in (5.102), r is random number,

μ_+, μ_- : $\mu_+ = \frac{1}{m^+} \sum_{i=1}^{m^+} p_i^+$ which is mean of positive class features,

$\mu_- = \frac{1}{m^-} \sum_{j=1}^{m^-} p_j^-$ which is mean of negative class features,

$\beta_1, \beta_2, \gamma_1, \gamma_2$: scaling parameters (scalars),

u_1, u_2, v_1, v_2 : translational parameters (vectors)

with a huge amount of data samples and when the learning model is to be updated frequently, the batch mode learning which requires the entire model to be re-trained for each arriving bulk of samples could be impractical. The recursive learning is also known as online learning and sometimes called incremental learning. In the following, the conventional recursive least squares will be first presented. Subsequently, it is extended to TER- and AUC-based learning.

5.4.1 Recursive Least Squares Learning

Suppose the training samples arrive one by one in the following sequence $x_1, x_2, \ldots,$ x_m, x_{m+1} where x_{m+1} denotes the newly arrived sample vector. These training samples are projected onto the feature space by linear projection giving column vectors $p(x_1), p(x_2), \ldots, p(x_m), p$ which are stacked as $\mathbf{P}^T = [p(x_1), \ldots, p(x_m), p]$.

Let

$$\mathbf{R}_{m+1} = \mathbf{P}^T \mathbf{P} + b\mathbf{I}, \tag{5.117}$$

and

$$\mathbf{Q}_{m+1} = \mathbf{P}^T \mathbf{y}. \tag{5.118}$$

By separating the covariance of the newly arrived sample p from the remaining stack, we can write

$$\mathbf{P}^T \mathbf{P} = \sum_{i=1}^{m+1} p(x_i) p(x_i)^T$$

$$= \sum_{i=1}^{m} p(x_i)p(x_i)^T + pp^T$$

$$= \mathbf{P}_m^T\mathbf{P}_m + pp^T. \tag{5.119}$$

Hence

$$\mathbf{R}_{m+1} = \mathbf{R}_m + pp^T, \tag{5.120}$$

and

$$\mathbf{Q}_{m+1} = \mathbf{Q}_m + py_{m+1}. \tag{5.121}$$

If the system is to forget the old training samples, then (5.120) and (5.121) can be modified as

$$\mathbf{R}_{m+1} = \lambda(t)\mathbf{R}_m + pp^T, \tag{5.122}$$

and

$$\mathbf{Q}_{m+1} = \lambda(t)\mathbf{Q}_m + py_{m+1}, \tag{5.123}$$

where $\lambda(t) \in (0, 1)$ is called a forgetting factor which can vary with time.

Let $\mathbf{A} = \mathbf{R}_m, \mathbf{B} = p(x_{m+1}) = p, \mathbf{C} = \mathbf{I} = 1$ (scalar), $\mathbf{D} = p(x_{m+1})^T = p^T$, then based on the *matrix inversion lemma*

$$(\mathbf{A} + \mathbf{BCD})^{-1} = \mathbf{A}^{-1} - \mathbf{A}^{-1}\mathbf{B}\left(\mathbf{C}^{-1} + \mathbf{DA}^{-1}\mathbf{B}\right)^{-1}\mathbf{DA}^{-1} \tag{5.124}$$

we have

$$\mathbf{R}_{m+1}^{-1} = \left[\lambda(t)\mathbf{R}_m + pp^T\right]^{-1}$$

$$= \frac{1}{\lambda(t)}\mathbf{R}_m^{-1} - \frac{1}{\lambda(t)}\mathbf{R}_m^{-1}\lambda p\left(1 + p^T\frac{1}{\lambda(t)}\mathbf{R}_m^{-1}p\right)^{-1}p^T\frac{1}{\lambda(t)}\mathbf{R}_m^{-1}$$

$$= \frac{1}{\lambda(t)}\mathbf{R}_m^{-1} - \frac{1}{(\lambda(t))^2}\mathbf{R}_m^{-1}pp^T\mathbf{R}_m^{-1}\left/\left(\frac{\lambda(t)}{\lambda(t)} + \frac{1}{\lambda(t)}p^T\mathbf{R}_m^{-1}p\right)\right.$$

$$\tag{5.125}$$

Substituting (5.125) and (5.123) into

$$w_{m+1} = (\mathbf{P}^T\mathbf{P} + b\mathbf{I})^{-1}\mathbf{P}^T\mathbf{y} = \mathbf{R}_{m+1}^{-1}\mathbf{Q}_{m+1}, \tag{5.126}$$

and simplifying the multiplied-out expressions, we have the recursive solution given by

$$w_{m+1} = w_m + \frac{\mathbf{R}_m^{-1}p\left(y_{m+1} - p^T w_m\right)}{\left(\lambda(t) + p^T\mathbf{R}_m^{-1}p\right)}. \tag{5.127}$$

If $\lambda(t) = \lambda$, and when $|\lambda| < 1$, then the memory of past data converges in the form of

$$\sum_{i=0}^{\infty} \lambda^i = \frac{1}{1 - \lambda}. \tag{5.128}$$

5.4.2 Recursive Classification TER Learning

Suppose the sequence of arriving data is indexed by $t = 1, 2, \ldots, t - 1, t$ where the total number of samples arrived is $m_t = m_t^+ + m_t^-$. In this formulation for recursive binary classification learning, a zero-one value of $y_i \in \{0, 1\}$ is used to indicate the class label of the ith sample, i.e., $y_i = 0$ if sample x_i (and its transformed feature $p := p(x_i)$) belongs to the negative class labeled by y_i^- and $y_i = 1$ if the sample belongs to the positive class labeled by y_i^+. This class indicator $y_i \in \{0, 1\}$ when used together with the sample feature p (such as $(1 - y_i)p$ or $y_i p$) can be used to sort out the two classes and relieves the need to label p by a $+$ or $-$ superscript.

Based on (5.74), we can write the estimation at time instant t as

$$
\hat{w}_t = \left[b\mathbf{I} + \frac{1}{m_t^-} \sum_{j=1}^{m_t^-} p_j^- p_j^{-T} + \frac{1}{m_t^+} \sum_{i=1}^{m_t^+} p_i^+ p_i^{+T} \right]^{-1} \left[\frac{(\tau - \eta)}{m_t^-} \sum_{j=1}^{m_t^-} p_j^- + \frac{(\tau + \eta)}{m_t^+} \sum_{i=1}^{m_t^+} p_i^+ \right]
$$

$$
= \left[b\mathbf{I} + \frac{1}{m_t^-} \sum_{j=1}^{m_t^-} p_j^- p_j^{-T} + \frac{1}{m_t^+} \sum_{i=1}^{m_t^+} p_i^- p_i^{+T} \right]^{-1} \left[\frac{1}{m_t^-} \sum_{j=1}^{m_t^-} p_j^- y_i^- + \frac{1}{m_t^+} \sum_{i=1}^{m_t^+} p_i^+ y_i^+ \right]
$$

$$
= \left[b\mathbf{I} + \underbrace{\sum_{j=1}^{m_t^-} \frac{1 - y_j}{m_t^-} p_j p_j^{T}}_{\mathbf{S}_t^-} + \underbrace{\sum_{i=1}^{m_t^+} \frac{y_i}{m_t^+} p_i p_i^{T}}_{\mathbf{S}_t^+} \right]^{-1} \left[\underbrace{\sum_{j=1}^{m_t^-} \frac{1 - y_j}{m_t^-} p_j y_i^-}_{\mathbf{y}_t^-} + \underbrace{\sum_{i=1}^{m_t^+} \frac{y_i}{m_t^+} p_i y_i^+}_{\mathbf{y}_t^+} \right]
$$

$$
= (b\mathbf{I} + \mathbf{S}_t^- + \mathbf{S}_t^+)^{-1} (\mathbf{y}_t^- + \mathbf{y}_t^+)
$$

$$
= (b\mathbf{I} + \mathbf{S}_t)^{-1} \mathbf{y}_t
$$

$$
= \mathbf{R}^{-1} \mathbf{y}_t \tag{5.129}
$$

The recursion of the projected target term $\mathbf{y}_t$ and the covariance term $\mathbf{R}_t^{-1}$ shall be treated separately below.

Recursion of $\mathbf{y}_t$: The newly arrived target at time instant t could either belong to the positive class or the negative class. Hence either the positive class or the negative class of the projected target should be updated utilizing the indicator y_t:

$$
\mathbf{y}_t^- = \mathbf{y}_{t-1}^- + \frac{(1 - y_t)}{m_t^-} (p_t^- y_t^- - \mathbf{y}_{t-1}^-), \tag{5.130}
$$

$$
\mathbf{y}_t^+ = \mathbf{y}_{t-1}^+ + \frac{y_t}{m_t^+} (p_t^+ y_t^+ - \mathbf{y}_{t-1}^+). \tag{5.131}
$$

Recursion of $\mathbf{R}_t^{-1}$: The recursion here requires to separate out the newly arrived data in $\mathbf{R}_t$ at time instant t from those past data before $t-1$. In other words, we need to express $\mathbf{R}_t$ in terms of $\mathbf{R}_{t-1}$ plus the newly arrived term. Based on those definitions in (5.129), the terms $\mathbf{S}_t^+$ and $\mathbf{S}_t^-$ within $\mathbf{R}_t$ can be processed as follows:

$$
\begin{aligned}
\mathbf{S}_t^+ &= \sum_{i=1}^{m_t^+} \frac{y_i}{m_t^+} \boldsymbol{p}_i \boldsymbol{p}_i{}^T \\[2mm]
&= \frac{y_t}{m_t^+} \boldsymbol{p}_t \boldsymbol{p}_t{}^T + \frac{m_{t-1}^+}{m_t^+} \sum_{i=1}^{m_{t-1}^+} \frac{y_i}{m_{t-1}^+} \boldsymbol{p}_i \boldsymbol{p}_i{}^T \\[2mm]
&= \frac{y_t}{m_t^+} \boldsymbol{p}_t \boldsymbol{p}_t{}^T + \frac{m_{t-1}^+}{m_t^+} \mathbf{S}_{t-1}^+
\end{aligned}
\tag{5.132}
$$

In a similar manner of separating the current sample for $\mathbf{S}_t^-$, we have

$$
\mathbf{S}_t^- = \frac{1-y_t}{m_t^-} \boldsymbol{p}_t \boldsymbol{p}_t{}^T + \frac{m_{t-1}^-}{m_t^-} \mathbf{S}_{t-1}^-
\tag{5.133}
$$

By putting them in $\mathbf{R}_t$, we have

$$
\begin{aligned}
\mathbf{R}_t &= \mathbf{S}_t + b\mathbf{I} \\[2mm]
&= \frac{y_t}{m_t^+} \boldsymbol{p}_t \boldsymbol{p}_t{}^T + \frac{m_{t-1}^+}{m_t^+} \mathbf{S}_{t-1}^+ + \frac{1-y_t}{m_t^-} \boldsymbol{p}_t \boldsymbol{p}_t{}^T + \frac{m_{t-1}^-}{m_t^-} \mathbf{S}_{t-1}^- + b\mathbf{I} \\[2mm]
&= \left(\frac{y_t}{m_t^+} + \frac{1-y_t}{m_t^-} \right) \boldsymbol{p}_t \boldsymbol{p}_t{}^T + \frac{m_{t-1}^+}{m_t^+} \mathbf{S}_{t-1}^+ + \frac{m_{t-1}^-}{m_t^-} \mathbf{S}_{t-1}^- + b\mathbf{I} \\[2mm]
&= \left(\frac{y_t}{m_t^+} + \frac{1-y_t}{m_t^-} \right) \boldsymbol{p}_t \boldsymbol{p}_t{}^T + (\mathbf{S}_{t-1}^+ + \mathbf{S}_{t-1}^- + b\mathbf{I}) \\[2mm]
&\qquad + \left(\frac{m_{t-1}^+}{m_t^+} - 1 \right) \mathbf{S}_{t-1}^+ + \left(\frac{m_{t-1}^-}{m_t^-} - 1 \right) \mathbf{S}_{t-1}^- \\[2mm]
&= \left(\frac{y_t}{m_t^+} + \frac{1-y_t}{m_t^-} \right) \boldsymbol{p}_t \boldsymbol{p}_t{}^T + \mathbf{R}_{t-1} \\[2mm]
&\qquad + \left(\frac{m_{t-1}^+ - m_t^+}{m_t^+} \right) \mathbf{S}_{t-1}^+ + \left(\frac{m_{t-1}^- - m_t^-}{m_t^-} \right) \mathbf{S}_{t-1}^- \\[2mm]
&= \left(\frac{y_t}{m_t^+} + \frac{1-y_t}{m_t^-} \right) \boldsymbol{p}_t \boldsymbol{p}_t{}^T + \mathbf{R}_{t-1} + \frac{-y_t}{m_t^+} \mathbf{S}_{t-1}^+ + \frac{-(1-y_t)}{m_t^-} \mathbf{S}_{t-1}^- .
\end{aligned}
\tag{5.134}
$$

The terms $m_{t-1}^+ - m_t^+ = -y_t$ and $m_{t-1}^- - m_t^- = -(1-y_t)$ indicate that the newly arrived sample belongs to one of the two classes labeled by y_t and $1-y_t$.

Let

$$\alpha_t = \left(\frac{y_t}{m_t^+} + \frac{1 - y_t}{m_t^-} \right) \tag{5.135}$$

then (5.134) can be rewritten as

$$\mathbf{R}_t = \mathbf{R}_{t-1} + \frac{-y_t}{m_t^+} \mathbf{S}_{t-1}^+ + \frac{-(1 - y_t)}{m_t^-} \mathbf{S}_{t-1}^- + \alpha_t \, \boldsymbol{p}_t \boldsymbol{p}_t^{\,T}. \tag{5.136}$$

To facilitate the recursion based on the Sherman-Morrison formula, we define

$$\mathbf{Q}_{t-1} = \mathbf{R}_{t-1} + \frac{-y_t}{m_t^+} \mathbf{S}_{t-1}^+ + \frac{-(1 - y_t)}{m_t^-} \mathbf{S}_{t-1}^-, \tag{5.137}$$

so that (5.136) can be written in two terms as shown below:

$$\mathbf{R}_t = \mathbf{Q}_{t-1} + \alpha_t \, \boldsymbol{p}_t \boldsymbol{p}_t^{\,T}. \tag{5.138}$$

Next, the Sherman-Morrison formula

$$(\mathbf{A} + \mathbf{b}\mathbf{c}^T)^{-1} = \mathbf{A}^{-1} - \frac{\mathbf{A}^{-1}\mathbf{b}\mathbf{c}^T\mathbf{A}^{-1}}{1 + \mathbf{c}^T\mathbf{A}^{-1}\mathbf{b}} \tag{5.139}$$

is applied for recursion of $\mathbf{R}_t$ with $\mathbf{A} := \mathbf{Q}_t$, $\mathbf{b} := \alpha_t \boldsymbol{p}_t$ and $\mathbf{c} := \boldsymbol{p}_t$ giving

$$\mathbf{R}_t^{-1} = (\mathbf{Q}_{t-1} + \alpha_t \boldsymbol{p}_t \boldsymbol{p}_t^T)^{-1} = \mathbf{Q}_{t-1}^{-1} - \frac{\alpha_t \mathbf{Q}_{t-1}^{-1} \boldsymbol{p}_t \boldsymbol{p}_t^T \mathbf{Q}_{t-1}^{-1}}{1 + \alpha_t \boldsymbol{p}_t^T \mathbf{Q}_{t-1}^{-1} \boldsymbol{p}_t}. \tag{5.140}$$

For recursion of $\mathbf{Q}_{t-1}$, the Kailath variant of the Woodbury identity is adopted as

$$(\mathbf{A} + \mathbf{B}\mathbf{C})^{-1} = \mathbf{A}^{-1} - \mathbf{A}^{-1}\mathbf{B}(\mathbf{I} + \mathbf{C}\mathbf{A}^{-1}\mathbf{B})^{-1}\mathbf{C}\mathbf{A}^{-1} \tag{5.141}$$

with $\mathbf{A} := \mathbf{R}_{t-1}$, $\mathbf{B} := \left(\frac{-y_t}{m_t^+} \mathbf{S}_{t-1}^+ + \frac{-(1-y_t)}{m_t^-} \mathbf{S}_{t-1}^- \right)$ and $\mathbf{C} := \mathbf{I}$ giving

$$\mathbf{Q}_{t-1}^{-1} = (\mathbf{R}_{t-1} + \mathbf{B})^{-1} = \mathbf{R}_{t-1}^{-1} - \mathbf{R}_{t-1}^{-1}\mathbf{B}(\mathbf{I} + \mathbf{R}_{t-1}^{-1}\mathbf{B})^{-1}\mathbf{R}_{t-1}^{-1}. \tag{5.142}$$

5.4.3   *Recursive AUC Learning*

The solution for binary classification in AUC optimization (5.90) contains a double summation of pairwise comparison between negative (such as class-1) and positive (such as class-2) samples. This means that in recursive formulation, each new incoming data sample needs to be paired with all the accumulated samples of the opposite class. The number of pairwise samples thus increases almost exponentially with the number of training samples.

In order to facilitate formulation in recursive form, the elementwise summation is expressed in terms of matrix-vector operation. The pairwise samples are stacked as a feature error (difference) column vector given by

$$\mathbf{e}_k = \boldsymbol{p}(\boldsymbol{x}_j^-) - \boldsymbol{p}(\boldsymbol{x}_i^+) \in \mathbb{R}^{D+1}, \tag{5.143}$$

where $k \in \{1, 2, \ldots, m^- \times m^+\}$ is an index for each pairwise sample for all pairing between $j \in \{1, 2, \ldots, m^-\}$ and $i \in \{1, 2, \ldots, m^+\}$ which are respectively the indices of negative and positive samples. A packed matrix of pairwise vectors can then be written as

$$\mathbf{E} = \begin{bmatrix} \mathbf{e}_1, \ \mathbf{e}_2, \ \ldots, \ \mathbf{e}_{(m^- \times m^+)} \end{bmatrix}^T \in \mathbb{R}^{(m^- \times m^+) \times (D+1)}. \tag{5.144}$$

Based on this matrix notation for pairwise samples, the solution (5.90) can be rewritten as

$$\hat{w} = \left(\frac{1}{m^- m^+} \mathbf{E}^T \mathbf{E} + b\mathbf{I} \right)^{-1} \left(\frac{-\eta}{m^+ m^-} \mathbf{E}^T \mathbf{1} \right), \tag{5.145}$$

where $\mathbf{1} = \begin{bmatrix} 1, \ 1, \ \ldots, \ 1 \end{bmatrix}^T \in \mathbb{R}^{m^- m^+ \times 1}$. With some algebraic manipulation, we have

$$
\begin{aligned}
\hat{w} &= \left(\tfrac{1}{m^- m^+} \right)^{-1} \left(\mathbf{E}^T \mathbf{E} + m^- m^+ b\mathbf{I} \right)^{-1} \tfrac{1}{m^- m^+} \left(-\eta \mathbf{E}^T \mathbf{1} \right) \\
&= \left(\mathbf{E}^T \mathbf{E} + c\mathbf{I} \right)^{-1} \mathbf{E}^T \boldsymbol{\eta} \\
&= \mathbf{R}^{-1} \mathbf{Q}
\end{aligned}
\tag{5.146}
$$

where $c = m^- m^+ b$, $\mathbf{R} = \mathbf{E}^T \mathbf{E} + c\mathbf{I}$, $\mathbf{Q} = \mathbf{E}^T \boldsymbol{\eta}$, and $\boldsymbol{\eta} = -\eta \begin{bmatrix} 1, \ 1, \ \ldots, \ 1 \end{bmatrix}^T \in \mathbb{R}^{m^- m^+ \times 1}$.

Consider the case where the training samples arrive one by one in a sequence such as $\boldsymbol{x}_1, \boldsymbol{x}_2, \ldots, \boldsymbol{x}_m, \boldsymbol{x}_{m+1}$. Here, $\boldsymbol{x}_{m+1}$ is a newly arrived sample vector at the $(m+1)^{th}$ instance.

For each of the new $(m+1)^{th}$ incoming sample for training, a set of pairwise error vectors $\boldsymbol{\varepsilon}_{m+1}$ can be generated by pairing the new sample with those stored samples of opposite label:

$$\boldsymbol{\varepsilon}_{m+1} = \begin{cases} \{ \boldsymbol{p}(\boldsymbol{x}_{m+1}) - \boldsymbol{p}(\boldsymbol{x}_i^+), \ i = 1, 2, \ldots, m^+ \} & \text{if } \boldsymbol{x}_{m+1} \in \text{class}^-, \\ \{ \boldsymbol{p}(\boldsymbol{x}_j^-) - \boldsymbol{p}(\boldsymbol{x}_{m+1}), \ j = 1, 2, \ldots, m^- \} & \text{if } \boldsymbol{x}_{m+1} \in \text{class}^+, \end{cases} \tag{5.147}$$

where m^- and m^+ are respectively the number of accumulated negative and positive samples just before the arrival of the $(m+1)^{th}$ sample. Since the subtraction is based upon whether the first item is larger than the second item which resulted in an error count, the newly arrived sample is placed before the positive sample if it belongs to the negative class and after the negative sample otherwise. The error features are subsequently packed as $\varepsilon_{m+1} = [\mathbf{e}_1, \mathbf{e}_2, \ldots, \mathbf{e}_{(m^- \text{ or } m^+)}]^T \in \mathbb{R}^{m^- \times (D+1)}$ or $\in \mathbb{R}^{m^+ \times (D+1)}$.

Let $\mathbf{P}_m^- \in \mathbb{R}^{m^- \times (D+1)}$ and $\mathbf{P}_m^+ \in \mathbb{R}^{m^+ \times (D+1)}$ denote respectively the accumulated negative and positive projected samples. Then, by stacking the comparisons of each pair of ε_{m+1} with the entire set of positive and negative samples in matrix notation, we have

$$\varepsilon_{m+1} = \begin{cases} \mathbf{1}_m^+ p(\mathbf{x}_{m+1})^T - \mathbf{P}_m^+ & \text{if } \mathbf{x}_{m+1} \in \text{class}^- \ (y_{m+1} = 0), \\ \mathbf{P}_m^- - \mathbf{1}_M^- p(\mathbf{x}_{m+1})^T & \text{if } \mathbf{x}_{m+1} \in \text{class}^+ \ (y_{m+1} = 1), \end{cases}$$
$$= (1 - 2y_{m+1}) \left(\mathbf{1} p(\mathbf{x}_{m+1})^T - \mathbf{P}_m^{\{y=1-y_{m+1}\}} \right), \tag{5.148}$$

where $(1 - 2y_{m+1})$ is a sign to indicate the order of subtraction, $\mathbf{P}_m^{\{y=1-y_{m+1}\}}$ consists of either negative samples ($\mathbf{P}_m^{\{y=0\}} = \mathbf{P}_m^-$) or positive samples ($\mathbf{P}_m^{\{y=1\}} = \mathbf{P}_m^+$) depending on the known class label (y_{m+1}) of the newly arrived training sample, $\mathbf{1} = [1, \ 1, \ldots, \ 1]^T$ where its length is the number of accumulated samples given by the number of rows of $\mathbf{P}_m^{\{y=1-y_{m+1}\}}$.

Let $\mathbf{E}_{m+1}$ denotes the accumulated pairwise matrix containing $(m^- + 1)m^+$ or $(m^+ + 1)m^-$ number of pairwise samples. With this inclusion of matrix size notation, (5.146) can be rewritten as

$$\hat{w}_{m+1} = \left(\mathbf{E}_{m+1}^T \mathbf{E}_{m+1} + c\mathbf{I} \right)^{-1} \mathbf{E}_{m+1}^T \boldsymbol{\eta}_{m+1} = \mathbf{R}_{m+1}^{-1} \mathbf{Q}_{m+1}, \tag{5.149}$$

where $\mathbf{R}_{m+1} = \mathbf{E}_{m+1}^T \mathbf{E}_{m+1} + c\mathbf{I}, \mathbf{Q}_{m+1} = \mathbf{E}_{m+1}^T \boldsymbol{\eta}_{m+1}$.

The next step is to express $\mathbf{E}_{m+1}$ in terms of the newly generated chunk of ε_{m+1} and the previously accumulated $\mathbf{E}_m$:

$$\mathbf{E}_{m+1} = \begin{bmatrix} \mathbf{E}_m \\ \varepsilon_{m+1} \end{bmatrix}, \tag{5.150}$$

and then rewrite the correlation matrix $\mathbf{R}_{m+1}$ and $\mathbf{Q}_{m+1}$ in (5.149) as

$$\mathbf{R}_{m+1} = (\mathbf{E}_m^T \mathbf{E}_m + c\mathbf{I}) + \varepsilon_{m+1}^T \varepsilon_{m+1} = \mathbf{R}_m + \varepsilon_{m+1}^T \varepsilon_{m+1}, \tag{5.151}$$

$$\mathbf{Q}_{m+1} = \mathbf{E}_m^T \boldsymbol{\eta}_m + \varepsilon_{m+1}^T \boldsymbol{\eta} = \mathbf{Q}_m + \varepsilon_{m+1}^T \boldsymbol{\eta}. \tag{5.152}$$

By substituting (5.151) and (5.152) into (5.149), the recursive AUC learning solution can be obtained as follows:

$$\begin{aligned} \hat{w}_{m+1} &= \mathbf{R}_{m+1}^{-1} \mathbf{Q}_{m+1} \\ &= \mathbf{R}_{m+1}^{-1} \left(\mathbf{Q}_m + \varepsilon_{m+1}^T \boldsymbol{\eta} \right) \\ &= \mathbf{R}_{m+1}^{-1} \left(\mathbf{R}_m \hat{w}_m + \varepsilon_{m+1}^T \boldsymbol{\eta} \right) \\ &= \mathbf{R}_{m+1}^{-1} \left[\left(\mathbf{R}_{m+1} - \varepsilon_{m+1}^T \varepsilon_{m+1} \right) \hat{w}_m + \varepsilon_{m+1}^T \boldsymbol{\eta} \right] \end{aligned} \tag{5.153}$$

$$= \mathbf{R}_{m+1}^{-1} \left[\mathbf{R}_{m+1} \hat{\boldsymbol{w}}_m + \varepsilon_{m+1}^T \left(\eta - \varepsilon_{m+1} \hat{\boldsymbol{w}}_m \right) \right]. \tag{5.154}$$

Therefore

$$\hat{\boldsymbol{w}}_{m+1} = \hat{\boldsymbol{w}}_m + \mathbf{R}_{m+1}^{-1} \varepsilon_{m+1}^T \left(\eta - \varepsilon_{m+1} \hat{\boldsymbol{w}}_m \right), \tag{5.155}$$

where

$$\mathbf{R}_{m+1}^{-1} = \mathbf{R}_m^{-1} - \mathbf{R}_m^{-1} \varepsilon_{m+1}^T \left(\mathbf{I} + \varepsilon_{m+1} \mathbf{R}_m^{-1} \varepsilon_{m+1}^T \right)^{-1} \varepsilon_{m+1} \mathbf{R}_m^{-1}. \tag{5.156}$$

Here, (5.156) has been obtained based on the *matrix inversion lemma* (Woodbury 1950; Sherman and Morrison 1950),

$$(\mathbf{A} + \mathbf{BCD})^{-1} = \mathbf{A}^{-1} - \mathbf{A}^{-1} \mathbf{B} \left(\mathbf{C}^{-1} + \mathbf{D} \mathbf{A}^{-1} \mathbf{B} \right)^{-1} \mathbf{D} \mathbf{A}^{-1} \tag{5.157}$$

by letting $\mathbf{A} = \mathbf{R}_m$, $\mathbf{B} = \varepsilon_{m+1}^T$, $\mathbf{C} = \mathbf{I}$, $\mathbf{D} = \varepsilon_{m+1}$.

Equation (5.156) can also be written as

$$\mathbf{R}_{m+1}^{-1} = \mathbf{R}_m^{-1} - \mathbf{R}_m^{-1} \left(\mathbf{I} + \varepsilon_{m+1}^T \varepsilon_{m+1} \mathbf{R}_m^{-1} \right)^{-1} \varepsilon_{m+1}^T \varepsilon_{m+1} \mathbf{R}_m^{-1}, \tag{5.158}$$

based on Woodbury (1950), Sherman and Morrison (1950):

$$(\mathbf{A} + \mathbf{BCD})^{-1} = \mathbf{A}^{-1} - \mathbf{A}^{-1} \left(\mathbf{C}^{-1} + \mathbf{BCD} \mathbf{A}^{-1} \right)^{-1} \mathbf{BCD} \mathbf{A}^{-1} \tag{5.159}$$

The main difference between (5.156) and (5.158) is the dimension of matrix operation for updating $\mathbf{R}_{m+1}^{-1}$. In (5.156), the dimension of $\left(\mathbf{I} + \varepsilon_{m+1} \mathbf{R}_m^{-1} \varepsilon_{m+1}^T \right)^{-1}$ is similar to that of $\varepsilon_{m+1} \varepsilon_{m+1}^T$ which is either $\mathbb{R}^{m^+ \times m^+}$ or $\mathbb{R}^{m^- \times m^-}$ according to the class label of the newly arrived data. If the newly arrived data belongs to positive category, then the dimension of $\varepsilon_{m+1} \varepsilon_{m+1}^T$ is $\mathbb{R}^{m^- \times m^-}$ since this newly arrived data is paired with those accumulated data of negative category to generate ε_{m+1}. The converse applies if the newly arrived data belongs to negative category. In (5.158), the dimension of $\left(\mathbf{I} + \varepsilon_{m+1}^T \varepsilon_{m+1} \mathbf{R}_m^{-1} \right)^{-1}$ is $\mathbb{R}^{(D+1) \times (D+1)}$ where $D + 1$ is the input feature dimension derived from $p(x)$.

Between the two alternatives ($D + 1$ or $\{m^-, m^+\}$) to compute the matrix inverse, the one with smaller dimension can be chosen in order to lower the computational cost during the recursive update. To further reduce the computational load, a subsampling can be applied to ε_{m+1} to obtain a reduced pairing set. To cater for possible data drift, the subsampling can be performed to include only recently accumulated data.

In summary, for each newly incoming sample, a pairwise difference vector ε_{m+1} is first generated. Then the learning solution is updated using the newly generated ε_{m+1} and the previous solution. Since the new solution is generated cumulatively based on the stored solution and the new data, the online computational cost is often much lower than that of the batch mode learning where a re-computation of the entire bulk of data is performed for every newly arrived data.

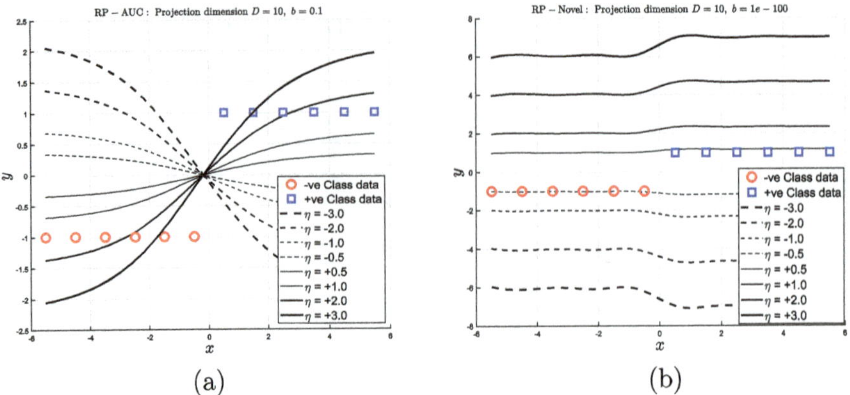

Fig. 5.6 Training data and outputs of a single-dimensional learning example with varying offset η values. Data used in training are marked in circles and squares; **a** Based on RP-AUC (5.90); **b** Based on RP-Novel (5.102)

5.5   Decision Landscapes

In the following, we observe the decision landscapes of the above learning methods using one- and two-dimensional examples. The Random Projection model given by (3.104) shall be used as the learning model. Firstly, the one-dimension example shall be used to observe the impact of the offset value η in the AUC and Novel learning outputs. Subsequently, the decision boundary for learning based on the Least Squares Error Ridge regression (LSE) given by (5.32), the Total Error Rate (TER) given by (5.75), the Area Under the ROC Curve (AUC) given by (5.90), the Fisher Linear Discriminant (FLD) given by (5.103), and the Novel given by (5.102) shall be studied. Finally, the trained weight coefficients of these learning methods shall be observed in terms of their L_2-norm magnitudes.

Effect of Offset on Output Mapping: In Fig. 5.6a–b, the learning outputs of RP-AUC and RP-Novel using a ten-dimensional random projection model at varying offset (η) values are shown. The notations RP-AUC and RP-Novel indicate the Random Projection model (3.104) has been adopted to learn the one-dimensional data using the AUC (5.90) and the Novel (5.102) methods respectively. Here, we recall that $p \in \mathbb{R}^{(D+1) \times d}$ (3.104), where $D = 10$ is the chosen projection dimension and $d = 1$ for the one-dimensional input x as shown in Fig. 5.6. It is seen that a high offset value results in a large output range, and this phenomenon applies to negative offset values but in a reverse sense. Except for numerical scaling and offset, similar trends can be observed for varying $b \in [10^{-100}, 1]$. These plots show that the distributions of learning outputs provide relevant mapping of the training data which distinguishes between the positive and negative classes where an appropriate threshold setting can be applied.

Effect of Projection Dimension: Fig. 5.7 show the decision boundaries obtained from training the Random Projection model using the Linear regression, LSE, TER, AUC, FLD, and Novel methods based on the given set of two-dimensional data marked by blue squares and red circles. In the figure, several dimensions of projection models ($D = 2, 10, 50$, and 200) have been trained under a regularization setting of $b = 10^{-4}$. The thresholds for RP-LSE and RP-TER have been set at zero whereas thresholds for RP-FLD, RP-AUC, and Novel classifiers have been set by maximizing the Total Error Rate. The learning outcome shows that as the projection dimension D goes higher (higher model complexity), all the predictors (except for the linear regression) show a higher order output behavior (noticeable from the more curved decision boundaries at high projection dimensions). While RP-LSE (5.32) attempts a zero training error fit at high projection dimension $D = 200$, RP-FLD, RP-TER, and RP-AUC appear to minimize the training error count without compromising much on model complexity. The decision boundary of RP-Novel is seen to fall between that of RP-LSE and RP-AUC at high projection dimensions. These results show that RP-Novel, due to its random pairing weights setting, gives a different behavior from all other predictors in terms of decision boundary. Such flexibility can be utilized in a validation-based tuning procedure to locate the best classifier for any intended application.

Impact on Weight Coefficients: In order to observe the resultant weight coefficients after training, the L_2-norm of the trained coefficient vector is observed over different projection dimensions based on the two-dimensional overlapping data set used in Fig. 5.7. Figure 5.8 shows the values of $\| \hat{\boldsymbol{w}} \|_2$ (L_2-norm of the projection weight coefficient vector) being plotted over projection dimension $D \in \{1, 11, 21, \ldots, 251\}$. Here, RP-TER, RP-AUC, and RP-FLD show a consistently lower parameter norm than that of RP-LSE in all projection dimensions. The RP-Novel, attributed to its random model settings, shows a varying parameter norm around that of RP-AUC and RP-FLD. Here, we note that a low value of parameter norm can be related to good generalization in the spirit of weight-decay regularization (Neter et al. 1996; Bishop 1995).

5.6 Computational Complexity

AUC using the original data set, (5.90): For simplicity, consider a D-dimensional projection (instead of $D + 1$-dimension), the covariance computation within the inverse term of (5.90) requires $m^+ \times m^-$ summations of D^2 operations. This is followed by an $O(D^3)$ operation of the inverse assuming a Gaussian elimination process (Duda et al. 2001). Considering both the covariance and the inverse operations, we obtain a computational complexity of $O(D^2 m^+ m^-)$. Since $(m^+ m^-)$ is bounded between $(m - 1)$ and $m^2/4$ when a non-empty set for each pattern category is assumed, the above computational complexity of $O(D^2 m^+ m^-)$ is bounded between $O(D^2(m - 1))$ and $O(D^2 m^2/4)$. This shows an almost always higher

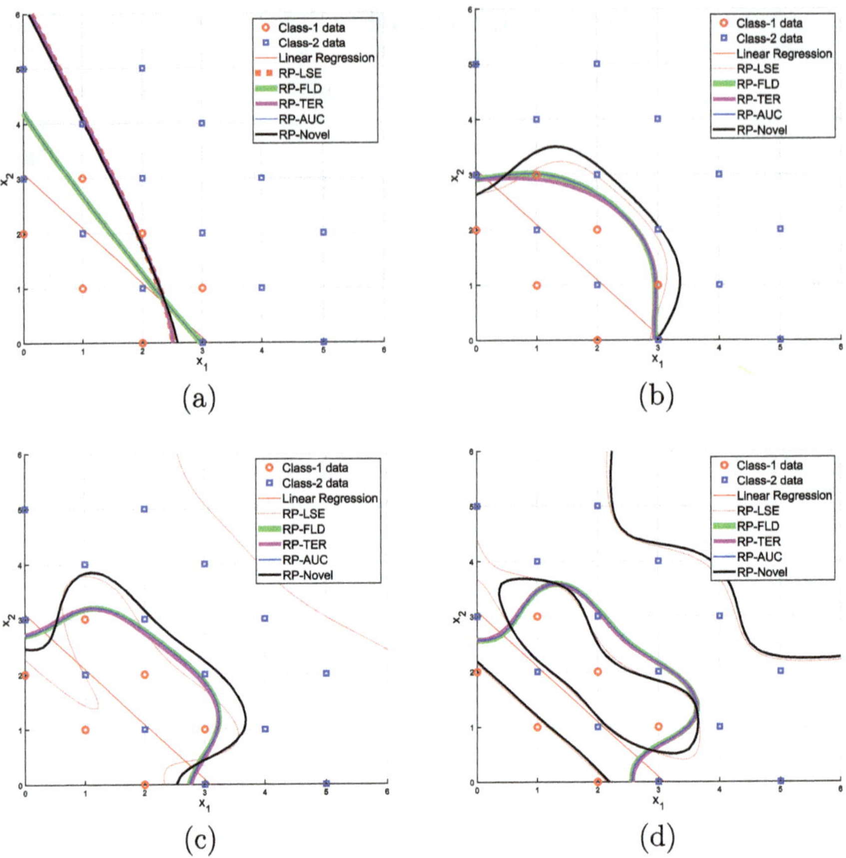

Fig. 5.7 Decision boundaries at different projection dimensions: **a** $D = 2$, **b** $D = 10$, **c** $D = 50$, **d** $D = 200$

computational complexity for AUC as compared with LSE estimation which is dominated by $O(D^2 m)$ for $m \geqslant D$.

TAUC (AUC using transformed data, (5.98)**):** When K number of data transformations is applied to the original data (containing m samples) for training, an additional summation loop is seen for the covariance of (5.98) as compared with that of (5.90). This results in a dominant computational complexity of $O(D^2 K m^+ m^-)$ for (5.98) which is bounded between $O(D^2 K(m - 1))$ and $O(D^2 K m^2 / 4)$.

TER (AUC in scaling space, (5.75)**):** For the total error rate solution given by (5.74), the covariance terms within the inverse function requires, respectively, m^- and m^+ summations of D^2 operations. This is followed by an $O(D^3)$ inverse operation assuming a Gaussian elimination process. The dominating computational complexity for this inverse term is thus either $O(D^3)$ if $D > \max(m^-, m^+)$ or $O(D^2 m^*)$

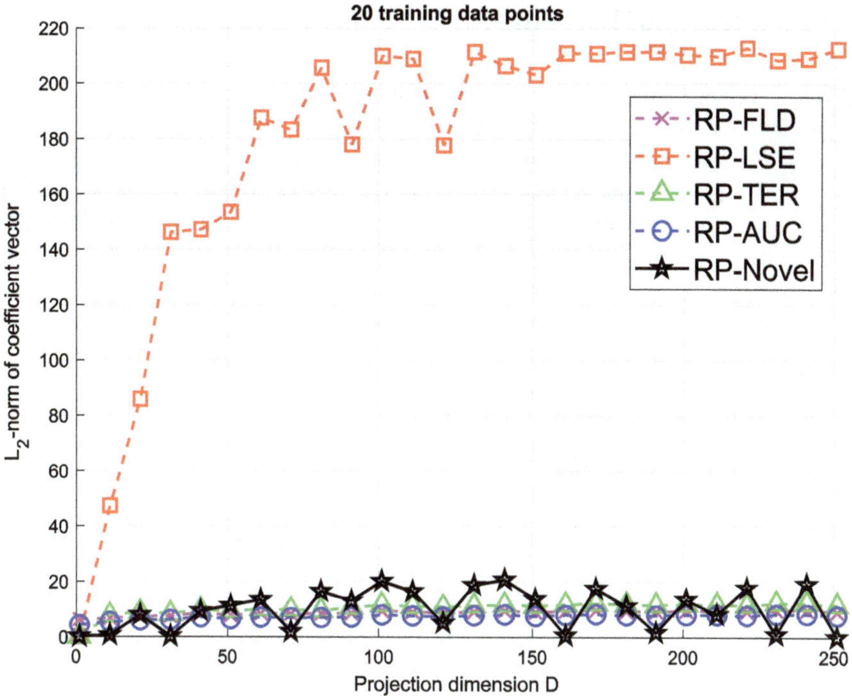

Fig. 5.8 L_2-norm of projection weight coefficient vector ($\|\hat{w}\|_2$) versus projection dimension (D) for the 20 data points

if $D < \max(m^-, m^+)$, where $m^* = \max(m^-, m^+)$. For an over-determined system with $D < m$ and for the extreme case of having a one class classification problem, the dominating complexity of (5.74) would be $O(D^2 m)$. This shows that the dominating complexity of (5.74) approaches that of least squares error learning.

FLD (AUC in translated scaling space, (5.103)): Apart from an offset from the mean vector, the covariance terms within the inverse function of (5.103) appear similar to that of (5.74). The computational complexity is thus either $O(D^3)$ or $O(D^2 m)$ depending on whether $m > D$. Again, the dominating complexity of (5.74) approaches that of least squares errorlearning.

Novel classifier (5.102): Consider $K = 2$ number of transformations and based on the above results, the Novel classifier (5.102) has a computational complexity bounded between $O(2D^2(m - 1))$ and $O(D^2 m^2/2)$.

5.7   Summary of Learning Algorithms

Given a training data set which consists of m examples $(\boldsymbol{x}_i, \boldsymbol{y}_i)$, $i = 1, \ldots, m$, where $\boldsymbol{x}_i \in \mathbb{R}^d$ denotes the ith feature sample, and $\boldsymbol{y}_i = [0, \ldots, 1_k, \ldots, 0]_i^T$ ($k \in \{1, \ldots, C\}$) denotes the corresponding target label using an indicator vector (i.e., an entry of "1" is put at the kth column position corresponding to the kth-class with other entries "0"). Both the feature samples (either the original features $\boldsymbol{x}_i \in \mathbb{R}^d$ or the expanded features $\boldsymbol{p}(\boldsymbol{x}_i) = [1, p_1(\boldsymbol{x}_i), \ldots, p_D(\boldsymbol{x}_i)]^T \in \mathbb{R}^{D+1}$ for $i = 1, \ldots, m$) and the corresponding indicator target vectors $\boldsymbol{y}_i$, $i = 1, \ldots, m$ are stacked row-by-row to form matrices as follows.

$$
\mathbf{P} = \left.\begin{bmatrix} 1 & p_1(\boldsymbol{x}_1) & \cdots & p_D(\boldsymbol{x}_1) \\ 1 & p_1(\boldsymbol{x}_2) & \cdots & p_D(\boldsymbol{x}_2) \\ \vdots & \vdots & \ddots & \vdots \\ 1 & p_1(\boldsymbol{x}_m) & \cdots & p_D(\boldsymbol{x}_m) \end{bmatrix}\right\} m, \tag{5.160}
$$

$$
\mathbf{Y} = \begin{bmatrix} \mathbf{Y}_1 \\ \mathbf{Y}_2 \\ \vdots \\ \mathbf{Y}_C \end{bmatrix} = \begin{bmatrix} 1 & 0 & \cdots & 0 \\ 1 & 0 & \cdots & 0 \\ \vdots & \vdots & \cdots & \vdots \\ 0 & 1 & \cdots & 0 \\ 0 & 1 & \cdots & 0 \\ \vdots & & & \vdots \\ 0 & 0 & \ddots & 1 \\ 0 & 0 & \cdots & 1 \end{bmatrix} m, \tag{5.161}
$$

where each row of $\mathbf{Y}$ and $\mathbf{P}$ corresponds to a data sample with respective class label. In other words, each row of $\mathbf{P}$ must be aligned to its corresponding row of target class indicator label $\boldsymbol{y}_i$ in $\mathbf{Y}$. A linear estimation model is written as

$$
\mathbf{G} = \mathbf{PW}, \tag{5.162}
$$

where $\mathbf{W}$ is the parameter matrix to be estimated which is given by

$$
\mathbf{W} = [\boldsymbol{w}_1, \boldsymbol{w}_2, \ldots, \boldsymbol{w}_C]
$$

$$
= \overbrace{
\begin{bmatrix}
w_{1,1} & w_{2,1} & \cdots & w_{C,1} \\
w_{1,2} & w_{2,2} & \cdots & w_{C,2} \\
\vdots & & \ddots & \vdots \\
w_{1,D} & w_{2,D} & \cdots & w_{C,D}
\end{bmatrix}}^{C}
\left.\rule{0pt}{42pt}\right\} D+1,
\tag{5.163}
$$

with each column of $\mathbf{W}$ corresponds to a parameter vector for each category. At testing stage, the given n unseen test data are expanded and packed as

$$
\mathbf{P}_t =
\overbrace{
\begin{bmatrix}
1 & p_2(\boldsymbol{x}_1) & \cdots & p_D(\boldsymbol{x}_1) \\
1 & p_2(\boldsymbol{x}_2) & \cdots & p_D(\boldsymbol{x}_2) \\
\vdots & \vdots & \ddots & \vdots \\
1 & p_2(\boldsymbol{x}_n) & \cdots & p_D(\boldsymbol{x}_n)
\end{bmatrix}}^{D+1}
\left.\rule{0pt}{42pt}\right\} n.
\tag{5.164}
$$

The learning and testing steps for each algorithm are summarized in the sequel.

5.7.1 Algorithm 1: LSE Primal Learning ($m \geqslant D + 1$)

—— **Training stage** ——

Step 1: Form matrices $\mathbf{P} \in \mathbb{R}^{m \times (D+1)}$ and $\mathbf{Y} \in \mathbb{R}^{m \times C}$ based on the given training data as mentioned in the chapter. Select a value for λ such as 0.01.

Step 2: Compute predictor parameter matrix $\mathbf{W} \in \mathbb{R}^{(D+1) \times C}$ using

$$
\hat{\mathbf{W}} = (\mathbf{P}^T \mathbf{P} + \lambda \mathbf{I})^{-1} \mathbf{P}^T \mathbf{Y}.
\tag{5.165}
$$

—— **Testing Stage** ——

Step 3: Compute prediction values $\hat{\mathbf{G}}_t \in \mathbb{R}^{n \times C}$ for n test samples packed in data matrix $\mathbf{P}_t$ using

$$
\hat{\mathbf{G}}_t = \mathbf{P}_t \hat{\mathbf{W}}.
\tag{5.166}
$$

Step 4: Classify each jth ($j = 1, \ldots, n$) row of prediction data based on

$$
cls(\hat{\mathbf{G}}_t(j, :)) = \arg\max_k g(\hat{\boldsymbol{w}}_k, \boldsymbol{x}_j), \ \forall k = 1, 2, \ldots, C.
\tag{5.167}
$$

—— **End** ——

5.7.2   Algorithm 2: LSE-Dual Learning ($m < D + 1$)

—— **Training stage** ——

Step 1: Form matrices $\mathbf{P} \in \mathbb{R}^{m \times (D+1)}$ and $\mathbf{Y} \in \mathbb{R}^{m \times C}$ based on the given training data as mentioned in the chapter.

Step 2: Compute predictor parameter matrix $\mathbf{A} \in \mathbb{R}^{(D+1) \times C}$ using

$$\hat{\mathbf{A}} = (\mathbf{P}\mathbf{P}^T + \lambda\mathbf{I})^{-1}\mathbf{Y}, \tag{5.168}$$

—— **Testing Stage** ——

Step 3: Compute prediction values $\hat{\mathbf{G}}_t \in \mathbb{R}^{n \times C}$ for n test samples packed in data matrix $\mathbf{P}_t$ using

$$\hat{\mathbf{G}}_t = \mathbf{P}_t \mathbf{P}^T \hat{\mathbf{A}}. \tag{5.169}$$

Step 4: Classify each jth ($j = 1, \ldots, n$) row of prediction data based on

$$cls(\hat{\mathbf{G}}_t(j, :)) = \arg \max_k g(\hat{\mathbf{w}}_k, \mathbf{x}_j), \ \forall k = 1, 2, \ldots, C. \tag{5.170}$$

—— **End** ——

5.7.3   Algorithm 3: TER Primal Learning ($m \geqslant D + 1$)

—— **Training stage** ——

Step 1: Form matrices $\mathbf{P} \in \mathbb{R}^{m \times (D+1)}$ and $\mathbf{Y} \in \mathbb{R}^{m \times C}$ based on the given training data as mentioned in the chapter. Set $\tau = 0.5$ and $\eta = 0.5$ without loss of generality (this choice is synchronized with those $\{0, 1\}$ labels within $\mathbf{Y}$).

Step 2: Loop for $k = 1, \ldots, C$,

(i) Extract each column of $\mathbf{Y}$ as $\mathbf{y}_k$. Based on the label of each element in the column vector $\mathbf{y}_k$, sort out each corresponding row of $\mathbf{P}$ which belongs to the positive category (labeled as "1" in $\mathbf{y}_k$) or the negative category (labeled as "0" in $\mathbf{y}_k$) and pack them respectively into matrices $\mathbf{P}_+$ and $\mathbf{P}_-$ (indexing by k is omitted here for simplicity).

(ii) Obtain the sample lengths (number of rows) of $\mathbf{P}_+$ and $\mathbf{P}_-$ as m^+ and m^- respectively.

(iii) Compute predictor parameter vector

$$\hat{\mathbf{w}}_k = \left(b\mathbf{I} + \frac{1}{m_k^-}\mathbf{P}_-^T\mathbf{P}_- + \frac{1}{m_k^+}\mathbf{P}_+^T\mathbf{P}_+ \right)^{-1}$$

$$\times \left(\frac{1}{m_k^-} \mathbf{P}_-^T (\tau - \eta) \mathbf{1}_- + \frac{1}{m_k^+} \mathbf{P}_+^T (\tau + \eta) \mathbf{1}_+ \right), \qquad (5.171)$$

where $\mathbf{1}_+$ is a column vector comprising m^+ number of 1's, and $\mathbf{1}_-$ is a column vector comprising m^- number of 1's.
End loop

Step 3: Pack the estimated parameter vectors in matrix form $\hat{\mathbf{W}} = [\hat{\boldsymbol{w}}_1, \ldots, \hat{\boldsymbol{w}}_C] \in \mathbb{R}^{(D+1) \times C}$.

——— **Testing Stage** ———
Step 4: Compute prediction values $\hat{\mathbf{G}}_t \in \mathbb{R}^{n \times C}$ for n test samples packed in data matrix $\mathbf{P}_t$ using

$$\hat{\mathbf{G}}_t = \mathbf{P}_t \hat{\mathbf{W}}. \qquad (5.172)$$

Step 5: Classify each jth $(j = 1, \ldots, n)$ row of prediction data based on

$$cls(\hat{\mathbf{G}}_t(j, :)) = \arg\max_k g(\hat{\boldsymbol{w}}_k, \boldsymbol{x}_j), \ \forall k = 1, 2, \ldots, C. \qquad (5.173)$$

——— **End** ———

5.7.4 Algorithm 4: TER Dual Learning ($m < D + 1$)

——— **Training stage** ———
Step 1: Form matrices $\mathbf{P} \in \mathbb{R}^{m \times (D+1)}$ and $\mathbf{Y} \in \mathbb{R}^{m \times C}$ based on the given training data as mentioned in the chapter.
Step 2: Loop for $k = 1, \ldots, C$,

(i) Extract each column of $\mathbf{Y}$ as $\mathbf{y}_k$.
(ii) Setup diagonal weight matrix $\mathbf{M}_k$ (consisting of $\mathbf{M}_-$ and $\mathbf{M}_+$ matrices with corresponding $1/m_k^+$ and $1/m_k^-$ normalization) for $k = 1, \ldots, C$ using (5.78) and (5.79) which are aligned to the class labels of $\mathbf{y}_k$ where $\mathbf{y}_k$ is the kth column of target $\mathbf{Y}$. Similarly, each row of the matrix $\mathbf{P}_k$ has been aligned to the class labels of $\mathbf{y}_k$.
(iii) Compute predictor parameter vector for each category

$$\hat{\boldsymbol{w}}_k = \mathbf{P}_k^T \mathbf{M}_k \left(b\mathbf{I} + \mathbf{P}_k^T \mathbf{M}_k \mathbf{P}_k \right)^{-1} \mathbf{y}_k, \qquad (5.174)$$

End loop

Step 3: Pack the estimated parameter vectors in matrix form $\hat{\mathbf{W}} = [\hat{\boldsymbol{w}}_1, \ldots, \hat{\boldsymbol{w}}_C] \in \mathbb{R}^{(D+1) \times C}$

—— **Testing Stage** ——

Step 4: Compute prediction values $\hat{\mathbf{G}}_t \in \mathbb{R}^{n \times C}$ for n test samples packed in data matrix $\mathbf{P}_t$ using

$$\hat{\mathbf{G}}_t = \mathbf{P}_t \hat{\mathbf{W}}. \tag{5.175}$$

Step 5: Classify each jth ($j = 1, \ldots, n$) row of prediction data based on

$$cls(\hat{\mathbf{G}}_t(j, :)) = \arg\max_k g(\hat{\boldsymbol{w}}_k, \boldsymbol{x}_j), \ \forall k = 1, 2, \ldots, C. \tag{5.176}$$

—— **End** ——

5.7.5 *Algorithm 5: Weighted-LS Learning ($m \geqslant D + 1$)*

—— **Training stage** ——

Step 1: Form matrices $\mathbf{P} \in \mathbb{R}^{m \times (D+1)}$ and $\mathbf{Y} \in \mathbb{R}^{m \times C}$ based on the given training data as mentioned in the chapter. Depending on the definition of $\mathbf{Y}$, this algorithm applies to both regression and classification.

Step 2: Loop for $k = 1, \ldots, C$,

(i)   Extract each column of $\mathbf{Y}$ as $\mathbf{y}_k$.
(ii) Setup diagonal weight matrix $\mathbf{M}_k$ (i.e., it is permitted to have different weights for each kth output).
(iii) Compute predictor parameter vector

$$\hat{\boldsymbol{w}}_k = \left(b\mathbf{I} + \mathbf{P}^T \mathbf{M}_k \mathbf{P}\right)^{-1} \mathbf{P}^T \mathbf{M}_k \mathbf{y}_k, \tag{5.177}$$

End loop

Step 3: Pack the estimated parameter vectors in matrix form $\hat{\mathbf{W}} = [\hat{\boldsymbol{w}}_1, \ldots, \hat{\boldsymbol{w}}_C] \in \mathbb{R}^{(D+1) \times C}$

—— **Testing Stage** ——

Step 4: Compute prediction values $\hat{\mathbf{G}}_t \in \mathbb{R}^{n \times C}$ for n test samples packed in data matrix $\mathbf{P}_t$ using

$$\hat{\mathbf{G}}_t = \mathbf{P}_t \hat{\mathbf{W}}. \tag{5.178}$$

Step 5: Classify each jth ($j = 1, \ldots, n$) row of prediction data based on

$$cls(\hat{\mathbf{G}}_t(j, :)) = \arg\max_k g(\hat{\boldsymbol{w}}_k, \boldsymbol{x}_j), \ \forall k = 1, 2, \ldots, C. \tag{5.179}$$

—— **End** ——

5.7.6 Algorithm 6: AUC Learning

—— **Training stage (for 2-category problems)**——
Step 1: Form matrix $\mathbf{P} \in \mathbb{R}^{m \times D}$ based on the given training data as mentioned in the chapter. Label each row of $\mathbf{P}$ as column vector $\boldsymbol{p}^+$ if it belongs to positive class and $\boldsymbol{p}^-$ otherwise.
Step 2: Compute predictor parameter

$$\hat{\boldsymbol{w}} = \left[b\mathbf{I} + \frac{1}{m^+ m^-} \sum_{i=1}^{m^+} \sum_{j=1}^{m^-} \left(\boldsymbol{p}_j^- - \boldsymbol{p}_i^+ \right) \left(\boldsymbol{p}_j^- - \boldsymbol{p}_i^+ \right)^T \right]^{-1}$$

$$\times \left[\frac{-\eta}{m^+ m^-} \sum_{i=1}^{m^+} \sum_{j=1}^{m^-} \left(\boldsymbol{p}_j^- - \boldsymbol{p}_i^+ \right) \right]. \tag{5.180}$$

Step 4: Choose a threshold τ according to

$$\tau = \frac{1}{2m^-} \sum_{j=1}^{m^-} \hat{\boldsymbol{w}}^T \boldsymbol{p}_j^- + \frac{1}{2m^+} \sum_{i=1}^{m^+} \hat{\boldsymbol{w}}^T \boldsymbol{p}_i^+. \tag{5.181}$$

—— **Testing Stage** ——
Step 4: Compute prediction values $\hat{\mathbf{g}}_t = [g(\hat{\boldsymbol{w}}, \boldsymbol{x}_1), \ldots, g(\hat{\boldsymbol{w}}, \boldsymbol{x}_n)]^T \in \mathbb{R}^n$ for n test samples packed in data matrix $\mathbf{P}_t$ using

$$\hat{\mathbf{g}}_t = \mathbf{P}_t \hat{\boldsymbol{w}}. \tag{5.182}$$

Step 5: Classify each jth $(j = 1, \ldots, n)$ element of prediction data based on

$$cls(g(\hat{\boldsymbol{w}}, \boldsymbol{x})) = \begin{cases} 1 \text{ if } g(\hat{\boldsymbol{w}}, \boldsymbol{x}_j) \geqslant \tau \\ 0 \text{ if } g(\hat{\boldsymbol{w}}, \boldsymbol{x}_j) < \tau \end{cases}. \tag{5.183}$$

—— **End** ——

5.7.7 Algorithm 7: Novel Learning

—— **Training stage (for 2-category problems)**——
Step 1: Set the parameter values for $\beta_1, \beta_2, \gamma_1, \gamma_2, \boldsymbol{u}_1, \boldsymbol{u}_2, \boldsymbol{v}_1, \boldsymbol{v}_1, b = 10^{-4}, \eta = 0.5$ (see Table 5.1).
Step 2: Form matrix $\mathbf{P} \in \mathbb{R}^{m \times D}$ based on the given training data as mentioned above. Label each row of $\mathbf{P}$ as $\boldsymbol{p}^+$ if it belongs to positive class and $\boldsymbol{p}^-$ otherwise.
Step 3: Compute predictor parameter

$$\tilde{w} = \left[b\mathbf{I} + \frac{1}{m^+m^-} \sum_{i=1}^{m^+} \sum_{j=1}^{m^-} \left((\beta_1 p_j^- - u_1) - (\gamma_1 p_i^+ - v_1) \right) \left((\beta_1 p_j^- - u_1) - (\gamma_1 p_i^+ - v_1) \right)^T \right.$$

$$\left. + \frac{1}{m^+m^-} \sum_{i=1}^{m^+} \sum_{j=1}^{m^-} \left((\beta_2 p_j^- - u_2) - (\gamma_2 p_i^+ - v_2) \right) \left((\beta_2 p_j^- - u_2) - (\gamma_2 p_i^+ - v_2) \right)^T \right]^{-1}$$

$$\times \left[\frac{-\eta}{m^+m^-} \sum_{i=1}^{m^+} \sum_{j=1}^{m^-} \left((\beta_1 p_j^- - u_1) - (\gamma_1 p_i^+ - v_1) \right) + \frac{-\eta}{m^+m^-} \sum_{i=1}^{m^+} \sum_{j=1}^{m^-} \left((\beta_2 p_j^- - u_2) - (\gamma_2 p_i^+ - v_2) \right) \right].$$

$$(5.184)$$

—— Testing Stage ——
Step 4: Compute prediction values $\hat{\mathbf{g}}_t = [g(\hat{w}, x_1), \ldots, g(\hat{w}, x_n)]^T \in \mathbb{R}^n$ for n test samples packed in data matrix $\mathbf{P}_t$ using

$$\hat{\mathbf{g}}_t = \mathbf{P}_t \hat{w}. \tag{5.185}$$

Step 5: Classify each jth ($j = 1, \ldots, n$) element of prediction data based on

$$cls(g(\hat{w}, x)) = \begin{cases} 1 \text{ if } g(\hat{w}, x_j) \geqslant \tau \\ 0 \text{ if } g(\hat{w}, x_j) < \tau \end{cases}. \tag{5.186}$$

—— End ——

5.7.8 Algorithm 8: Recursive Least Squares (RLS) for Binary Learning

—— Initialization ——
Step 0: Initialize $w_0 = random\ vector$, $\mathbf{R}_0^{-1} = \frac{1}{c}\mathbf{I}$, $c > 0$ and select a value for $0 < \lambda \leqslant 1$.

—— Training stage ——
Step 1: At each iteration $i + 1$, form projection column vector $p(x_{i+1})$
Step 2: Update

$$\mathbf{R}_i^{-1} = \frac{1}{\lambda}\mathbf{R}_{i-1}^{-1} - \frac{\frac{1}{\lambda^2}\mathbf{R}_{i-1}^{-1} p(x_i) p(x_i)^T \mathbf{R}_{i-1}^{-1}}{1 + \frac{1}{\lambda} p(x_i)^T \mathbf{R}_{i-1}^{-1} p(x_i)}, \tag{5.187}$$

and

$$\hat{w}_{i+1} = \hat{w}_i + \frac{\mathbf{R}_i^{-1} p(x_{i+1})(y_{i+1} - p(x_{i+1})^T \hat{w}_i)}{\lambda + p(x_{i+1})^T \mathbf{R}_i^{-1} p(x_{i+1})}. \tag{5.188}$$

Loop back to Step 1 until end of training sample.
—— Testing Stage ——
Step 3: Compute prediction value for test data x_t

$$\hat{g}_t = p(x_t)^T \hat{w}_{i+1}.$$ (5.189)

Step 4: Classify the prediction based on the decision threshold

$$cls(\hat{g}_t) = \begin{cases} 1 \text{ if } \hat{g}_t \geqslant \tau \\ 0 \text{ if } \hat{g}_t < \tau \end{cases}.$$ (5.190)

—— **End** ——

5.7.9 Algorithm 9: Recursive TER (RTER) for Binary Learning

—— **Initialization** ——
Step 0: Initialize $w_0 = random\ vector$ and $\mathbf{R}_0 = \mathbf{S}_i^+ = \mathbf{S}_i^- = \mathbf{I}$.
—— **Training stage** ——
Step 1: At each iteration $i + 1$, form projection column vector $p(x_{i+1})$
Step 2: Update

$$\mathbf{S}_i^+ = \sum_{k=1}^{m_i^+} \frac{y_i}{m_i^+} p(x_k) p(x_k)^T, \quad \mathbf{S}_i^- = \sum_{k=1}^{m_i^-} \frac{1 - y_i}{m_i^-} p(x_k) p(x_k)^T,$$ (5.191)

$$\mathbf{B} := \left(\frac{-y_{i+1}}{m_{i+1}^+} \mathbf{S}_i^+ + \frac{-(1 - y_{i+1})}{m_{i+1}^-} \mathbf{S}_i^- \right),$$ (5.192)

$$\mathbf{Q}_i^{-1} = \mathbf{R}_i^{-1} - \mathbf{R}_i^{-1} \mathbf{B} (\mathbf{I} + \mathbf{R}_i^{-1} \mathbf{B})^{-1} \mathbf{R}_i^{-1},$$ (5.193)

$$\alpha_i = \left(\frac{y_i}{m_i^+} + \frac{1 - y_i}{m_i^-} \right),$$ (5.194)

$$\mathbf{R}_i^{-1} = \mathbf{Q}_{i-1}^{-1} - \frac{\alpha_i \mathbf{Q}_{i-1}^{-1} p(x_i) p(x_i)^T \mathbf{Q}_{i-1}^{-1}}{1 + \alpha_i p(x_i)^T \mathbf{Q}_{i-1}^{-1} p(x_i)},$$ (5.195)

and

$$\hat{w}_{i+1} = \hat{w}_i + \frac{\mathbf{R}_i^{-1} p(x_{i+1})(y_{i+1} - p(x_{i+1})^T \hat{w}_i)}{p(x_{i+1})^T \mathbf{R}_i^{-1} p(x_{i+1}) + 1}.$$ (5.196)

Loop back to Step 1 until end of training sample.
—— **Testing Stage** ——
Step 3: Compute prediction value for test data x_t

$$\hat{g}_t = \boldsymbol{p}(\boldsymbol{x}_t)^T \hat{\boldsymbol{w}}_{i+1}. \tag{5.197}$$

Step 4: Classify the prediction based on the decision threshold

$$cls(\hat{g}_t) = \begin{cases} 1 \text{ if } \hat{g}_t \geqslant \tau \\ 0 \text{ if } \hat{g}_t < \tau \end{cases}. \tag{5.198}$$

—— **End** ——

5.7.10   *Algorithm 10: Recursive AUC for Binary Learning*

—— **Initialization** ——
Step 0: Initialize $\boldsymbol{w}_0 = random\ vector \in \mathbb{R}^{(D+1)}$ and $\mathbf{R}_0^{-1} = \frac{1}{c}\mathbf{I} \in \mathbb{R}^{(D+1)\times(D+1)}$, $c > 0$.
—— **Training stage** ——
Step 1: At each iteration $i + 1$, based on the class label of newly arrived $(i + 1)^{th}$ sample $\boldsymbol{x}_{i+1}$, form column vector $\boldsymbol{p}(\boldsymbol{x}_{i+1})$ and then a set of sample-difference pairs ε_{i+1} (of either $(m^- \times (D + 1))$ or $(m^+ \times (D + 1))$ dimension) by pairing it with those accumulate samples of opposite class:

$$\varepsilon_{i+1} = (1 - 2y_{i+1})\left(\mathbf{1}\boldsymbol{p}(\boldsymbol{x}_{i+1})^T - \mathbf{P}_i^{\{y=1-y_{i+1}\}}\right). \tag{5.199}$$

Step 2: Update

$$\mathbf{R}_{i+1}^{-1} = \mathbf{R}_i^{-1} - \mathbf{R}_i^{-1}\varepsilon_{i+1}^T\left(\mathbf{I} + \varepsilon_{i+1}\mathbf{R}_i^{-1}\varepsilon_{i+1}^T\right)^{-1}\varepsilon_{i+1}\mathbf{R}_i^{-1}, \tag{5.200}$$

or

$$\mathbf{R}_{i+1}^{-1} = \mathbf{R}_i^{-1} - \mathbf{R}_i^{-1}\left(\mathbf{I} + \varepsilon_{i+1}^T\varepsilon_{i+1}\mathbf{R}_i^{-1}\right)^{-1}\varepsilon_{i+1}^T\varepsilon_{i+1}\mathbf{R}_i^{-1}, \tag{5.201}$$

and

$$\hat{\boldsymbol{w}}_{i+1} = \hat{\boldsymbol{w}}_i + \mathbf{R}_{i+1}^{-1}\varepsilon_{i+1}^T\left(\boldsymbol{\eta} - \varepsilon_{i+1}\boldsymbol{w}_i\right). \tag{5.202}$$

Loop back to Step 1 until end of training sample.
—— **Testing Stage** ——
Step 3: Compute prediction value for test data $\boldsymbol{x}_t$

$$\hat{g}_t = \boldsymbol{p}(\boldsymbol{x}_t)^T \hat{\boldsymbol{w}}_{i+1}. \tag{5.203}$$

Step 4: Classify the prediction based on the decision threshold

$$cls(\hat{g}_t) = \begin{cases} 1 \text{ if } \hat{g}_t \geqslant \tau \\ 0 \text{ if } \hat{g}_t < \tau \end{cases}. \tag{5.204}$$

—— **End** ——

5.8 Bibliographic Notes

Least Squares

Carl Friedrich Gauss is often credited with the development of fundamentals of the basis for least squares analysis while Adrien-Marie Legendre is credited for its priority. The original memoirs on least squares that Gauss published in 1820s was in Latin with German Notices. The translation by G. W. Stewart in 1995 had largely made the articles accessible to the English-speaking community (Gauss and Stewart 1995). The historical applications of least squares to astronomy and geodesy are well summarized in Nievergelt (2000). In terms of the foundations of linear algebra, a very comprehensive account can be found in Strang (2016), Boyd and Vandenberghe (2018).

Dual Ridge Regression

The dual ridge regression (Saunders et al. 1998) is also called the kernel ridge regression in the literature of pattern analysis and machine learning (Shawe-Taylor and Cristianini 2004; Bishop 2006; Welling 2013) where it is also known as the *kernel trick* in Support Vector Machines (SVM) (Boser et al. 1992; Vapnik 1998). Essentially, the following matrix identity is adopted to traverse between the two possibilities of dimensions for summing the matrix $\mathbf{B}$:

$$(\mathbf{A}^{-1} + \mathbf{B}^T\mathbf{C}^{-1}\mathbf{B})^{-1}\mathbf{B}^T\mathbf{C}^{-1} = \mathbf{A}\mathbf{B}^T(\mathbf{B}\mathbf{A}\mathbf{B}^T + \mathbf{C})^{-1}. \tag{5.205}$$

If $\mathbf{B}$ is not a square matrix (say, $\mathbb{R}^{m \times n}$), the inverse in the LHS and the RHS of (5.205) is performed in spaces of different dimensionality (i.e., $\mathbb{R}^{m \times m}$ or $\mathbb{R}^{n \times n}$). The solution in (5.35) can be obtained from (5.32) by putting $\mathbf{B} = \mathbf{P}$, $\mathbf{A} = \mathbf{I} \in \mathbb{R}^{n \times n}$, $\mathbf{C} = \mathbf{I} \in \mathbb{R}^{m \times m}$.

The Step Loss Function and Its Approximation

The zero-one step function from the error view as shown in Fig. 5.1 can also be interpreted from the margin view as shown in Fig. 5.9 (see Sect. 4.3.4 of this book and the book of Hastie et al. (2017)). For binary target $y \in \{-1, +1\}$ response, the margin given by $y \cdot g$ with $g = g(\boldsymbol{w}, \boldsymbol{x})$ being the predictor response plays a role analogous to the error $y - g(\boldsymbol{x}, \boldsymbol{w})$ in regression but with an implicit zero value threshold to give a positive value for correct classification and a negative value for misclassification. Due to the non-differentiable nature of the step loss, several schemes have been developed in the literature to obtain a smooth function for optimization search.

Using sigmoid or logistic-like function: The sigmoid function or logistic-like functions such as *tanh* and *arctan* functions have been a natural choice to approximate the step function (Yan et al. 2003; McDermott and Katagiri 2004; Toh 2006a). Denoted by σ the sigmoid function, the minimization function for TER can be written as

$$\min_{\boldsymbol{w}} \text{TER} \approx \min_{\boldsymbol{w}} \left\{ \frac{1}{m^-} \sum_{j=1}^{m^-} \sigma(\epsilon_j) + \frac{1}{m^+} \sum_{i=1}^{m^+} \sigma(\varepsilon_i) \right\}. \tag{5.206}$$

where

$$\sigma(x) = \frac{1}{1 + e^{-\gamma x}}, \quad \gamma > 0 , \tag{5.207}$$

and $\epsilon_j = g(\boldsymbol{x}_j^-, \boldsymbol{w}) - \tau$ and $\varepsilon_i = \tau - g(\boldsymbol{x}_i^+, \boldsymbol{w})$. From the margin-loss perspective, the error counting based on sigmoid approximation is a flipped version in the x-axis (i.e., using $\sigma(x) = \frac{1}{1+e^{\gamma x}}$) as depicted in the dashed-dotted curve in Fig. 5.9. Such a sigmoidal approximation to the loss facilitates a differentiable cost function for optimization. However, it also introduces local plateaus resulting from summing the flat regions of the sigmoid function. This poses difficulty in the optimization since much search effort may be spent in making little progress at these locally flat regions (Toh 2006a).

Using power and piecewise functions: Since the goal of classification is to maximize accuracy, the loss function should penalize negative margins ($yg < 0$, indicating inaccuracy) more heavily than positive margins. This necessitates a monotonically decreasing function for discrimination (Hastie et al. 2017). However, the flat regions in the sigmoid function require special attention regarding search efficiency. Two piecewise power functions were adopted in Yan et al. (2003) to approximate the step function under the Wilcoxon-Mann-Whitney (WMW) statistic formulation:

$$R_1(x) = \begin{cases} [-(x - \eta)]^r & : x < \eta \\ 0 & : \text{otherwise} \end{cases}, \tag{5.208}$$

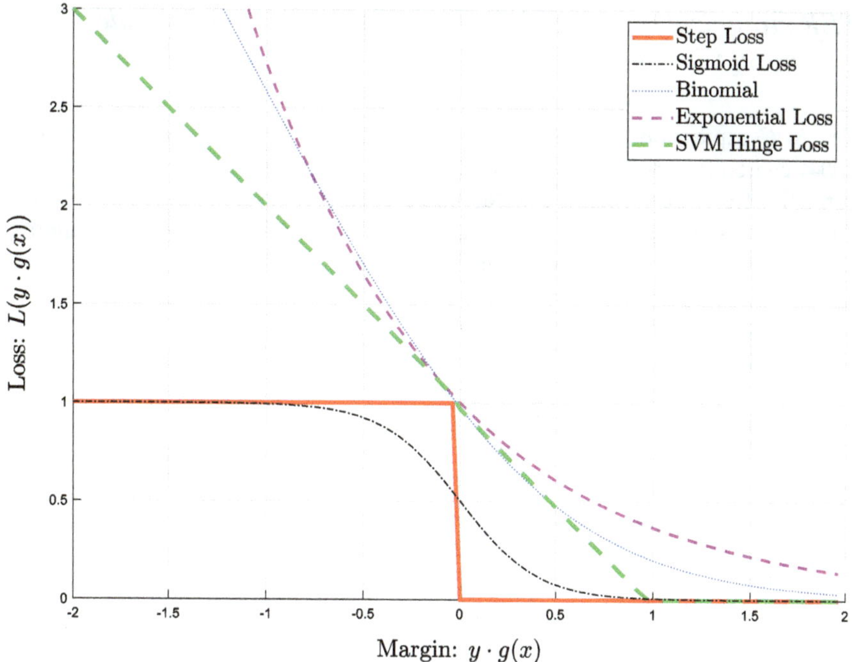

Fig. 5.9 Loss functions for binary classification problems. The target is $y = \pm 1$; the prediction is g with class prediction sign(g). Common loss functions include: misclassification given by $L(y \cdot g < 0)$ where L denotes a zero-one step loss (red solid line); sigmoid given by $1/(1 + exp(\gamma y \cdot g))$, $\gamma = 5$; exponential loss given by $exp(-y \cdot g)$; and support vector hinge loss given by $(1 - y \cdot g) \cdot L(y \cdot g < 1)$

where $0 < \eta \leqslant 1$ and $r > 1$.

$$R_2(x) = \begin{cases} (x - \eta)^r & : x > \eta \\ 0 & : \text{otherwise} \end{cases}, \tag{5.209}$$

where $0 < \eta \leqslant 1$ and $r > 1$. Here, $R_1(x)$ can be considered an approximation to the step loss $L(x)$ shown in Fig. 5.9, and $R_2(x)$ approximates $L(-x)$. According to Yan et al. (2003), the performance of these R-based approximations was observed to be superior to that of sigmoid-based approximation.

Apart from the above, other smooth functions such as *binomial* and *exponential* functions have also been explored to approximate the step loss function in a monotonic manner (Hastie et al. 2017; Yan et al. 2006). Moreover, piecewise functions such as the *support vector* hinge function have been adopted in place of step loss (Hastie et al. 2017; Vapnik 1998; Boser et al. 1992; Bobrowski and Sklansky 1995; Bronshtein et al. 2003) as shown in Fig. 5.9. Due to either the nonlinear nature or piecewise nature of the formulations, these monotonic losses require an iterative procedure to locate the solution.

TER minimization, AUC, Wilcoxon-Mann-Whitney Statistic and Novel Classifier

The AUC expression in (5.87) (Hanley and McNeil 1982; Herschtal and Raskutti 2004) can be interpreted as the fraction of positive-negative pairs that are ranked correctly. This value is also known as Wilcoxon-Mann-Whitney statistic (Yan et al. 2003) where some approaches (Yan et al. 2003; Rakotomamonjy 2004) have approximated the computation of the step function using $L(\xi > 0)$ instead of the Heaviside function.

The analytic form for TER and AAC $(1-\text{AUC})$ minimization was first seen in Toh (2006a) where their feasibility was illustrated through biometric scores fusion (Toh et al. 2008a, b). Apart from biometrics, the fundamental idea was also studied for a relatively large group of classification data sets from the UCI machine learning repository based on a reduced polynomial model (Toh & Eng 2008) and a random projection network model (Toh 2008). The deterministic and single learning step of AAC (5.90) is differentiated from those search-based and boosting-based algorithms (Gurwitz and Overton 1989; Kam and Kopec 1996; Yang et al. 2010).

In Toh and Tan (2014), the relationship among several binary classifier solutions such as those based on the Least Squares Errors (LSE), the Total Error Rate (TER) minimization, the Area Under the ROC Curve (AUC) optimization, and the Fisher's Linear Discriminant (FLD) analysis has been uncovered through simple data transformation. This establishes the ground for the development of the Novel classifier which exploits the combinations beyond those known transformation forms.

Recursive Learning

Recursive algorithms are also known as online algorithms where the learning parameters are updated sample-wise. Here, we can treat the sample-wise update in a more generic sense as either sample-by-sample or chunk-by-chunk (a small-sample subset or block of training data) processing where the gradient computation is gradually accumulated. Although such accumulation does not change the total computational cost with respect to the entire training data, computation of gradient using a small subset of training data relieves the instantaneous need for large computational memory/facility for large training data size. The early formulation of sample-wise computation can be traced back to Plackett (1950), Gauss and Stewart (1995) for the *recursive least squares* formulation, Rosenblatt (1958) for the *perceptron* learning, Kalman (1960) for the well-known *Kalman filtering* algorithm, and Widrow and Holf (1960) for the well-known *least-mean-squares* (LMS) algorithm which found wide applications in control (see, e.g., Åström and Wittenmark (1994)) and signal processing (see, e.g., Haykin (2013)) areas. A summary of differences among these formulations can be found in Kim et al. (2012) apart from optimizing the AUC in recursive form. The recursive form for TER minimization can be found in Jang et al. (2020) with application to online face recognition.

5.9 Exercises

5.1 The polynomial function given by $g(x, w) = w_0 + w_1 x_1 + w_2 x_2 + w_3 x_1 x_2 + w_4 x_1^2 + w_5 x_2^2$ is linear with respect to the input feature vector $x = [x_1, x_2]^T$, and hence the learning of its parameters can be analytic. True or False?

5.2 The contours of the decision surface in the following plot have been produced by learning a second-order polynomial model based on four data points drawn from two categories. Which of the following statement(s) is/are true?

(a) The system is an over-determined one
(b) The system is an under-determined one
(c) The system is an even-determined one
(d) The learned solution is exact with no approximation
(e) None of these

5.3 Consider the optimization problem given by

$$\min_{w} J(w) = \min_{w} \left\{ \sum_{i=1}^{m} (f(w, x) - y_i)^2 + \lambda w^T w \right\}$$

where $f(w, x_i) = x_i^T w$ with each x_i being non-zero independent entry, $w \in \mathbb{R}^{(d+1) \times 1}$ is the parameter, each $y_i \in \mathbb{R}$ is the learning target. When the covariance $X^T X$ is invertible, a least squares solution exists in the form of $\hat{w} = (X^T X)^{-1} X^T y$. Which of the following statement(s) is/are correct?

(a) The first derivative of $f(w, x_i)$ with respect to w at $\hat{w}$ must be zero.
(b) The first derivative of $f(w, x_i)$ with respect to w at $\hat{w}$ need not be zero.
(c) The second derivative of $f(w, x_i)$ with respect to w at $\hat{w}$ must be zero.
(d) The second derivative of $f(w, x_i)$ with respect to w at $\hat{w}$ must be positive definite.
(e) Non of these.

5.4 Consider the following data samples with x_1 and x_2 being the data features, and y being the learning target output class label.

x_1	0	1	1	0
x_2	0	1	0	1
y	-1	-1	$+1$	$+1$

Without regularization, construct a polynomial model of second-order to learn these data and then predict the test output labels of $(x_1, x_2) \in \{(0.1, 0.1), (0.9, 0.9), (0.9, 0.1), (0.1, 0.9)\}$.

5.5 Consider the following data samples with x_1 and x_2 being the data features, and y being the learning target output value.

x_1	0	0	1	2	0	1	3	2	4	5	3	6
x_2	0	1	0	1	2	2	1	3	5	3	6	8
y	-2.1	-1.3	-1.5	1.0	0.5	0.8	2.1	1.8	3.3	2.8	3.8	5.0

First construct a polynomial model of third order without regularization to learn these data and then predict the test outputs of $(x_1, x_2) \in \{(0.5, 0.3), (0.8, 0.8)\}$. Next learn the same third-order polynomial model with regularization at $\lambda = 0.0001$ for predicting the same test points.

5.6 You are given a collection of 6 training data points of two features ($x = [x_1, x_2]^T$) and their class labels ($y \in \{-1, +1\}$) as follows:

$$y = -1 : \quad x \in \left\{ \begin{bmatrix} 0 \\ 0 \end{bmatrix}, \begin{bmatrix} 0 \\ 0.2 \end{bmatrix}, \begin{bmatrix} 0.2 \\ 0.2 \end{bmatrix}, \begin{bmatrix} 0.2 \\ 0 \end{bmatrix} \right\},$$

$$y = +1 : \quad x \in \left\{ \begin{bmatrix} 0.2 \\ 0.1 \end{bmatrix}, \begin{bmatrix} 0.3 \\ 0.1 \end{bmatrix} \right\}.$$

These data have been used for training a partial polynomial model of third-order given by equation (1) utilizing the criterion function given by equation (2) below.

$$g(w, x) = w_0 + \sum_{i=1}^{2} \sum_{j=1}^{2} w_{ij} x_i x_j + \sum_{i=1}^{2} \sum_{j=1}^{2} \sum_{k=1}^{2} w_{ijk} x_i x_j x_k, \quad (1)$$

$$\min_{w} J(w) = \min_{w} \sum_{i=1}^{m} (g(w, x) - y_i)^2 . \quad (2)$$

How many of the above samples are incorrectly classified by the partial polynomial model incorporating a signum threshold?

5.7 Generate the data points using the MATLAB codes below:

```
x = [0.1:.1:10];
rand('state',8);
y = [0.6+0.05*rand(1,length(x)/2),0.3+0.05*rand
(1,length(x)/2)]';
```

Based on the knowledge that a kernel measures some kind of distance between two data points, construct the kernel matrix corresponding to $K(i, j) = e^{\frac{1}{2\sigma^2} \|x_i - x_j\|_2^2}$ for two σ values given by 0.1 and 1. Learn the above data by kernel ridge regression with the two given σ values and use the two learning results to predict test samples given by

```
x_ts = [0.1:.01:10.05];
```

What do you observe regarding the difference between the two predictions?

5.8 Consider the following training data points for two categories,

$$\begin{bmatrix} x_1 \\ x_2 \end{bmatrix} : \begin{bmatrix} 3 \\ 1 \end{bmatrix}, \begin{bmatrix} 2 \\ 3 \end{bmatrix}, \begin{bmatrix} 0 \\ 1 \end{bmatrix}, \begin{bmatrix} 2 \\ 1 \end{bmatrix}, \begin{bmatrix} 1 \\ 3 \end{bmatrix},$$ which correspond to labels y : $\{0, 0, 0, 1, 1\}$ respectively.

(a) The first goal is to predict the class label of $\mathbf{x}_{t1} = \begin{bmatrix} 3 \\ 3 \end{bmatrix}$ and $\mathbf{x}_{t2} = \begin{bmatrix} 1 \\ 1 \end{bmatrix}$ based on the primal TER method utilizing a linear model with a single output (i.e., utilize $y : \{0, 0, 0, 1, 1\}$ as the single column of learning target).

(b) The second goal is to predict the class label of $\mathbf{x}_{t1} = \begin{bmatrix} 3 \\ 3 \end{bmatrix}$ and $\mathbf{x}_{t2} = \begin{bmatrix} 1 \\ 1 \end{bmatrix}$ based on the dual TER method utilizing a third-order polynomial model.

5.9 Solve the above question based on one-hot encoding of the target instead of utilizing a single target output.

5.10 Given the data points of two categories which are labeled by circles and squares as shown in the following figure. The decision boundaries of Classifier A, Classifier B, and Classifier C are shown as dotted line, dashed line, and solid line respectively.

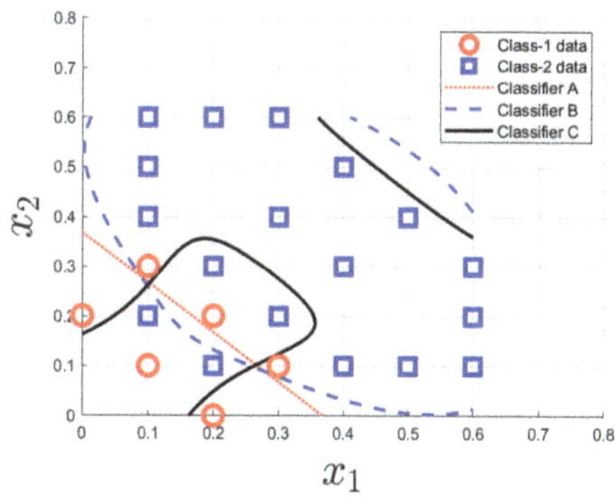

(i)   Write out the confusion matrix for each classifier.
(ii) Calculate the Total Error Rate for each Classifier.
(iii) Which classifier can predict better for unseen data? Why?

Chapter 6
Penalized Learning

While the main goal of supervised learning is to minimize an error cost function toward fitting the training data to the target values or labels, an overdoing of the fitting may result in a phenomenon called overfitting. In other words, an over-trained model might not provide accurate predictions for unseen data. This is particularly true when only a small number of training samples is given where only a partial picture of the unknown data distribution can be inferred. In this chapter, we focus on learning with coefficient shrinkage, which is also known as penalized learning or regularization. While Sect. 6.1 provides an overview on penalty learning, Sect. 6.2 focuses on ridge regression and related methods. Section 6.3 then introduces bridge regression, providing deterministic solutions. Following this, Sect. 6.4 examines coefficient profiles by adjusting the penalty strength. Finally, Sect. 6.5 presents some bibliographic notes.

6.1 An Overview of Coefficient Shrinkage

The complexity of a predictor model has a direct impact on its predictivity in regression as well as generalization in classification. Apart from the model structure, which includes the system order and nonlinearity, the magnitudes of estimated coefficients or parameters play a critical role in system stability. Shrinkage of estimation coefficients during parametric learning is thus an attractive research topic in machine learning.

Essentially, coefficient shrinkage can be formulated either as a constraint to the main minimization problem of an error objective or as the main optimization goal itself subject to an error constraint. The former is also known as *penalized learning* or *regularization* while the later is considered a parametric norm minimization with the goal of having the *least norm* solution.

6.1.1 *Regularization Approach*

Consider a given set of training data containing m samples and the predictor $g(x_i, w) = p_i^T w$ in (5.2), a weight-decay or L_2-norm related penalty term $\|w\|_2^2 := \sum_{j=0}^{D} w_j^2 = w^T w$ can be included along with the minimization of a squared error objective such as that in (5.31) which is reproduced here with a scaling by half in view of the squaring:

$$J_{\text{SSEr}}(w) = \frac{1}{2} \sum_{i=1}^{m} \left(y_i - p_i^T w \right)^2 + \frac{\lambda}{2} \|w\|_2^2. \tag{6.1}$$

The scalar λ is called the regularization factor which controls the weighting between the error term and the weight coefficient term. Alternatively, (6.1) can be written in constrained optimization form as minimizing the Sum of Squared Errors (SSE) subject to $\|w\|_2^2 \le t$, $t \in \mathbb{R}^+$. An analytical closed-form solution for (6.1) (or (5.31)) is reproduced in (6.9) below (or (5.32)). This is known as *ridge regression* in the literature (see Chap. 5).

Instead of adopting an L_2-norm penalty, LASSO (Least Absolute Shrinkage and Selection Operator) adopts an L_1-norm penalty term in the regression problem which finds w that minimizes

$$J_{\text{lasso}}(w) = \sum_{i=1}^{m} \left(y_i - p_i^T w \right)^2 + \lambda \|w\|_1, \tag{6.2}$$

where $\|w\|_1 := \sum_{j=0}^{D} |w_j|$. This problem is equivalent to minimizing the SSE subject to a constraint of the form $\|w\|_1 \le c$, $c \in \mathbb{R}^+$.

By generalizing the penalty norm term to arbitrary power $p \ge 0$, as

$$\|w\|_p := \left(\sum_{j=0}^{D} |w_j|^p \right)^{1/p}, \tag{6.3}$$

the problem of minimizing

$$\text{SSE}_{\text{bridge}}(w) = \sum_{i=1}^{m} \left(y_i - p_i^T w \right)^2 + \lambda \|w\|_p^p, \tag{6.4}$$

is called *bridge regression*. Here $0 \le p < 2$ is of particular interest due to its parametric selection and compression properties. When $0 \le p \le 1$, a parametric subset selection with some coefficients compressed to zero is expected while for $1 < p < 2$, a skewed parametric compression is expected.

6.1.2 Least Norm Approach

As discussed in Chap. 5, by putting the parametric norm as the main minimization objective with target fitting as constraint, we can write

$$\text{minimizes} \quad \|\boldsymbol{w}\|_2^2 \tag{6.5}$$
$$\text{subject to} \quad \mathbf{y} - \mathbf{P}\boldsymbol{w} = \mathbf{0},$$

where an analytical solution can be obtained as

$$\hat{\boldsymbol{w}} = \mathbf{P}^T(\mathbf{P}\mathbf{P}^T)^{-1}\mathbf{y}, \tag{6.6}$$

when $\mathbf{P}\mathbf{P}^T$ is non-singular. Here, we see the *right pseudoinverse* term $\mathbf{P}^T(\mathbf{P}\mathbf{P}^T)^{-1}$ replaces the *left pseudoinverse* term $(\mathbf{P}^T\mathbf{P})^{-1}\mathbf{P}^T$ of LSE solution (5.6). This is also known as the *minimum-norm* or *least norm* solution which applies particularly well to an under-determined system when $m < d + 1$ for the original features or $m < D + 1$ for the transformed features. The constrained optimization problem in (6.5) can, again, be generalized to minimization of L_p-norm as

$$\text{minimize} \quad \|\boldsymbol{w}\|_p^p \tag{6.7}$$
$$\text{subject to} \quad \mathbf{y} - \mathbf{P}\boldsymbol{w} = \mathbf{0},$$

where the corresponding unconstrained minimization problem is given by

$$\min_{\boldsymbol{w}} \|\boldsymbol{w}\|_p^p + \boldsymbol{\alpha}^T(\mathbf{y} - \mathbf{P}\boldsymbol{w}). \tag{6.8}$$

Here, the elements in vector $\boldsymbol{\alpha}$ are known as Lagrange multipliers, each corresponding to a given data sample. This topic shall be discussed in greater detail in Sect. 6.3.

6.2 Ridge Regression

In Chap. 5, Sects. 5.1.3 and 5.1.4, we have learning based on primal (5.32) and dual (5.36) ridge regressions. The solutions for binary classification using a single target vector $\mathbf{y}$ are reproduced as follows:

$$\text{Primal ridge}: \quad \hat{\boldsymbol{w}} = (\mathbf{P}^T\mathbf{P} + \lambda\mathbf{I})^{-1}\mathbf{P}^T\mathbf{y}, \quad \lambda > 0 \tag{6.9}$$

$$\text{Dual ridge}: \quad \hat{\boldsymbol{w}} = \mathbf{P}^T(\mathbf{P}\mathbf{P}^T + \lambda\mathbf{I})^{-1}\mathbf{y}, \quad \lambda > 0. \tag{6.10}$$

Here, we shall show the derivation of the dual ridge form based on the constrained optimization approach. Apart from using the penalty-based weight-decay formulation (5.31), an equivalent way to write the primal ridge regression minimization objective is based on constrained optimization:

$$\min_{w} \|\mathbf{y} - \mathbf{P}w\|_2^2 \tag{6.11}$$

$$\text{subject to } \|w\|_2^2 \le c,$$

where the parameter size constraint is explicitly expressed via a positive real constant c. Alternatively, the minimization problem of dual ridge regression can begin with

$$\min_{w} \|w\|_2^2 \tag{6.12}$$

$$\text{subject to } \mathbf{y} = \mathbf{P}w,$$

in view of parameters shrinkage. Consequently, with scaling consideration, (6.12) can be rewritten in unconstrained form via the method of Lagrange:

$$\min_{w} \frac{1}{2} \|w\|_2^2 + \boldsymbol{\alpha}^T (\mathbf{y} - \mathbf{P}w), \tag{6.13}$$

where $\boldsymbol{\alpha}$ denotes the Lagrange multipliers. Differentiation of the above equation with respect to w and put it zero gives

$$w = \mathbf{P}^T \boldsymbol{\alpha}, \tag{6.14}$$

which can be substituted into (6.13) to obtain a reparameterized minimization objective given by

$$\min_{\alpha} \frac{1}{2} \boldsymbol{\alpha}^T \mathbf{P}\mathbf{P}^T \boldsymbol{\alpha} + \boldsymbol{\alpha}^T (\mathbf{y} - \mathbf{P}\mathbf{P}^T \boldsymbol{\alpha}). \tag{6.15}$$

Next, the minimization can include a norm term of $\boldsymbol{\alpha}$:

$$\min_{\alpha} \frac{1}{2} \boldsymbol{\alpha}^T \mathbf{P}\mathbf{P}^T \boldsymbol{\alpha} + \boldsymbol{\alpha}^T (\mathbf{y} - \mathbf{P}\mathbf{P}^T \boldsymbol{\alpha}) - \frac{\lambda}{2} \boldsymbol{\alpha}^T \boldsymbol{\alpha}. \tag{6.16}$$

Here the sign of the norm term of $\boldsymbol{\alpha}$ can be arbitrary since the equality constraint can be written either as $\mathbf{y} - \mathbf{P}w = 0$ or as $\mathbf{P}w - \mathbf{y} = 0$ which causes sign change. Differentiate (6.16) with respect to $\boldsymbol{\alpha}$ and put it to zero gives

$$\mathbf{P}\mathbf{P}^T \boldsymbol{\alpha} + \mathbf{y} - 2\mathbf{P}\mathbf{P}^T \boldsymbol{\alpha} - \lambda \boldsymbol{\alpha} = 0$$

$$\mathbf{y} = \mathbf{P}\mathbf{P}^T \boldsymbol{\alpha} + \lambda \boldsymbol{\alpha}$$

$$\mathbf{y} = (\mathbf{P}\mathbf{P}^T + \lambda \mathbf{I})\boldsymbol{\alpha}$$

$$\hat{\boldsymbol{\alpha}} = (\mathbf{P}\mathbf{P}^T + \lambda \mathbf{I})^{-1} \mathbf{y}, \ \lambda > 0. \tag{6.17}$$

Substitute (6.17) back into (6.14) gives the dual solution in (6.10).

Apart from the arbitrary sign issue in the Lagrange formulation, another possible concern for (6.12) is that it requires the data to fit exactly according to the linear constraint term $\mathbf{y} = \mathbf{P}\mathbf{w}$. It is arguable that such constraint might pose a stringent requirement in practice. To get away from such an argument, the problem can be reformulated to include an error term as follows,

$$\min_{\mathbf{w}} \ \|\mathbf{w}\|_2^2 + \gamma \|\boldsymbol{\epsilon}\|_2^2$$
$$\text{subject to} \quad \mathbf{y} = \mathbf{P}\mathbf{w} + \boldsymbol{\epsilon}, \quad \gamma > 0 \tag{6.18}$$

where its unconstrained Lagrange function is

$$J(\mathbf{w}, \boldsymbol{\alpha}, \boldsymbol{\epsilon}) = \mathbf{w}^T \mathbf{w} + \gamma \boldsymbol{\epsilon}^T \boldsymbol{\epsilon} + \boldsymbol{\alpha}^T (\mathbf{y} - \mathbf{P}\mathbf{w} - \boldsymbol{\epsilon}). \tag{6.19}$$

Here the error term $\boldsymbol{\epsilon}$ can absorb the sign concerned in $\mathbf{y} - \mathbf{P}\mathbf{w} - \boldsymbol{\epsilon}$. Differentiate (6.19) with respect to $\mathbf{w}$ and put it to zero gives

$$2\mathbf{w} = \mathbf{P}^T \boldsymbol{\alpha}$$
$$\Rightarrow \quad \mathbf{w} = \frac{1}{2}\mathbf{P}^T \boldsymbol{\alpha}. \tag{6.20}$$

This result (6.20) is substituted into (6.19) to give

$$l(\boldsymbol{\alpha}, \boldsymbol{\epsilon}) = \left(\frac{1}{2}\right)^2 \boldsymbol{\alpha}^T \mathbf{P}\mathbf{P}^T \boldsymbol{\alpha} + \gamma \boldsymbol{\epsilon}^T \boldsymbol{\epsilon} + \boldsymbol{\alpha}^T \left(\mathbf{y} - \mathbf{P}(\frac{1}{2}\mathbf{P}^T \boldsymbol{\alpha}) - \boldsymbol{\epsilon}\right)$$
$$= -\frac{1}{4}\boldsymbol{\alpha}^T \mathbf{P}\mathbf{P}^T \boldsymbol{\alpha} + \gamma \boldsymbol{\epsilon}^T \boldsymbol{\epsilon} + \boldsymbol{\alpha}^T \mathbf{y} - \boldsymbol{\alpha}^T \boldsymbol{\epsilon}, \quad \gamma > 0. \tag{6.21}$$

Next, differentiate (6.21) with respect to $\boldsymbol{\epsilon}$ and put it to zero gives

$$2\gamma \boldsymbol{\epsilon} - \boldsymbol{\alpha} = \mathbf{0}$$
$$\Rightarrow \quad \boldsymbol{\epsilon} = \frac{1}{2\gamma}\boldsymbol{\alpha}. \tag{6.22}$$

Then, substitute (6.22) into (6.21) gives

$$l(\boldsymbol{\alpha}) = -\frac{1}{4}\boldsymbol{\alpha}^T \mathbf{P}\mathbf{P}^T \boldsymbol{\alpha} + \frac{\gamma}{4\gamma^2}\boldsymbol{\alpha}^T \boldsymbol{\alpha} + \boldsymbol{\alpha}^T \mathbf{y} - \frac{1}{2\gamma}\boldsymbol{\alpha}^T \boldsymbol{\alpha}$$
$$= -\frac{1}{4}\boldsymbol{\alpha}^T \mathbf{P}\mathbf{P}^T \boldsymbol{\alpha} - \frac{2-1}{4\gamma}\boldsymbol{\alpha}^T \boldsymbol{\alpha} + \boldsymbol{\alpha}^T \mathbf{y}, \quad \gamma > 0. \tag{6.23}$$

Finally, differentiate (6.23) with respect to $\boldsymbol{\alpha}$ and put it to zero gives

$$-\frac{1}{2}\mathbf{P}\mathbf{P}^T\alpha - \frac{1}{2\gamma}\alpha + \mathbf{y} = \mathbf{0}$$

$$\mathbf{P}\mathbf{P}^T\alpha + \frac{1}{\gamma}\alpha = 2\mathbf{y}$$

$$\left(\mathbf{P}\mathbf{P}^T + \frac{1}{\gamma}\mathbf{I}\right)\alpha = 2\mathbf{y}$$

$$\Rightarrow \quad \hat{\alpha} = 2(\mathbf{P}\mathbf{P}^T + \lambda\mathbf{I})^{-1}\mathbf{y}, \quad \lambda = \frac{1}{\gamma} \tag{6.24}$$

where (6.24) is substituted into (6.20) to give the ridge solution in dual form:

$$\hat{w} = \mathbf{P}^T(\mathbf{P}\mathbf{P}^T + \lambda\mathbf{I})^{-1}\mathbf{y}, \quad \lambda = \frac{1}{\gamma}, \gamma > 0. \tag{6.25}$$

In summary, both primal and dual ridge regressions attempt to shrink the size of parameters in order to avoid overfitting.

6.3 Bridge Regression

The previous section discussed about inclusion of a parametric penalty with the well-known L_2-norm type which has an analytic solution when minimized together with the least squares loss. In this section, we present a particular range of bridge regression which can be solved analytically.

6.3.1 Mathematical Preliminary

Consider a parameter vector w of dimension $w \in \mathbb{R}^{D+1}$. The L_p-norm of this parameter vector is defined as

$$\|w\|_p := \left(\sum_{j=0}^{D} |w_j|^p\right)^{1/p}, \quad p \geq 1. \tag{6.26}$$

To facilitate a differentiable norm function for minimization, the absolute operator can be approximated by $f(w_i) = \sqrt{w_i^2 + \epsilon}$ (see Ramirez et al. (2014)) for an arbitrarily small $\epsilon > 0$ (see Fig. 6.1 for a plot with different ϵ values). This turns the L_p-norm into a smooth function as follows:

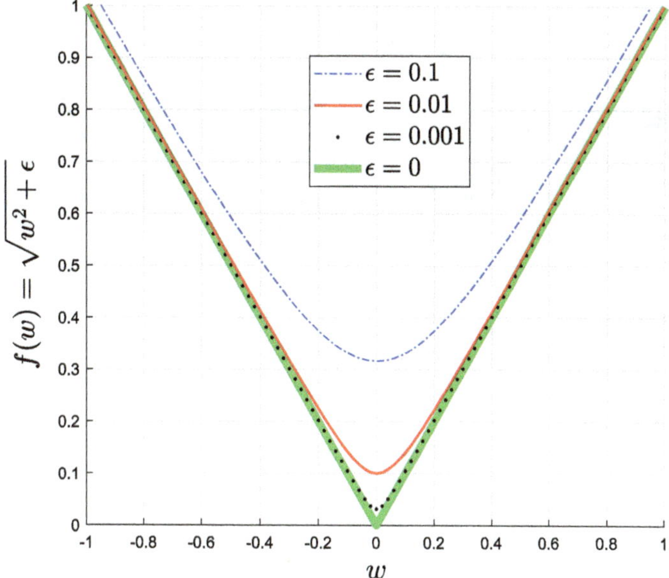

Fig. 6.1 Plot of $f(w) = \sqrt{w^2 + \epsilon}$ at several ϵ values

$$\wr w \wr_k := \left(\sum_{j=0}^{D-1} f(\boldsymbol{w}_j)^k \right)^{1/k}. \tag{6.27}$$

Here, we observed that $\lim_{\epsilon \to 0} f(w_i) = |w_i|$ and that the loosely defined $\wr \boldsymbol{w} \wr_k$ does not constitute a *normed vector space* for $\boldsymbol{w}$ since it violates the absolute homogeneity axiom. We shall call $\wr \cdot \wr_k$ defined in (6.27) a *k-measure* operator for convenience hereon.

Another approximation to the L_p-norm can be achieved based on a Local Quadratic Approximation (LQA, Park and Yoon (2011)) such as

$$|w_j|_q^q \approx |w_{0j}|^q + \frac{q}{2} \frac{|w_{0j}|^{q-1}}{|w_{0j}|} (w_j^2 - w_{0j}^2), \tag{6.28}$$

where the minimization problem of an approximated bridge regression can be written as

$$\min_{\boldsymbol{w}} \left\{ (\mathbf{y} - \mathbf{P}\boldsymbol{w})^T (\mathbf{y} - \mathbf{P}\boldsymbol{w}) + \frac{q}{2} \sum_{j=0}^{D} |w_{0j}|^{q-2} w_j^2 \right\}. \tag{6.29}$$

Figure 6.2 shows the plot of utilizing an one-dimensional ($p = 1$) LQA to approximation the absolute function at several q values and at $\lambda = 1$ with $w_{0j} = w_j - 0.01$.

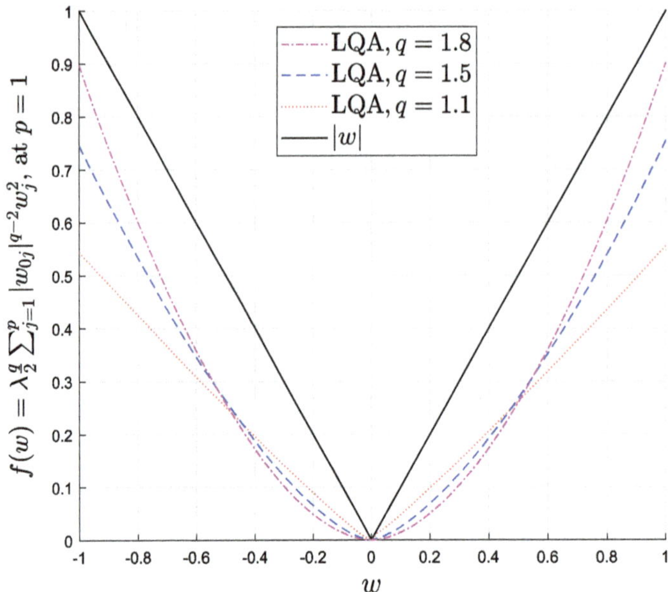

Fig. 6.2 Plot of one-dimensional ($p = 1$) Linear Quadratic Approximation (LQA) $f(w) = \lambda \frac{q}{2} \sum_{j=1}^{p} |w_{0j}|^{q-2} w_j^2$ being used to approximate the absolute function at several q values and at $\lambda = 1$ with $w_{0j} = w_j - 0.01$

Figure 6.3 plots the contours of the L_p-norm metric for $1 < p \leq 2$ and the corresponding k-measure ($\wr w \wr_k$, (6.27)) together with its k-powered form ($\wr w \wr_k^k$). From the second and the third rows of plots in Fig. 6.3, we see that except for the difference in curvature, the entire k-measure and its k-powered form approximate well to the solution p-norm for the plotted range of $1 < \{p, k\} \leq 2$. A similar observation also applies to the LQA as shown in the bottom row of the plots in Fig. 6.3. This observation suggests vertices of the vector space being feasible solutions for the desired constrained solution search. Such an observation shall be exploited in the following development for compressive solution when $1 < k \leq 2$.

In the following, the raised power form of k-measure is shown (Toh et al., 2023) to be convex when the approximation function $f_\epsilon(\cdot)$ is convex.

Lemma 6.1 $\wr w \wr_k^k$ *is convex on* w *when* f_ϵ *is convex for all* $k \geq 1$.

This result states the convexity of $\wr w \wr_k^k$ for a general convex approximation function f_ϵ. For our specific case of $f(w_i) = \sqrt{w_i^2 + \epsilon}$, it is easy to verify its convexity because the second derivative of $f(w)$ for each $w \in \{w_0, w_1, \ldots, w_{D-1}\}$ is

$$\frac{d^2 f}{d w^2} = \left(w^2 + \epsilon\right)^{-\frac{1}{2}} - w^2 \left(w^2 + \epsilon\right)^{-\frac{3}{2}}, \tag{6.30}$$

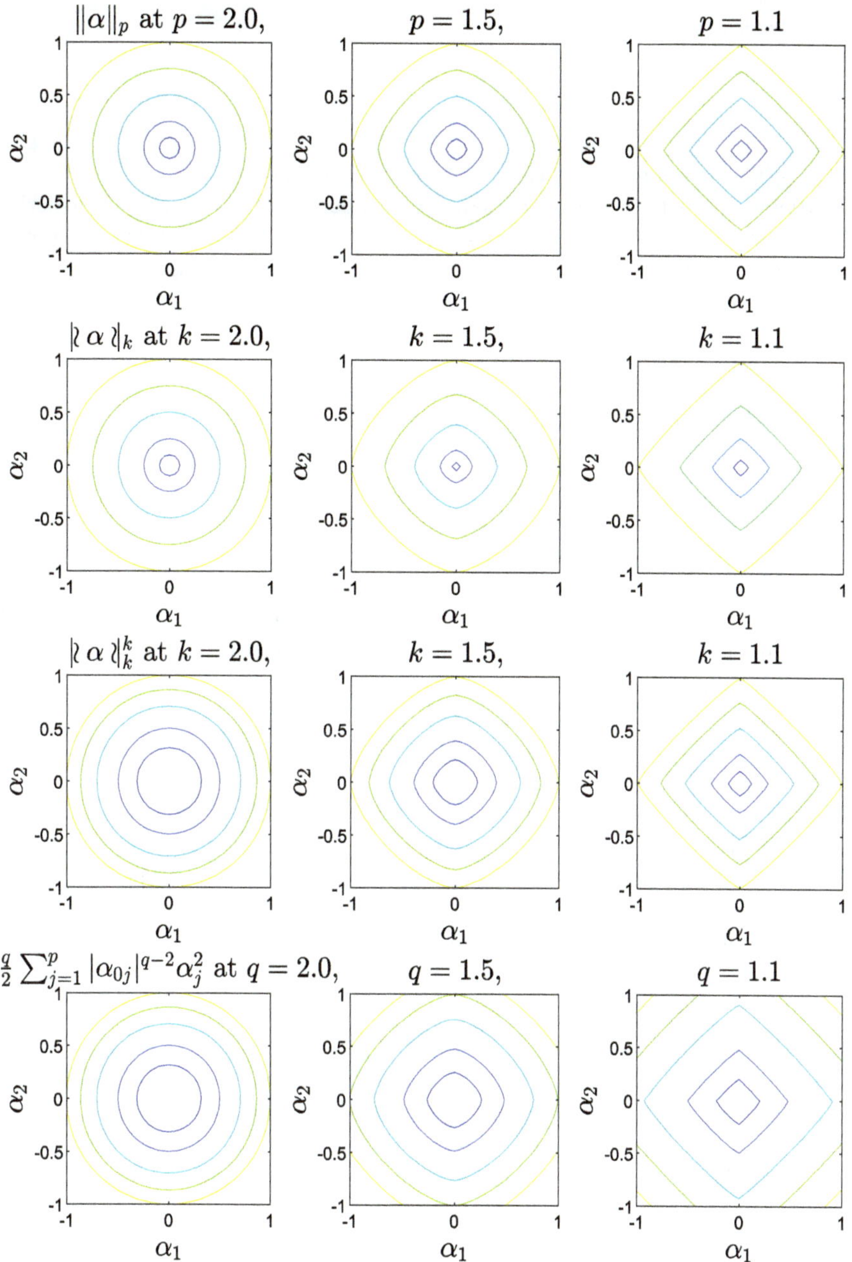

Fig. 6.3 Contour plots at levels [0.1, 0.25, 0.5, 0.75, 1]. Top row: $\|\boldsymbol{w}\|_p$ for $p \in \{2.0, 1.5, 1.1\}$. Second row: $\wr \boldsymbol{w} \wr_k$ for $k \in \{2.0, 1.5, 1.1\}$ at $\epsilon = 0.0001$. Third row: $\wr \boldsymbol{w} \wr_k^k$ for $k \in \{2.0, 1.5, 1.1\}$ at $\epsilon = 0.0001$. Bottom row: $\frac{q}{2} \sum_{j=1}^{p} |w_{0j}|^{q-2} w_j^2$ for $q \in \{2.0, 1.5, 1.1\}$ with $w_{0j} = w_j - 0.01$

where we see that

$$\frac{\left(w^2 + \epsilon\right)^{-\frac{1}{2}}}{w^2 \left(w^2 + \epsilon\right)^{-\frac{3}{2}}} = \frac{w^2 + \epsilon}{w^2} > 1, \tag{6.31}$$

for $\epsilon > 0$, which implies convexity of f on w based on $d^2 f / d w^2 > 0$.

After verifying the convexity of the k-measure penalty term, we are ready to derive the solution for the penalized learning. This is grounded on the fact that the sum of two convex functions, namely the squared error term and the k-measure penalty term, is also convex.

6.3.2 Proximal Bridge Regression in Primal Form

Given the training data $\{x_i, y_i\}$ with samples indexed by $i = 1, \ldots, m$ where $x_i = [x_{i,0}, x_{i,1}, \ldots, x_{i,D}]^T$ and y_i are respectively the regressors and the response for the ith observation. Consider the linear regression model using transformed features given by $\mathbf{P}w$ with parameter vector $w \in \mathbb{R}^{D+1}$ and regressor matrix $\mathbf{P} = [p(x_1), \ldots, p(x_m)]^T$ (see also (3.31)). For over-determined systems with $m > D + 1$, the minimization objective is the sum of squared errors with a k-measure penalty as follows:

$$J(w) = (\mathbf{y} - \mathbf{P}w)^T (\mathbf{y} - \mathbf{P}w) + \lambda \wr w \wr_k^k. \tag{6.32}$$

This minimization will be referred to as *proximal bridge regression in primal form* and in brief *primal p-bridge*. In the forthcoming derivation, $\circ$ denotes an elementwise operator. For example, $\mathbf{A}^{\circ k}$ indicates raising each element of $\mathbf{A}$ to the power k. Also, $\mathrm{diag}(a)$ denotes a diagonal matrix with its diagonal elements given by vector a.

The minimization (Toh et al., 2023) begins by defining $\bar{w} := [(w_0^2 + \epsilon)^{k/4}, \ldots, (w_{D-1}^2 + \epsilon)^{k/4}]^T$ where we can write $\wr w \wr_k = (\bar{w}^T \bar{w})^{1/k}$ and $\wr w \wr_k^k = (\bar{w}^T \bar{w})$ according to the definition of k-measure in (6.27). Next, by taking the first derivative of (6.32) with respect to w and set it to zero, we have

$$\frac{\partial}{\partial w} \left((\mathbf{y} - \mathbf{P}w)^T (\mathbf{y} - \mathbf{P}w) + \lambda \bar{w}^T \bar{w} \right) = 0$$

$$-2\mathbf{P}^T (\mathbf{y} - \mathbf{P}w) + \lambda \frac{k}{4} \cdot 2w \circ \left(w^{\circ 2} + \epsilon\right)^{\circ\left(\frac{k}{4} - 1\right)} \circ 2\bar{w} = 0$$

$$-2\mathbf{P}^T (\mathbf{y} - \mathbf{P}w) + \lambda k\, w \circ \left(w^{\circ 2} + \epsilon\right)^{\circ\left(\frac{k}{4} - 1\right)} \circ \left(w^{\circ 2} + \epsilon\right)^{\circ\frac{k}{4}} = 0$$

$$\Rightarrow \quad \lambda k\, w \circ \left(w^{\circ 2} + \epsilon\right)^{\circ\left(\frac{k}{2} - 1\right)} = 2\mathbf{P}^T (\mathbf{y} - \mathbf{P}w). \tag{6.33}$$

The limiting case of $\epsilon \to 0$ leads to

$$\lim_{\epsilon \to 0} k\, \boldsymbol{w} \circ \left(\boldsymbol{w}^{\circ 2} + \boldsymbol{\epsilon}\right)^{\circ\left(\frac{k}{2} - 1\right)} = k\, \boldsymbol{w} \circ \left(\boldsymbol{w}^{\circ 2}\right)^{\circ\left(\frac{k}{2} - 1\right)}$$

$$= k\, \mathrm{sgn}(\boldsymbol{w}) \circ (\boldsymbol{w}^{\circ 2})^{\circ\frac{1}{2}} \circ \left(\boldsymbol{w}^{\circ 2}\right)^{\circ\frac{k-2}{2}}$$

$$= k\, \mathrm{sgn}(\boldsymbol{w}) \circ \left(\boldsymbol{w}^{\circ 2}\right)^{\circ\frac{k-1}{2}} . \tag{6.34}$$

Substitute (6.34) into the LHS of Eq. (6.33), we have

$$\lambda k\, \mathrm{sgn}(\boldsymbol{w}) \circ \left(\boldsymbol{w}^{\circ 2}\right)^{\circ\frac{k-1}{2}} = 2\mathbf{P}^T(\mathbf{y} - \mathbf{P}\boldsymbol{w})$$

$$\Rightarrow \frac{\lambda k}{2}\, \mathrm{sgn}(\boldsymbol{w}) \circ |\boldsymbol{w}|^{\circ(k-1)} = \mathbf{P}^T(\mathbf{y} - \mathbf{P}\boldsymbol{w}).$$

$$\frac{\lambda k}{2}\, \mathrm{sgn}(\boldsymbol{w}) \circ |\boldsymbol{w}|^{\circ(k-1)} + \mathbf{P}^T\mathbf{P}\boldsymbol{w} = \mathbf{P}^T\mathbf{y},$$

$$\frac{\lambda k}{2}\, \boldsymbol{w} \circ |\boldsymbol{w}|^{\circ(k-2)} + \mathbf{P}^T\mathbf{P}\boldsymbol{w} = \mathbf{P}^T\mathbf{y},$$

$$\left(\frac{\lambda k}{2}\, \mathrm{diag}\{|\boldsymbol{w}|^{\circ(k-2)}\} + \mathbf{P}^T\mathbf{P}\right)\boldsymbol{w} = \mathbf{P}^T\mathbf{y}. \tag{6.35}$$

This leads to the solution

$$\hat{\boldsymbol{w}} = \left(\frac{\lambda k}{2}\, \mathrm{diag}\{|\boldsymbol{w}|^{\circ(k-2)}\} + \mathbf{P}^T\mathbf{P}\right)^{-1}\mathbf{P}^T\mathbf{y} \tag{6.36}$$

when $\left(\frac{\lambda k}{2}\, \mathrm{diag}\{|\boldsymbol{w}|^{\circ(k-2)}\} + \mathbf{P}^T\mathbf{P}\right)$ is non-singular.

To summarize, (6.36) is the solution to the minimization problem (6.32) for $m > D + 1$. This solution is expressed in recursive form and hence an initialization of the parameter $\boldsymbol{w}$ is needed. A good candidate for initialization is the solution to ridge regression given by $\boldsymbol{w}_o = (\mathbf{P}^T\mathbf{P} + \lambda\mathbf{I})^{-1}\mathbf{P}^T\mathbf{y}$ and an iteration of two to five cycles gives reasonably good results.

6.3.3 Proximal Bridge Regression in Dual Form

For under-determined systems with $m < D + 1$, the minimization is formulated as

$$\min_{\boldsymbol{w}} \| \boldsymbol{w} \|_k^k \quad \text{subject to} \quad \mathbf{y} = \mathbf{P}\boldsymbol{w}. \tag{6.37}$$

This minimization shall be called *proximal bridge regression in dual form* or *dual p-bridge* in brief. Similar to the primal p-bridge regression, the goal here is to have a skewed compressive estimate for $1 < k \le 2$.

The minimization (Toh et al., 2023) begins by writing (6.37) in its Lagrangian form,

$$\min_{\boldsymbol{w}} \wr \boldsymbol{w} \wr_k^k + \boldsymbol{\beta}^T (\mathbf{y} - \mathbf{P}\boldsymbol{w}), \tag{6.38}$$

where $\boldsymbol{\beta}$ corresponds to the Lagrange multiplier. Similar to the primal case, let $\bar{\boldsymbol{w}} := [(w_0^2 + \epsilon)^{k/4}, \ldots, (w_D^2 + \epsilon)^{k/4}]^T$ where we can write $\wr \boldsymbol{w} \wr_k = (\bar{\boldsymbol{w}}^T \bar{\boldsymbol{w}})^{1/k}$ and $\wr \boldsymbol{w} \wr_k^k = (\bar{\boldsymbol{w}}^T \bar{\boldsymbol{w}})$. Then, by taking the first derivative of the Lagrange function in (6.38) and setting it to zero, we have

$$\frac{\partial}{\partial \boldsymbol{w}} \left(\bar{\boldsymbol{w}}^T \bar{\boldsymbol{w}} + \boldsymbol{\beta}^T (\mathbf{y} - \mathbf{P}\boldsymbol{w}) \right) = \mathbf{0}$$

$$\frac{k}{4} \cdot 2\boldsymbol{w} \circ \left(\boldsymbol{w}^{\circ 2} + \boldsymbol{\epsilon} \right)^{\circ(\frac{k}{4} - 1)} \circ 2\bar{\boldsymbol{w}} - \mathbf{P}^T \boldsymbol{\beta} = \mathbf{0}$$

$$k\,\boldsymbol{w} \circ \left(\boldsymbol{w}^{\circ 2} + \boldsymbol{\epsilon} \right)^{\circ(\frac{k}{4} - 1)} \circ \left(\boldsymbol{w}^{\circ 2} + \boldsymbol{\epsilon} \right)^{\circ \frac{k}{4}} - \mathbf{P}^T \boldsymbol{\beta} = \mathbf{0}$$

$$\Rightarrow \quad k\,\boldsymbol{w} \circ \left(\boldsymbol{w}^{\circ 2} + \boldsymbol{\epsilon} \right)^{\circ(\frac{k}{2} - 1)} = \mathbf{P}^T \boldsymbol{\beta}. \tag{6.39}$$

For the limiting case of $\epsilon \to 0$, we have

$$\lim_{\epsilon \to 0} k\,\boldsymbol{w} \circ \left(\boldsymbol{w}^{\circ 2} + \boldsymbol{\epsilon} \right)^{\circ(\frac{k}{2} - 1)} = k\,\boldsymbol{w} \circ \left(\boldsymbol{w}^{\circ 2} \right)^{\circ(\frac{k}{2} - 1)}, \tag{6.40}$$

which implies

$$k\,\boldsymbol{w} \circ \left(\boldsymbol{w}^{\circ 2} \right)^{\circ \frac{k-2}{2}} = \mathbf{P}^T \boldsymbol{\beta}$$

$$k\,\mathrm{sgn}(\boldsymbol{w}) \circ (\boldsymbol{w}^{\circ 2})^{\frac{1}{2}} \circ \left(\boldsymbol{w}^{\circ 2} \right)^{\circ \frac{k-2}{2}} = \mathbf{P}^T \boldsymbol{\beta}$$

$$k\,\mathrm{sgn}(\boldsymbol{w}) \circ \left(\boldsymbol{w}^{\circ 2} \right)^{\circ \frac{k-1}{2}} = \mathbf{P}^T \boldsymbol{\beta}$$

$$\Rightarrow \quad \left(\boldsymbol{w}^{\circ 2} \right)^{\circ \frac{k-1}{2}} = \mathrm{sgn}(\boldsymbol{w}) \circ \left(\frac{1}{k} \mathbf{P}^T \boldsymbol{\beta} \right). \tag{6.41}$$

Taking square elementwise for both sides of (6.41), we have

$$\left(\boldsymbol{w}^{\circ 2} \right)^{\circ(k - 1)} = \left(\frac{1}{k} \mathbf{P}^T \boldsymbol{\beta} \right)^{\circ 2}. \tag{6.42}$$

Since the vector $\lim_{\epsilon \to 0}(\boldsymbol{w}^{\circ 2} + \boldsymbol{\epsilon})$ has nonnegative elements, $\lim_{\epsilon \to 0} \left(\boldsymbol{w}^{\circ 2} + \boldsymbol{\epsilon} \right)^{\circ(\frac{k}{2} - 1)}$ has nonnegative elements. Therefore, we can infer from (6.39) that $\mathrm{sgn}(\boldsymbol{w}) = \mathrm{sgn}(\mathbf{P}^T \boldsymbol{\beta})$, and

$$\boldsymbol{w} = \mathrm{sgn}(\mathbf{P}^T \boldsymbol{\beta}) \circ \left| \mathbf{P}^T \left\{ \frac{1}{k} \boldsymbol{\beta} \right\} \right|^{\circ \frac{1}{k-1}}. \tag{6.43}$$

Next, suppose that

$$\text{sgn}(\mathbf{P}^T\boldsymbol{\beta}) \circ \left| \mathbf{P}^T \left\{ \frac{1}{k}\boldsymbol{\beta} \right\} \right|^{\circ \frac{1}{k-1}} = |\mathbf{P}^T|^{\circ \frac{1}{k-1}} \boldsymbol{\gamma} \tag{6.44}$$

for some $\boldsymbol{\gamma}$, then premultiply $\mathbf{P}$ to both sides of (6.43) gives

$$\mathbf{P}w = \mathbf{P}\,\text{sgn}(\mathbf{P}^T\boldsymbol{\beta}) \circ \left| \mathbf{P}^T \left\{ \frac{1}{k}\boldsymbol{\beta} \right\} \right|^{\circ \frac{1}{k-1}}$$

$$\Rightarrow \mathbf{P}w = \mathbf{P}|\mathbf{P}^T|^{\circ \frac{1}{k-1}}\boldsymbol{\gamma}, \quad \text{according to (6.47)}$$

$$\Rightarrow \mathbf{y} = \mathbf{P}|\mathbf{P}^T|^{\circ \frac{1}{k-1}}\boldsymbol{\gamma}, \quad \text{since } \mathbf{y} = \mathbf{P}w$$

$$\Rightarrow \boldsymbol{\gamma} = \left[\mathbf{P}|\mathbf{P}^T|^{\circ \frac{1}{k-1}} \right]^{-1} \mathbf{y}, \tag{6.45}$$

since $\left[\mathbf{P}|\mathbf{P}^T|^{\circ \frac{1}{k-1}} \right]^{-1}$ is invertible. Knowing also that $\mathbf{PP}^T$ is invertible, we substitute (6.45) into (6.44) and get

$$\text{sgn}(\mathbf{P}^T\boldsymbol{\beta}) \circ \left| \mathbf{P}^T \left\{ \frac{1}{k}\boldsymbol{\beta} \right\} \right|^{\circ \frac{1}{k-1}} = |\mathbf{P}^T|^{\circ \frac{1}{k-1}} \left[\mathbf{P}|\mathbf{P}^T|^{\circ \frac{1}{k-1}} \right]^{-1} \mathbf{y}$$

$$\mathbf{P}^T \left\{ \frac{1}{k}\boldsymbol{\beta} \right\} = \text{sgn}(\mathbf{P}^T\boldsymbol{\beta}) \circ \left| |\mathbf{P}^T|^{\circ \frac{1}{k-1}} \left[\mathbf{P}|\mathbf{P}^T|^{\circ \frac{1}{k-1}} \right]^{-1} \mathbf{y} \right|^{\circ(k-1)}$$

$$\mathbf{PP}^T \left\{ \frac{1}{k}\boldsymbol{\beta} \right\} = \text{sgn}(\mathbf{P}^T\boldsymbol{\beta}) \circ \mathbf{P} \left| |\mathbf{P}^T|^{\circ \frac{1}{k-1}} \left[\mathbf{P}|\mathbf{P}^T|^{\circ \frac{1}{k-1}} \right]^{-1} \mathbf{y} \right|^{\circ(k-1)}$$

$$\left\{ \frac{1}{k}\boldsymbol{\beta} \right\} = \text{sgn}(\mathbf{P}^T\boldsymbol{\beta}) \circ (\mathbf{PP}^T)^{-1} \mathbf{P} \left| |\mathbf{P}^T|^{\circ \frac{1}{k-1}} \left[\mathbf{P}|\mathbf{P}^T|^{\circ \frac{1}{k-1}} \right]^{-1} \mathbf{y} \right|^{\circ(k-1)}.$$

$$\tag{6.46}$$

Subsequently, substitute (6.46) into (6.43) and we have

$$\hat{w} = \text{sgn}(\mathbf{P}^T\boldsymbol{\beta}) \circ \left| \mathbf{P}^T (\mathbf{PP}^T)^{-1} \mathbf{P} \left(|\mathbf{P}^T|^{\circ \frac{1}{k-1}} \left[\mathbf{P}|\mathbf{P}^T|^{\circ \frac{1}{k-1}} \right]^{-1} \mathbf{y} \right)^{\circ(k-1)} \right|^{\circ \frac{1}{k-1}}$$

$$= \text{sgn}(\boldsymbol{\theta}) \circ \left| \mathbf{P}^T (\mathbf{PP}^T)^{-1} \mathbf{P}\boldsymbol{\theta}^{\circ(k-1)} \right|^{\circ \frac{1}{k-1}}, \tag{6.47}$$

where

$$\boldsymbol{\theta} = |\mathbf{P}^T|^{\circ \frac{1}{k-1}} \left[\mathbf{P}|\mathbf{P}^T|^{\circ \frac{1}{k-1}} \right]^{-1} \mathbf{y}. \tag{6.48}$$

The sign of $\text{sgn}(\mathbf{P}^T\boldsymbol{\beta}) = \text{sgn}(\boldsymbol{\theta})$ has been deduced from the top row of (6.46). Equations (6.47)–(6.48) hold well without singularity for all $\boldsymbol{\theta} \in \mathbb{R}^{D+1}$ and all $\mathbf{P} \in \mathbb{R}^{m \times (D+1)}$ for $k > 1$. Finally, since $\| w \|_k^k$ is convex according to Lemma 6.1

and the linear constraint function $(\mathbf{y} = \mathbf{P}\mathbf{w})$ is also convex, the Lagrange function (6.38) is convex. Hence the minimizer.

In summary, (6.47) with (6.48) is the solution to the minimization problem (6.37) when $m < D + 1$. This solution is global since the formulation is convex. A restriction to this solution is that k cannot be equal to 1.

6.3.4 Learning with Multiple Outputs

When the outputs are independent, the estimations for proximal bridge in both the primal and the dual cases can be stacked utilizing the same regressor matrix (such as $\mathbf{P}$ in (3.31)) in a similar manner to the least squares method in (5.18). Suppose $\hat{\mathbf{W}}$ stacks the multiple columns of estimated coefficient vectors $[\hat{\mathbf{w}}_1, \ldots, \hat{\mathbf{w}}_C] \in \mathbb{R}^{(D+1)\times C}$ of the learning model $\mathbf{G} = \mathbf{P}\mathbf{W}$, then the prediction can be computed as $\hat{\mathbf{G}} = \mathbf{P}\hat{\mathbf{W}}$.

Given the data $\{\mathbf{x}_i, y_{i,l}\}, i = 1, \ldots, m, \ l = 1, \ldots, C$ where $\mathbf{x}_i = [x_{i,0}, \ldots, x_{i,D}]^T$ and $y_{i,l}$ are respectively the inputs and the response for the ith observation of the lth output. The transformed regressor matrix $\mathbf{P}$ (see also (3.31)) has been generated based on $\mathbf{x}_i, i = 1, \ldots, m$, by a mapping such as the polynomials with D expansion terms. Suppose $\mathbf{P}^T\mathbf{P}$ is of full rank for the over-determined case where $m > D + 1$. Then, under the limiting case of $\epsilon_l \to \mathbf{0}$ and for $k \geq 1$, $\hat{\mathbf{W}} = [\hat{\mathbf{w}}_1, \ldots, \hat{\mathbf{w}}_C]$ where

$$\hat{\mathbf{w}}_l = \left(\frac{\lambda k}{2} \operatorname{diag}\{|\mathbf{w}_l|^{\circ(k-2)}\} + \mathbf{P}^T\mathbf{P} \right)^{-1} \mathbf{P}^T\mathbf{y}_l, \quad l = 1, \ldots, C, \qquad (6.49)$$

when the matrix $(\frac{\lambda k}{2} \operatorname{diag}\{|\mathbf{w}_l|^{\circ(k-2)}\} + \mathbf{P}^T\mathbf{P})$ is invertible.

For an under-determined system $\mathbf{y}_l = \mathbf{P}\mathbf{w}_l, l = 1, \ldots, C$ with $\mathbf{y}_l \in \mathbb{R}^m$ being the given target vector for each output, $\mathbf{P} \in \mathbb{R}^{m \times D+1}$ being the regressor matrix and $\mathbf{w} \in \mathbb{R}^{D+1}$ being the parameter vector, with number of samples $m < D + 1$ regressor dimensions, the estimation is given by $\hat{\mathbf{W}} = [\hat{\mathbf{w}}_1, ..., \hat{\mathbf{w}}_C]$ where

$$\hat{\mathbf{w}}_l = \operatorname{sgn}(\boldsymbol{\theta}_l) \circ \left| \mathbf{P}^T \left(\mathbf{P}\mathbf{P}^T \right)^{-1} \mathbf{P}\hat{\boldsymbol{\theta}}_l^{\circ(k-1)} \right|^{\circ\frac{1}{k-1}}, \quad l = 1, \ldots, C, \qquad (6.50)$$

with

$$\hat{\boldsymbol{\theta}}_l = |\mathbf{P}^T|^{\circ\frac{1}{k-1}} \left[\mathbf{P}|\mathbf{P}^T|^{\circ\frac{1}{k-1}} \right]^{-1} \mathbf{y}_l. \qquad (6.51)$$

In summary, when an indicator label matrix $\mathbf{Y} = [\mathbf{y}_1, \ldots, \mathbf{y}_C]$ (see (5.15)) has been adopted for classification, each column $\mathbf{y}_l, l = 1, \ldots, C$ of the indicator label matrix $\mathbf{Y}$ is utilized to learn separately using the same regressor matrix $\mathbf{P}$. The solutions $\hat{\mathbf{w}}_l, l = 1, \ldots, C$ are stacked to form the solution $\hat{\mathbf{W}} = [\hat{\mathbf{w}}_1, \ldots, \hat{\mathbf{w}}_C]$ for multiple outputs.

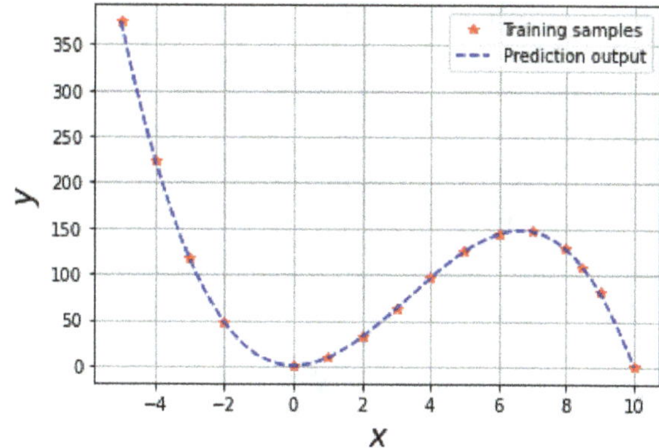

Fig. 6.4 The learning outputs of the 10th-order polynomial model at $k = 1.1$

Example 6.1 This example contains 16 training samples of single input dimension given by $x \in \{-5, -4, -3, -2, 0, 1, 2, 3, 4, 5, 6, 7, 8, 8.5, 9, 10\}$. The corresponding target outputs have been generated based on $y = 10x^2 - x^3$. To fit these data, a polynomial model of 10th-order is adopted with intercept. Including the intercept term, this polynomial model has 11 parameters. For testing, the input feature x has been generated for the range $x \in [-5 : 0.1 : 10]$. The p-bridge has been trained with $\lambda = 1$ and $k = 1.1$ where the estimated weight vector is given in four decimal places by

$$
w \approx \begin{bmatrix}
0.0169 \\
0.0000 \\
9.9439 \\
-0.9778 \\
0.0089 \\
-0.0007 \\
-0.0004 \\
0.0001 \\
0.0000 \\
-0.0000 \\
0.0000
\end{bmatrix}.
$$

These weights correspond to the polynomial terms given by $[1, x, x^2, x^3, x^4, x^5, x^6, x^7, x^8, x^9, x^{10}]$. This result shows compression of weights except for those corresponding to x^2 and x^3. Figure 6.4 shows the test prediction result plotted along with the training data points. The codes in Python are given below.

```python
import matplotlib.pyplot as plt
#--- Training data ---#
x = np.array([[-5],[-4],[-3],[-2],[0],[1],[2],[3],[4],[5],[6],[7],[8],[8.5],[9],[10]])
y = 10 * (x ** 2) - (x ** 3)
P = np.hstack([x ** i for i in range(11)])
#--- Test data ---#
xx = np.arange(-5, 10.1, 0.1).reshape(-1, 1)
PP = np.hstack([xx ** i for i in range(11)])
#--- p-bridge setting ---#
Primal = 1
k = 1.1
lambda_ = 1
#--- perform p-bridge learning ---#
w = pbridge(P,y,k,lambda_,Primal)
y_est = PP @ w
plt.plot(x, y, '*r', label='Training samples')
plt.plot(xx, y_est, '--b', label='Prediction output')
plt.legend(loc='upper right')
plt.grid()
plt.show()
print(w)
 [[ 1.69440773e-02]
  [-3.71654188e-07]
  [ 9.94394383e+00]
  [-9.97782217e-01]
  [ 8.85072149e-03]
  [-7.23873553e-04]
  [-4.35928118e-04]
  [ 5.59929816e-05]
  [ 5.61191057e-06]
  [-1.20831438e-06]
  [ 5.10376067e-08]]
```

6.4 Coefficient Profiles

In this section, we shall observe the profile of coefficient shrinkage by controlling the strength of penalty λ. The tic-tac-toe database[1] from the UCI machine learning

---

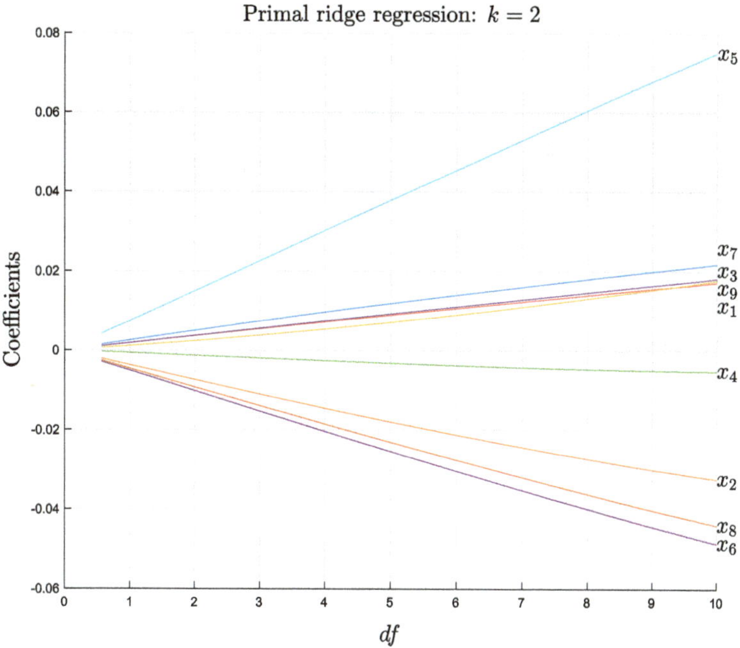

Fig. 6.5 Profiles of ridge coefficients for the over-determined tic-tac-toe example

repository will be utilized for studying the over-determined case and the Exclusive-OR (XOR) problem will be used for the under-determined case.

6.4.1   Coefficient Profile of the Over-Determined Case

In this example, a linear regression model is set to learn 600 training samples of the tic-tac-toe data set. The data for test prediction consists of the remaining 358 samples. This data set has a binary response with 9 input variables. The input features together with an intercept term give rise to 10 estimation coefficients for the linear regression model and this forms an over-determined regression system.

The estimation coefficient profile is observed according to variation of shrinkage setting controlled by the penalty λ value. Apart from the proximal bridge (p-bridge) solution, the ridge and lasso coefficient estimates are also shown for comparison purpose. These coefficient profiles have been plotted over $df(\lambda) = \text{tr}[\mathbf{P}(\mathbf{P}^T\mathbf{P} + \lambda\mathbf{I})\mathbf{P}^T]$ which defines the *effective degrees of freedom* implied by the penalty λ (see Hastie et al. (2017), Sect. 3.4).

In Figs. 6.5, 6.6a and b, the ridge, p-bridge and lasso regressions have been plotted respectively. The p-bridge has been plotted at $k = 1$ in order to observe its trace of parametric compression comparing with that of lasso. Figure 6.6a and b show certain

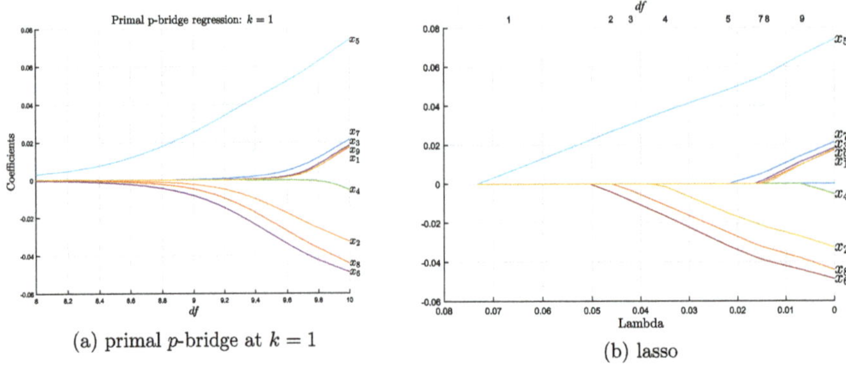

(a) primal p-bridge at $k = 1$

(b) lasso

Fig. 6.6 Profiles of primal p-bridge (at $k = 1$) and lasso coefficients for the over-determined tic-tac-toe example

Table 6.1 Tic-tac-toe example: test mean squared error (MSE) and variables (excluding the intercept) selected

Method	Tuned parameter(s)	Test MSE	Variables selected
Ols	–	0.2150 (0.009)	All
Ridge regression	$\lambda = 20$	0.2152 (0.009)	All
Lasso at (Alpha = 1)	Lambda = 0	0.2150 (0.009)	All
Elastic net	Lambda = 0.23, Alpha = 0.01	0.2159 (0.008)	All
p-bridge at ($k = 1$)	$\lambda = 0$	0.2150 (0.009)	All
p-bridge	$\lambda = 20, k = 1.58$	0.2153 (0.009)	All

Compared methods: ols (5.6), ridge regression (Tikhonov, 1963; Hoerl & Kennard, 1970a), lasso (Tibshirani, 1996), elastic-net (Zou & Hastie, 2005)

resemblance between the behaviors of primal p-bridge at $k = 1$ and lasso in terms of their order of shrinking sequences. In particular, the sequential shrinking order of x_4, x_9, x_1, x_3, and x_7 to zero show similarity for the two methods when lowering the df (raising the λ) value. However, for ridge regression as shown in Fig. 6.5, all the coefficients do not appear to shrink to zero sequentially as that of lasso and p-bridge. **Prediction results**: The prediction results of p-bridge, the ordinary least squares regression (ols), the ridge regression (ridge), the lasso, and the elastic-net are listed in Table 6.1 for comparison. For illustration purpose, this table shows the regression mean squared errors (MSE) between the target output and the measured output of the compared methods based on the 358 test samples. The chosen models have been obtained based on tenfold cross-validation utilizing the 600 training observations. The estimated coefficients are shown in Table 6.2 for each method. These results show comparable performance of the primal p-bridge with those of the well-known state of the art. Moreover, all selected models based on cross-validation show no preference for variable selection.

Table 6.2 Tic-tac-toe example: estimated coefficients for the chosen setting based cross-validation on the training set

Predictor	Ols	Ridge	Lasso	Elastic net	p-bridge at $k = 1$	p-bridge
0. intcpt	0.343	0.332	0.345	0.345	0.345	0.331
1. x_1	0.017	0.017	0.017	0.012	0.017	0.015
2. x_2	−0.033	−0.032	−0.033	−0.026	−0.033	−0.030
3. x_3	0.018	0.018	0.018	0.013	0.018	0.015
4. x_4	−0.005	−0.005	−0.005	−0.003	−0.005	−0.005
5. x_5	0.075	0.072	0.074	0.058	0.074	0.068
6. x_6	−0.049	−0.047	−0.049	−0.039	−0.049	−0.045
7. x_7	0.022	0.021	0.021	0.016	0.021	0.019
8. x_8	−0.044	−0.043	−0.044	−0.035	−0.044	−0.040
8. x_9	0.018	0.017	0.018	0.011	0.018	0.015

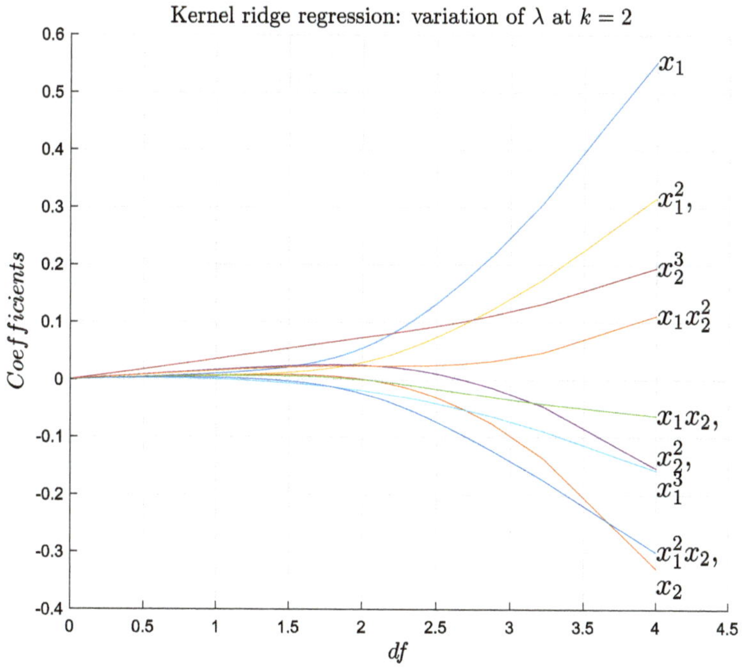

Fig. 6.7 Profiles of kernel ridge coefficients for the under-determined XOR example

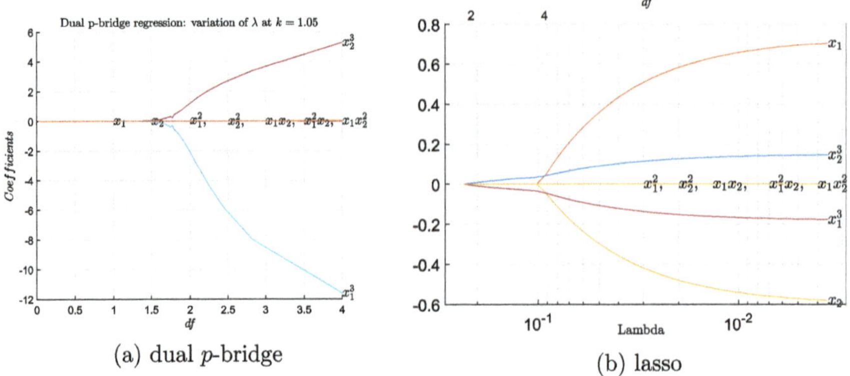

(a) dual p-bridge   (b) lasso

Fig. 6.8 Profiles of dual p-bridge and lasso coefficients for the under-determined XOR problem

6.4.2 Coefficient Profile of the Under-Determined Case

For the under-determined systems, the well-known XOR problem with four training data samples is evaluated. The inputs to the XOR problem are $(x_1, x_2) \in \{(0, 1), (2, 1), (1, 0), (1, 2)\}$ with their corresponding target outputs given by $y \in \{0, 0, 1, 1\}$, respectively. A third-order polynomial model given by

$$p(x_1, x_2) = w_0 + w_1 x_1 + w_2 x_2 + w_3 x_1^2 + w_4 x_2^2 + w_5 x_1 x_2 + w_6 x_1^3 + w_7 x_2^3$$
$$+ w_8 x_1^2 x_2 + w_9 x_1 x_2^2 \tag{6.52}$$

is deployed to learn the XOR data. Since the number of parametric coefficients $(w_0, \ldots, w_9)$ is larger than the four learning samples, this constitutes an under-determined learning system. For testing, a total of 200 test samples for this XOR problem have been generated for mapping evaluation. These test samples have been generated by a bivariate Gaussian random number generator with centers located at the four training points, each center corresponds to 50 samples with an identity covariance matrix scaled by 0.3. The responses of these test data follow the labels of the four centers respectively.

The coefficient profile of the kernel ridge regression ($k = 2$) is shown in Fig. 6.7 as baseline. Figure 6.8 shows the coefficient profiles of the dual p-bridge and the lasso. This plot shows a highly sparse estimation for p-bridge at low k-value ($k = 1.05$) comparing with lasso. Particularly, when the shrinkage penalty is low (at small λ value), p-bridge suppresses all coefficients except those of x_1^3 and x_2^3 whereas lasso emphasizes the coefficients of x_1 and x_2 more than that of x_1^3 and x_2^3 while suppressing all other coefficients. The decision boundaries for both the dual p-bridge regression and the lasso in Fig. 6.9 show much resemblance in view of the high contribution of the coefficients of x_1^3 and x_2^3.

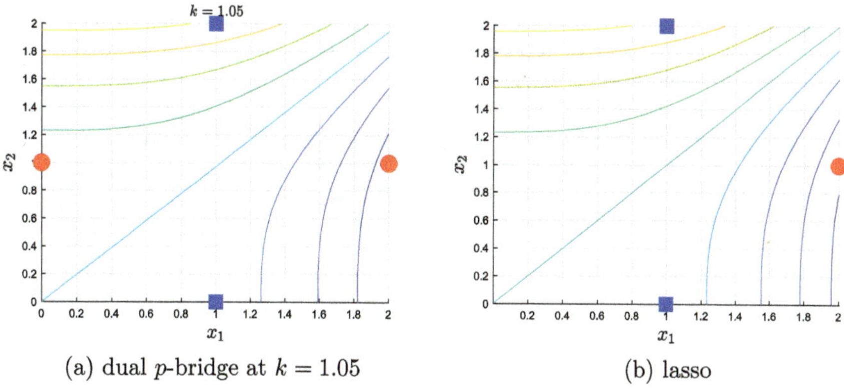

(a) dual p-bridge at $k = 1.05$ (b) lasso

Fig. 6.9 Decision contours of dual p-bridge and lasso

Table 6.3 XOR: test mean squared error (MSE) and variables (w_0, ..., w_9) selected based on the four training points

Method	Tuned parameter(s)	Test MSE	Variables selected
ols	–	0.513 (0.089)	All
ridge regression	$\lambda = 6$	0.503 (0.047)	(0, 1, 2, 3, 4, 6, 7, 8, 9)
lasso (`Alpha = 1`,`Lambda = 0.1`)	–	0.225 (0.011)	(0, 6, 7)
lasso at (`Alpha = 1`)	`Lambda = 0`	0.799 (0.189)	(0, 1, 2, 3, 4, 6, 7, 8, 9)
elastic-net	`Lambda = 0`, `Alpha = 0.01`	0.799 (0.189)	(0, 1, 2, 3, 4, 6, 7, 8, 9)
p-bridge at ($k = 1.05$)	$\lambda = 30$	0.504 (0.040)	(6, 7)
p-bridge	$\lambda = 0, k = 2$	0.513 (0.089)	All

Prediction results: The prediction accuracy is shown in Table 6.3 in terms of the mean squared errors between the generated and the estimated outputs of the compared methods for the 200 test observations. These results show comparable mapping performance of the dual p-bridge with that of the state of the art. The estimated coefficients as seen from Table 6.4 show sparseness for lasso at `Lambda = 0.1` (`Alpha = 1`) and dual p-bridge at $k = 1.05$ and $\lambda = 30$ (Fig. 6.10).

Example 6.2 Learning of the XOR problem by a third-order polynomial model can be implemented in Python as follows.

```python
import matplotlib.pyplot as plt
# XOR training data
X = np.array([[0, 1], [2, 1], [1, 0], [1, 2]]) # XOR inputs (training data)
y = np.array([0, 0, 1, 1])                      # XOR outputs (training data)
#--- 3rd-order Polynomial features ---%
P = np.column_stack((np.ones((4,1)), X[:,0], X[:,1], X[:,0]**2, X[:,1]**2,
X[:,0]*X[:,1], X[:,0]**3, X[:,1]**3, X[:,0]**2*X[:,1], X[:,0]*X[:,1]**2))
# generate test data
X1, X2 = np.meshgrid(np.arange(0, 2.05, 0.05), np.arange(0, 2.05, 0.05))
nrow, ncol = X1.shape
```

Table 6.4 XOR: estimated coefficients for the chosen setting based on the four training points (parameters within parenthesis have been prefixed, parameters without parenthesis have been determined based on training MSE)

Predictor	State-of-arts				p-bridge regression	
	Ols	Ridge	Lasso (A1, L0.1)	Elastic-net	($k = 1.05$) $\lambda = 30$	$k = 2$, $\lambda = 0$
0. intcpt	0.288	0.200	0.500	0.644	0.000	0.288
1. x_1	0.554	0.040	0	0.002	0.000	0.554
2. x_2	-0.329	-0.040	0	-0.002	-0.000	-0.329
3. x_1^2	0.316	-0.038	0	0.760	0.000	0.316
4. x_2^2	-0.154	0.038	0	-0.914	-0.000	-0.154
5. $x_1 x_2$	-0.063	-0.000	0	0.000	0.000	-0.063
6. x_1^3	-0.159	-0.071	-0.034	0.406	-0.050	-0.159
7. x_2^3	0.195	0.071	0.034	0.272	0.054	0.195
8. $x_1^2 x_2$	-0.301	-0.051	0	-0.179	-0.000	-0.301
9. $x_1 x_2^2$	0.111	0.051	0	0.460	0.000	0.111

```python
Xt = np.column_stack((X1.ravel(), X2.ravel()))
# generate P matrix for test data
Pt = np.column_stack((np.ones(nrow * ncol), Xt[:, 0], Xt[:, 1], Xt[:, 0] ** 2,
Xt[:, 1] ** 2, Xt[:, 0] * Xt[:, 1], Xt[:, 0] ** 3, Xt[:, 1] ** 3,
Xt[:, 0] ** 2 * Xt[:, 1], Xt[:, 0] * Xt[:, 1] ** 2))
# use training set {P,y} to learn w
k = 1.05
Lam = 30
w = pbridge(P, y, k, Lam)
# predict test data
y_test = Pt @ w
y_mesh = y_test.reshape(nrow, ncol)

fig, ax = plt.subplots()

ax.contour(X1, X2, y_mesh)

ax.plot(X[0:2, 0], X[0:2, 1], 'ro', markersize=12, linewidth=3)

ax.plot(X[2:4, 0], X[2:4, 1], 'bs', markersize=12, linewidth=3)

ax.set_title(f'$k = {k:.2f}$')

plt.grid()

plt.show()
```

6.5　Bibliographic Notes

Coefficient Shrinkage and Regularization

Being invented independently under different contexts, *ridge regression* is also known as weight decay or Tikhonov regularization which is named after Andrey

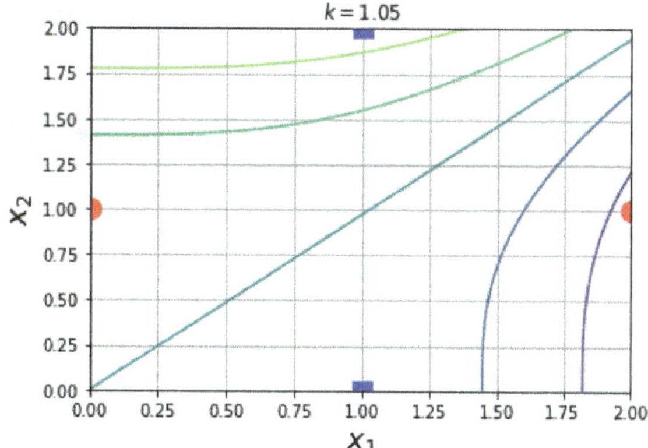

Fig. 6.10 Decision contours of dual p-bridge at $k = 1.05$ and $\lambda = 30$

Tikhonov (Tikhonov, 1963; Hoerl & Kennard, 1970b; Hastie et al., 2017). Due to its simplicity and viable solution, it is the most commonly adopted method of regularization for *ill-posed* problems.[2] The term *regularization* essentially refers to a trade-off between fitting the data and reducing a norm of the solution. It is thus also called *weight-decay regularization* particularly when referring to inclusion of a penalty term of L_2-norm of parameters in the minimization of sum of squared errors. In the space of dual variables, our derivation of ridge regression learning follows that in Saunders et al. (1998).

Subsequently, the *Least Absolute Shrinkage and Selection Operator* (LASSO) method was proposed by Tibshirani (1996) to adopt an L_1-norm for parameters in the penalty term during minimization of sum of squared errors. A least angle regression (LAR) was successively proposed for solving LASSO in Tibshirani (2011). When the penalty norm term is generalized to L_p-norm for $1 \leq p < 2$, the minimization is called *bridge regression* (Frank & Friedman, 1993; Fu, 1998). Since then, numerous extensions to the regularization solution search can be found from different treatments of the penalty terms. A well-known example is the *elastic-net* proposed by Zou and Hastie (2005) where the penalty term included both L_1-norm and L_2-norm of parameters.

The Least Norm Problem

According to Boyd and Vandenberghe (2018), the *minimum-norm* or *least norm* problem is a special case of the *linearly constrained least squares problem* or just constrained least squares problem given by

---

[2] Based on differential equations, Jacques Hadamard (1902) raised the concept of well-posedness for such systems. The monograph by Tikhonov and Arsenin (1977) presented the state of theory developed since then. A problem is said to be well-posed in the sense of Hadamard if there exists (condition 1) a unique solution (condition 2) to this problem that depends continuously on its data (condition 3) (Lavrent'ev et al., 1986; Kabanikhin, (Kabanikhin, 2008)). Problems which are not well-posed in the sense of Hadamard are often called *ill-posed*.

$$\min \quad \|\mathbf{Ax} - \mathbf{b}\|^2$$
$$\text{subj. to} \quad \mathbf{Cx} = \mathbf{d}. \tag{6.53}$$

When $\mathbf{A} = \mathbf{I}$ and $\mathbf{b} = \mathbf{0}$, the problem becomes the *least norm problem* or the *minimum-norm problem* given by

$$\min \quad \|\mathbf{x}\|^2$$
$$\text{subj. to} \quad \mathbf{Cx} = \mathbf{d}, \tag{6.54}$$

which is the form discussed in the under-determined case (see, e.g., (6.5)).

Bridge Regression

Bridge regression generalizes ridge regression by replacing the L_2-norm penalty with an L_p-norm penalty. The range of interest for p is $0 < p \le 2$, where subset selection $(0 < p \le 1)$ and coefficient weighting compression $(1 < p < 2)$ occur. Bridge regression employs the following cost function for minimization:

$$\sum_{i=1}^{M}(y_i - \boldsymbol{x}_i^T \boldsymbol{w})^2 + \lambda \sum_{j=1}^{D}|\boldsymbol{w}_j|^p. \tag{6.55}$$

To the best of our knowledge (Toh et al., 2023), bridge regression was first known from the work of Frank and Friedman (1993) who explored various statistical tools for Chemometrics regression. In their study, the parameter p was interpreted as the degree to which the prior probability was concentrated along favored directions. A value of $p \to 0$ places the prior mass toward the coordinate axes, expressing the prior belief that only a few of the predictor variables are likely to have high relative influence on the response. In the work of Fu (1998), a general approach for $p \ge 1$ was developed to solve the bridge optimization problem. The algorithm, based on a modified Newton-Raphson method, iteratively searched for the unique solution for bridge for $p \ge 1$. According to the work of Fan and Li (2001), the solution for bridge is continuous only when $p \ge 1$. They employed a local quadratic approximation iteratively for the L_p penalized likelihood, reducing the minimization problem to a quadratic one, solvable using the Newton-Raphson algorithm (Fan & Li, 2001). To enhance computational efficiency, Zou and Li (2008) replaced the local quadratic approximation with a local linear approximation when solving the penalized likelihood. Both local linear and local quadratic approximations were studied by Park and Yoon (2011), who demonstrated that the bridge estimator is a robust choice under various conditions, comparing favorably with ridge, lasso, and elastic-net.

Our initial effort to find an analytic solution for bridge regression were documented in Toh (2014), subsequently in Toh et al. (2018b) with a conjecture toward a simpler version of global solution. However, subsequent finding in Toh et al. (2023) indicates that the concerned solution cannot be simplified further. From the application perspective, although the validity of k stretches beyond 2 for both the primal and the dual cases, the region of interest for parametric shrinking is $k < 2$. The focus is thus on $k \in [1, 2]$ for over-determined systems and $k \in (1, 2]$ for under-determined systems. The formulation for the under-determined scenario is particularly useful when the data is scarce. This was termed *small-sample size* (SSS) problem according to Jain and Zongker (1997), Lu et al. (2003) before the deep learning era, and is also known as *few-shot learning* (Wang et al., 2021) in the recent literature.

It is noteworthy that the solution given by (6.36) has a similar form (see (6.56) below) to that in Fan and Li (2001); Park and Yoon (2011). In their work, a different L_p-norm approximation was used for the penalty term based on the LQA (i.e., minimization of (6.29)) instead of the k-measure (i.e., minimization of (6.32)).

$$\hat{w}_j = \left(\frac{\lambda q}{2} \operatorname{diag}\{|w_{0j}|^{\circ(q-2)}\} + \mathbf{P}^T\mathbf{P} \right)^{-1} \mathbf{P}^T\mathbf{y}. \tag{6.56}$$

6.6 Exercises

6.1 The solution for a primal ridge regression problem with feature vectors in $\mathbf{X}$ and targets in $\mathbf{y}$ can be written as $\hat{w} = (\mathbf{X}^T\mathbf{X} + \lambda\mathbf{I})^{-1}\mathbf{X}^T\mathbf{y}$ for $\lambda > 0$. As λ increases, $\hat{w}^T\hat{w}$ increases. True or False?

6.2 Which technique in classifier learning involves minimizing an error cost function while incorporating an L_p-norm penalty on the model parameters?

  (a) Ridge regression
  (b) Lasso regression
  (c) Elastic net
  (d) Bridge regression
  (e) None of these.

6.3 Consider the penalized learning formulated by

$$\min_{w} J(x, w) = \min_{w} \left\{ \sum_{i=1}^{m} (f(x_i, w) - y_i)^2 + \lambda\|w\|_p^p, \right\}, \quad \lambda > 0$$

where $f(x, w)$ is a learning model with augmented input $x \in \mathbb{R}^{d+1}$, learning parameter $w \in \mathbb{R}^{d+1}$ and y_i, $i = 1, ..., m$ being the learning target. Which of the following statement(s) is/are true?

  (a) When $p \leq 1$, it performs both variable selection and regularization.
  (b) When $p > 2$, it does not perform coefficient shrinkage.
  (c) When $p < 2$, it performs coefficient shrinkage.
  (d) When $p = 2$, the solution is sparse.
  (e) None of these.

6.4 Consider the criterion

$$\hat{w} = \min_{w} \left\{ \sum_{i=1}^{m} \left(y_i - w_0 - \sum_{j=1}^{d} x_{ij} w_j \right)^2 + \lambda \sum_{j=1}^{d} |w_j|^q \right\}, \quad q \geq 0,$$

where x and y are the input and target values, w is the parameter vector and λ is a penalty factor. The contours of constant value of $\sum_j |w_j|^q$ are shown in figure below for the case of two inputs.

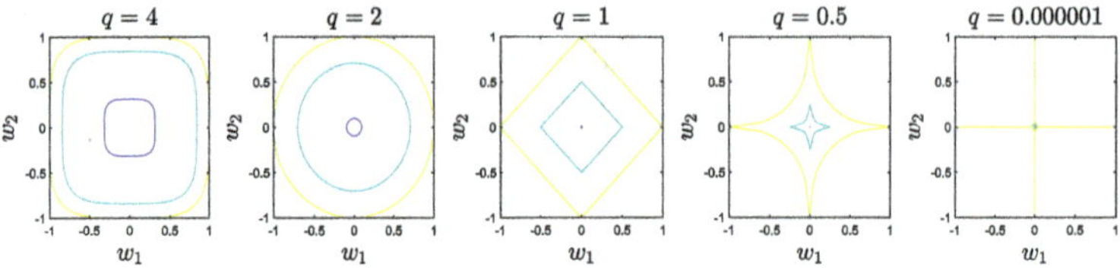

Write down the common name/term (for those items marked with "?") for each value of q and state whether there is coefficient shrinkage and sparseness for each estimation (by circling the answer) (Note: the answer for LASSO is already provided):

Value of q	Common names	Coefficient shrinkage?	Sparse?
$q = 0$	?	(Yes/No)?	(Yes/No)?
$q = 0.5$	–NA–	(Yes/No)?	(Yes/No)?
$q = 1$	Lasso	Yes	Yes
$q = 2$	?	(Yes/No)?	(Yes/No)?
$q = 4$	–NA–	(Yes/No)?	(Yes/No)?

6.5 Consider the XOR problem with four training data samples given by $(x_1, x_2) \in \{(0, 1), (2, 1), (1, 0), (1, 2)\}$ with their corresponding target outputs $y \in \{0, 0, 1, 1\}$. Use the following quadratic polynomial model to learn the XOR data based on the proximal bridge regression at $k = 1.1$ and $k = 2$ using $\lambda = 10$.

$$p(x_1, x_2) = w_0 + w_1 x_1 + w_2 x_2 + w_3 x_1^2 + w_4 x_2^2 + w_5 x_1 x_2$$

Rank the coefficients in terms of their magnitudes. Comment on any difference in the rankings.

6.6 Consider the simulated data with the true model given by $y = w^T x + \sigma \epsilon$ with $\epsilon \sim N(0, 1)$. Generate a data set consisting of 100/100/400 (training/validation/test) samples with $\sigma = 15$ and

$$w = [0, 3, 6, 8, 0, 0, 0, 0, 0, 0]^T.$$

The correlation between x_i and x_j is zero.

(Hint: the data can be generated using the following MATLAB code:

`Xall = mvnrnd(zeros(1,10),diag(ones(1,10)),600);`

which uses a fixed random seed such as `rng(8)`.)

Compute the Mean-Squared Error (MSE) of learning using ordinary least squares, ridge regression, and proximal bridge regression based on the generated data. Compare their learned parameters too.

(Hint: since the true model is known, the MSE can be computed using $\text{MSE} = (\hat{w} - w)^T \frac{X_t^T X_t}{n} (\hat{w} - w)$ where X_t and n are respectively the matrix containing the test samples and the test sample size.)

6.7 For the over-determined and under-determined solutions of the proximal bridge regression, are the estimations biased?

6.8 The penalty term in ridge regression can be considered as an augmentation data set on the ordinary least squares. By augmenting the data matrix $\mathbf{X}$ with d additional rows and the target $\mathbf{y}$ with d zeros, show that such a relationship between data augmentation and regularization can be established.

6.9 By utilizing the singular value decomposition of the centered regressor matrix $\mathbf{X}$, derive an expression to show that $\|\mathbf{w}\|$ increases as its tuning parameter $\lambda \to 0$.

6.10 Match among the types (L_1-norm, L_2-norm, $L_{1.5}$-norm) of penalized learning with their corresponding parameter profile below:

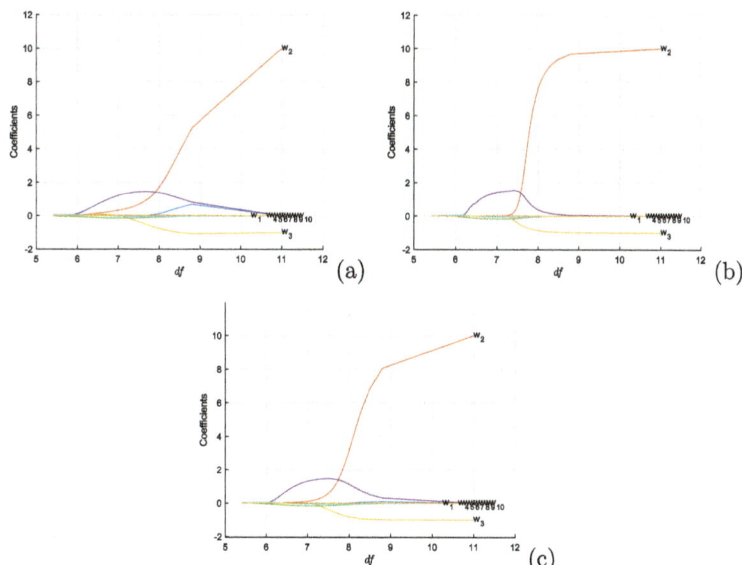

Chapter 7
Network Learning

In recent times, the multilayer neural networks have demonstrated their ability to learn from big data while maintaining good generalization. This is evidenced by the superior prediction accuracy of deep networks across various applications. The main ground for realizing the high learning capacity with good predictivity comes from several major advancements in the field. These advancements include the processing platform, the learning regimen, and the availability of big data size. However, despite their success in applications, a good understanding of the learning representation and generalization properties remains elusive. In this chapter, the mechanism of network learning is interpreted from the perspective of a linear system of equations in matrix form where the learning representation and generalization properties are investigated.

7.1 Mathematical Preliminaries

7.1.1 The Spaces of the Kernel and the Range

Consider the system of linear equations given by

$$\mathbf{X}w = \mathbf{y}, \tag{7.1}$$

where $\mathbf{X} \in \mathbb{R}^{m \times d}$ and $\mathbf{y} \in \mathbb{R}^{m \times 1}$ are the given system data and $w \in \mathbb{R}^{d \times 1}$ is the unknown parameter vector to be determined. In many application contexts, $\mathbf{X}$ constitutes the input of the system and $\mathbf{y}$ constitutes the output or response of the system. Strictly speaking, for an over-determined system with $m > d$, the m equations in (7.1) are generally unsolvable for a strict equality of each equation to hold (see, e.g., Strang (2016)). However, by multiplying $\mathbf{X}^T$ to both sides of (7.1), the resultant d equations

© The Author(s), under exclusive license to Springer Nature Singapore Pte Ltd. 2025
K.-A. Toh et al., *Analytic Learning Methods for Pattern Recognition*,
https://doi.org/10.1007/978-981-96-2151-4_7

$$\mathbf{X}^T\mathbf{X}\boldsymbol{w} = \mathbf{X}^T\mathbf{y} \tag{7.2}$$

give the least squares error solution (7.3) (Strang, 2016) under the *primal solution space* (see also Sect. 5.1.1) as follows:

$$\hat{\boldsymbol{w}} = (\mathbf{X}^T\mathbf{X})^{-1}\mathbf{X}^T\mathbf{y}. \tag{7.3}$$

Here, it is noted that this solution in the primal space does not fully fit Eq. (7.1) without error. In other words, the solution is just the best solution with the lowest fitting error (See Definition 7.1 below). To see this point, $\mathbf{y}$ can be split into the solvable part $\hat{\mathbf{y}}$ and the unsolvable part $\boldsymbol{e}$. The solvable part lies in the *range (column) space* of $\mathbf{X}$ and the unsolvable part lies in the *null (row) space* of $\mathbf{X}$. The null space is also known as the *kernel* of $\mathbf{X}$. Based on this splitting, (7.1) can be written (Albert, 1972) as

$$\mathbf{X}\boldsymbol{w} = \hat{\mathbf{y}} + \boldsymbol{e}. \tag{7.4}$$

Since $\mathbf{X}\boldsymbol{w}$ is on the linear manifold of the Euclidean space, then according to (7.1) and (7.4), we have

$$\|\mathbf{X}\boldsymbol{w} - \mathbf{y}\|^2 = \|\mathbf{X}\boldsymbol{w} - \hat{\mathbf{y}} - \boldsymbol{e}\|^2. \tag{7.5}$$

Because $\hat{\mathbf{y}}$ is the solvable part, we have $\mathbf{X}\boldsymbol{w} - \hat{\mathbf{y}} = \mathbf{0}$ and this results in

$$\|\mathbf{X}\boldsymbol{w} - \mathbf{y}\|^2 = \|\boldsymbol{e}\|^2. \tag{7.6}$$

Equation (7.2), which can also be obtained by setting the first partial derivative of (7.6) to zero (see Sect. 5.1.1), is known as the normal equation minimizing (7.6). Hence we see that solving (7.1) based on (7.2), which is an outcome of pre-multiplying $\mathbf{X}^T$ to both sides of the system equation, minimizes the least squares error given by $\|\boldsymbol{e}\|^2$. The process of pre-multiplying $\mathbf{X}^T$ to both sides of the system equation can be regarded as a manipulation under the *range* or *column space* of $\mathbf{X}$. Here, the size of $\mathbf{X}^T\mathbf{X}$ is determined by the column size of $\mathbf{X}$.

When the system is under-determined (i.e., $m < d$), then the number of equations is less than that of the unknowns. This gives rise to an infinite number of solutions. Nevertheless, we can constrain the solution space by projecting the unknown parameter vector $\boldsymbol{w} \in \mathbb{R}^{d \times 1}$ onto the $\mathbb{R}^m$ sample subspace (Madych, 1991) utilizing the row space of $\mathbf{X}$ (see also Sect. 5.1.2):

$$\boldsymbol{w} = \mathbf{X}^T\boldsymbol{\alpha}, \tag{7.7}$$

where $\boldsymbol{\alpha} \in \mathbb{R}^{m \times 1}$ is the projected parameter vector. Substituting (7.7) into (7.1) gives

$$\mathbf{X}\mathbf{X}^T\boldsymbol{\alpha} = \mathbf{y}. \tag{7.8}$$

Since $\mathbf{X}\mathbf{X}^T$ is now of full row rank when $\mathbf{X}$ has full row rank, $\boldsymbol{\alpha}$ has a unique solution given by

$$\hat{\boldsymbol{\alpha}} = (\mathbf{X}\mathbf{X}^T)^{-1}\mathbf{y}. \tag{7.9}$$

Substituting this projected parameter vector $\hat{\boldsymbol{\alpha}}$ back into (7.7) gives

$$\hat{\boldsymbol{w}} = \mathbf{X}^T(\mathbf{X}\mathbf{X}^T)^{-1}\mathbf{y}. \tag{7.10}$$

This process of constraining the parameters of d dimension (for $d > m$) to the subspace of m dimension via $\boldsymbol{w} = \mathbf{X}^T\boldsymbol{\alpha}$ can be regarded as a manipulation under the *kernel* or *row space* of $\mathbf{X}$.

As mentioned in Sect. 5.1.2, the solution given by (7.10) corresponds to minimizing $\|\boldsymbol{w}\|_2^2$ subject to $\mathbf{X}\boldsymbol{w} = \mathbf{y}$. This is also known as the *least norm solution* in the literature (see e.g., Boyd and Vandenberghe (2004)). Interestingly, this solution in the *dual space* (the m row space of $\mathbf{X}$) also minimizes the sum of squared errors (but this time to zero error)! This can be easily seen by substituting (7.7) and (7.9) into (7.6) which gives a minimized $(\mathbf{X}\hat{\boldsymbol{w}} - \mathbf{y})^T(\mathbf{X}\hat{\boldsymbol{w}} - \mathbf{y})$. With these observations, the following conclusion is obtained based on the conventional ground of linear algebra (see, e.g., Albert (1972), Ben-Israel and Greville (2003), Campbell and Meyer (2009), Strang (2016), Boyd and Vandenberghe (2018)): Solving for $\boldsymbol{w}$ in the system of linear equations of the form (7.1) in the column space (range) of $\mathbf{X}$ or in the row space (kernel) of $\mathbf{X}$ boils down to solving the least squares error of the form (7.6).

Moreover, the resultant solution $\hat{\boldsymbol{w}}$ is unique with a minimum-norm value in the sense that $\|\hat{\boldsymbol{w}}\|_2^2 \leq \|\boldsymbol{w}\|_2^2$ for all feasible $\boldsymbol{w}$. This can be observed from both the perspectives of the under-determined system ($m < d$) and the over-determined system ($m > d$). For the under-determined system, according to Boyd and Vandenberghe (2018) $\mathbf{X}(\boldsymbol{w} - \hat{\boldsymbol{w}}) = \mathbf{0}$ when $\mathbf{X}\boldsymbol{w} = \mathbf{y}$. This can be substituted into

$$\begin{aligned}
(\boldsymbol{w} - \hat{\boldsymbol{w}})^T\hat{\boldsymbol{w}} &= (\boldsymbol{w} - \hat{\boldsymbol{w}})^T\mathbf{X}^T(\mathbf{X}\mathbf{X}^T)^{-1}\mathbf{y} \\
&= (\mathbf{X}(\boldsymbol{w} - \hat{\boldsymbol{w}}))^T(\mathbf{X}\mathbf{X}^T)^{-1}\mathbf{y}
\end{aligned} \tag{7.11}$$

to obtain $(\boldsymbol{w} - \hat{\boldsymbol{w}})^T\hat{\boldsymbol{w}} = 0$. This means that $(\boldsymbol{w} - \hat{\boldsymbol{w}}) \perp \hat{\boldsymbol{w}}$ and so $\|\hat{\boldsymbol{w}} + \boldsymbol{w} - \hat{\boldsymbol{w}}\|_2^2 = \|\hat{\boldsymbol{w}}\|_2^2 + \|\boldsymbol{w} - \hat{\boldsymbol{w}}\|_2^2$ leads to

$$\begin{aligned}
\|\boldsymbol{w}\|_2^2 &= \|\hat{\boldsymbol{w}}\|_2^2 + \|\boldsymbol{w} - \hat{\boldsymbol{w}}\|_2^2 \\
&\geq \|\hat{\boldsymbol{w}}\|_2^2,
\end{aligned} \tag{7.12}$$

i.e., $\hat{\boldsymbol{w}}$ has the smallest norm among all feasible solutions.

On the other hand, for the over-determined system, the observation follows that in Albert (1972) where we have $\hat{\mathbf{y}} - \mathbf{X}\boldsymbol{w}$ in the range space of $\mathbf{A}$ and $\boldsymbol{e} \perp \hat{\mathbf{y}} - \mathbf{X}\boldsymbol{w}$. Therefore

$$\|\mathbf{y} - \mathbf{X}\boldsymbol{w}\|_2^2 = \|\hat{\mathbf{y}} - \mathbf{X}\boldsymbol{w} + e\|_2^2$$
$$= \|\hat{\mathbf{y}} - \mathbf{X}\boldsymbol{w}\|_2^2 + \|e\|_2^2$$
$$\geq \|e\|_2^2 \tag{7.13}$$

This lower bound is attained at $\hat{\boldsymbol{w}}$ only if $\hat{\boldsymbol{w}}$ is such that $\mathbf{X}\hat{\boldsymbol{w}} = \hat{\mathbf{y}}$. Similarly, any such $\hat{\boldsymbol{w}}$ can be decomposed into two orthogonal vectors $\hat{\boldsymbol{w}}^r + \hat{\boldsymbol{w}}^k$ where $\hat{\boldsymbol{w}}^r$ falls in the range space of $\mathbf{X}^T$ and $\hat{\boldsymbol{w}}^k$ falls in the kernel of $\mathbf{X}$. Thus $\mathbf{X}\hat{\boldsymbol{w}} = \mathbf{X}\hat{\boldsymbol{w}}^r$ so that $\|\mathbf{y} - \mathbf{X}\hat{\boldsymbol{w}}\|_2^2 = \|\mathbf{y} - \mathbf{X}\hat{\boldsymbol{w}}^r\|_2^2$ and $\|\hat{\boldsymbol{w}}\|_2^2 = \|\hat{\boldsymbol{w}}^r\|_2^2 + \|\hat{\boldsymbol{w}}^k\|_2^2 \geq \|\hat{\boldsymbol{w}}^r\|_2^2$.

7.1.2 The Moore-Penrose Inverse and the Least Squares Approximation

Consider again the system of linear equations given by $\mathbf{X}\boldsymbol{w} = \mathbf{y}$. A formal definition of the least squares error solution to this system is quoted as follows.

Definition 7.1 (see, e.g., Campbell and Meyer (2009), Albert (1972), Ben-Israel and Greville (2003), MacAusland (2014)) A least squares solution to the system $\mathbf{X}\boldsymbol{w} = \mathbf{y}$, where $\mathbf{X} \in \mathbb{R}^{m \times d}$, $\boldsymbol{w} \in \mathbb{R}^{d \times 1}$ and $\mathbf{y} \in \mathbb{R}^{m \times 1}$, is a vector $\hat{\boldsymbol{w}}$ such that

$$\|\hat{\mathbf{e}}\| = \|\mathbf{X}\hat{\boldsymbol{w}} - \mathbf{y}\| \leq \|\mathbf{X}\boldsymbol{w} - \mathbf{y}\| = \|\mathbf{e}\|, \tag{7.14}$$

where $\|\cdot\|$ denotes the ℓ^2-norm of "$\cdot$".

Here, we require another definition, commonly referred to as the pseudoinverse or Moore-Penrose inverse.

Definition 7.2 (see, e.g., Ben-Israel and Greville (2003), Campbell and Meyer (2009), Albert (1972), MacAusland (2014)) If $\mathbf{X}$ (square or rectangular) is a matrix of real or complex elements, then there exits a unique matrix $\mathbf{X}^\dagger$, known as the Moore-Penrose inverse or the pseudoinverse of $\mathbf{X}$, such that (i) $\mathbf{X}\mathbf{X}^\dagger\mathbf{X} = \mathbf{X}$, (ii) $\mathbf{X}^\dagger\mathbf{X}\mathbf{X}^\dagger = \mathbf{X}^\dagger$, (iii) $(\mathbf{X}\mathbf{X}^\dagger)^H = \mathbf{X}\mathbf{X}^\dagger$ and (iv) $(\mathbf{X}^\dagger\mathbf{X})^H = \mathbf{X}^\dagger\mathbf{X}$ where ^{H} denotes the conjugate transpose.

Based on these definitions, a well-known result from the literature is quoted as follows.

Lemma 7.1 (see, e.g., Albert (1972), Campbell and Meyer (2009), Ben-Israel and Greville (2003), MacAusland (2014)) $\hat{\boldsymbol{w}} = \mathbf{X}^\dagger\mathbf{y}$ *is the best approximate solution of* $\mathbf{X}\boldsymbol{w} = \mathbf{y}$.

When the columns of $\mathbf{X}$ are linearly independent in an over-determined system given by $\mathbf{X}\boldsymbol{w} = \mathbf{y}$, the Moore-Penrose inverse can be computed based on the *left inverse*, where $\mathbf{X}^\dagger = (\mathbf{X}^T\mathbf{X})^{-1}\mathbf{X}^T$ with $\mathbf{X}^\dagger\mathbf{X} = \mathbf{I}$. Pre-multiplying the system equation by $\mathbf{X}^\dagger$ on both sides yields the result stated in Lemma 7.1. On the other hand, when the rows of $\mathbf{X}$ are linearly independent in an under-determined system, the Moore-Penrose inverse can be computed based on the *right inverse*, where $\mathbf{X}^\dagger = \mathbf{X}^T(\mathbf{XX}^T)^{-1}$ with $\mathbf{XX}^\dagger = \mathbf{I}$. As mentioned, a least-norm solution can be obtained by projecting the d-space of the parameter vector $\boldsymbol{w}$ onto the m-space of samples using $\hat{\boldsymbol{w}} = \mathbf{X}^\dagger \hat{\boldsymbol{\alpha}}$. By substituting $\hat{\boldsymbol{w}}$ into $\mathbf{X}\boldsymbol{w} = \mathbf{y}$, we have $\hat{\boldsymbol{\alpha}} = \mathbf{y}$. The solution presented in Lemma 7.1 is obtained again by substituting $\hat{\boldsymbol{\alpha}} = \mathbf{y}$ back into $\hat{\boldsymbol{w}} = \mathbf{X}^\dagger \hat{\boldsymbol{\alpha}}$.

In practical scenarios, achieving full rank for high-dimensional data is often challenging without employing regularization techniques. However, this full rank condition is unnecessary when solving the linear system of equations using the Moore-Penrose inverse, as it is guaranteed to exist and be unique according to Definition 7.2. Specifically, for any given matrix $\mathbf{X}$, there is always a unique matrix $\mathbf{X}^\dagger$ that satisfies the four criteria outlined in Definition 7.2. Generally, if $\mathbf{X}$ can be factorized into $\mathbf{X} = \mathbf{FG}$ with full rank, then $\mathbf{X}^\dagger = \mathbf{G}^*(\mathbf{GG}^*)^{-1}(\mathbf{F}^*\mathbf{F})^{-1}\mathbf{F}^*$ will satisfy the Penrose conditions (Ben-Israel and Greville, 2003).

Collectively, the above results of single output nature can be packed into a system with multiple outputs as follows.

Lemma 7.2 *Solving for* $\mathbf{W}$ *in the system of linear equations of the form*

$$\mathbf{XW} = \mathbf{Y}, \quad \mathbf{X} \in \mathbb{R}^{m \times d}, \ \mathbf{W} \in \mathbb{R}^{d \times q}, \ \mathbf{Y} \in \mathbb{R}^{m \times q} \tag{7.15}$$

in the column space (range) of $\mathbf{X}$ *or in the row space (kernel) of* $\mathbf{X}$ *boils down to minimizing the sum of squared errors given by*

$$\mathrm{SSE} = trace\left((\mathbf{XW} - \mathbf{Y})^T(\mathbf{XW} - \mathbf{Y})\right). \tag{7.16}$$

Moreover, the resultant solution $\hat{\mathbf{W}} = \mathbf{X}^\dagger \mathbf{Y}$ *is unique with a minimum-norm value in the sense that* $\|\hat{\boldsymbol{w}}_i\|_2^2 \le \|\boldsymbol{w}_i\|_2^2$ *for all feasible* $\boldsymbol{w}_i \in \mathbb{R}^{d \times 1}$, $i = 1, \ldots, q$ *in* $\mathbf{W} = [\boldsymbol{w}_1, \ldots, \boldsymbol{w}_q]$.

Proof Equation (7.15) can be rewritten as a set of multiple linear systems as follows:

$$\mathbf{X}[\boldsymbol{w}_1, \ldots, \boldsymbol{w}_q] = [\mathbf{y}_1, \ldots, \mathbf{y}_q], \tag{7.17}$$

where the result of Lemma 7.1 can be utilized. Since the trace of $(\mathbf{XW} - \mathbf{Y})^T(\mathbf{XW} - \mathbf{Y})$ is equal to the sum of the squared lengths of the error vectors $\mathbf{X}\boldsymbol{w}_i - \mathbf{y}_i$, $i = 1, 2, \ldots, q$, the unique solution $\hat{\mathbf{W}} = (\mathbf{X}^T\mathbf{X})^{-1}\mathbf{X}^T\mathbf{Y} = \mathbf{X}^\dagger\mathbf{Y}$ in the column space of $\mathbf{X}$ or that $\hat{\mathbf{W}} = \mathbf{X}^T(\mathbf{XX}^T)^{-1}\mathbf{Y} = \mathbf{X}^\dagger\mathbf{Y}$ in the row space of $\mathbf{X}$, not only minimizes this sum, but also minimizes each term in the sum (Duda et al., 2001). Moreover, since the column and the row spaces are independent, the individually minimized norm (see Sect. 7.1.1) is also minimum. ∎

The algebraic relationships above indicate that learning a system of linear equations through minimizing the square of errors can be accomplished by exploiting its range or kernel space using either matrix projection (section 7.1.1) or the Moore-Penrose inverse (section 7.1.2). Essentially, multiplying the system matrix on both sides of the equation or applying the pseudoinverse operation inherently minimizes the least squares error during the algebraic manipulation process to find the solution. These observations suggest that algebraic manipulation of matrix products, along with the Moore-Penrose inverse, might be exploited to develop a gradient-free approximate solution for multilayer network learning, in the context of least squares error.

7.2   Network Structures

7.2.1   Fully Connected Network

In our context, a *fully connected* feedforward neural network of n layers is defined as one that consists of a chain of n replicas of multiplying a weight matrix to the input matrix and then passing the product through an activation function. This kind of network is also known as the *dense* network. Mathematically, a network with full bias connections can be written as

$$\mathbf{G} = f_n\left([\mathbf{1}, f_{n-1}(\cdots[\mathbf{1}, f_2([\mathbf{1}, f_1(\mathbf{XW}_1)]\mathbf{W}_2)]\cdots\mathbf{W}_{n-1})]\mathbf{W}_n\right), \tag{7.18}$$

where $\mathbf{X} \in \mathbb{R}^{m\times(d+1)}$ is the input matrix, $\mathbf{W}_1 \in \mathbb{R}^{(d+1)\times h_1}$, $\mathbf{W}_2 \in \mathbb{R}^{(h_1+1)\times h_2}$, ..., $\mathbf{W}_{n-1} \in \mathbb{R}^{(h_{n-2}+1)\times h_{n-1}}$, $\mathbf{W}_n \in \mathbb{R}^{(h_{n-1}+1)\times q}$ are the weight matrices, $\mathbf{1} = [1, \ldots, 1]^T \in \{\mathbb{R}^{m\times1}, \mathbb{R}^{h_k\times1}\}$, $k = 1, \ldots, h_{n-1}$ are the bias vectors, $f_j, j = 1, \ldots, n$ are the activation functions which operate elementwise on their matrix domain, and $\mathbf{G} \in \mathbb{R}^{m\times q}$ is the network output. $h_k, k \in \{1, \ldots, n-1\}$ correspond to the number of the hidden nodes in the kth layer. The number of hidden nodes in the output layer is equal to the number of outputs, i.e., $h_n = q$.

Figure 7.1 shows a fully connected feedforward network of n layers with bias connection in every layer. This network takes an input of dimension d and gives an output of dimension q. The number of data samples m is contained in the input matrix $\mathbf{X} \in \mathbb{R}^{m\times(d+1)}$ and is not reflected in this figure. When the functions $g_i, i = 1, ..., h_j$ in the h_jth layer share the same activation function, they are collectively denoted as f_j in Eq. (7.18). In this representation of the network structure, the bias term in each layer has been isolated from the inputs of the current layer (or the outputs of the previous layer) where the weight matrix can be partitioned as $\mathbf{W}_k = \begin{bmatrix} \mathbf{w}_k^T \\ \mathbf{W}_k \end{bmatrix}$ with $\mathbf{w}_k = [w_1, \ldots, w_{h_k}]^T \in \mathbb{R}^{h_k}$ being the weights corresponding to the bias vector, and $\mathbf{W}_k \in \mathbb{R}^{h_{k-1}\times h_k}$ being the weights for h_k number of hidden nodes at the kth layer.

The bias (also known as the *intercept* term) is an important component in multilayer networks to translate the entire fitting hyper-surface. In other words, the positive and negative translations of the entire network output are controlled by the bias term.

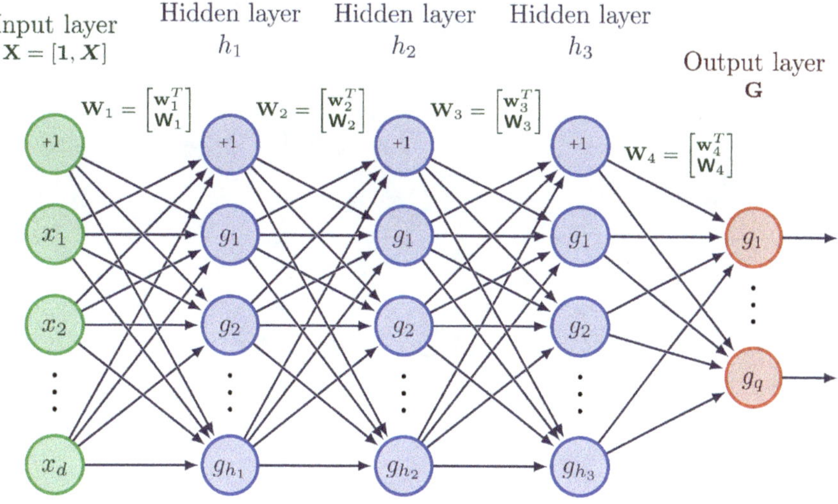

Fig. 7.1 A fully connected n-layer feedforward network. When the functions g_i, $i = 1, ..., h_j$ in each h_jth layer share the same activation function, they are collectively denoted as f_j in Eq. (7.18).

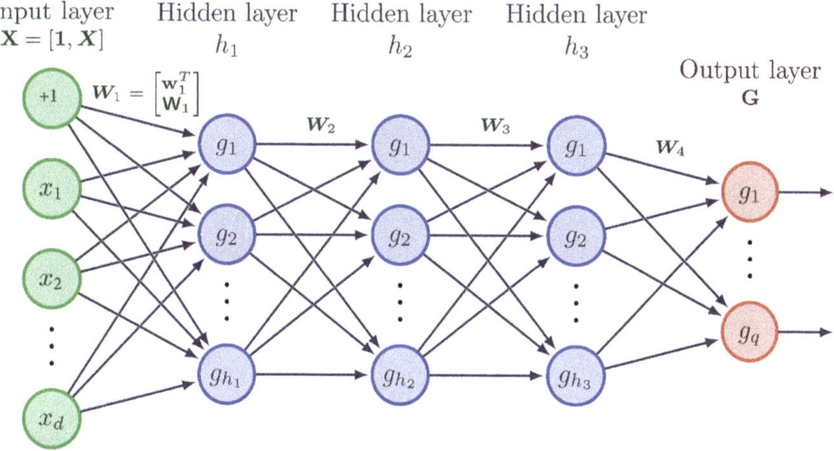

Fig. 7.2 A fully connected feedforward network with only bias at the input layer

While the network with full biases in the hidden nodes as shown in Fig. 7.1 has been conventionally adopted, a network with bias only at the input layer works equally well (see Fig. 7.2 for the network structure). Mathematically, the n-layer network model of h_1-h_1-$\cdots$-h_{n-1}-h_n structure can be written in matrix form as

$$\mathbf{G} = f_n\left(f_{n-1}(\cdots f_2(f_1(\mathbf{X}\mathbf{W}_1)\mathbf{W}_2)\cdots \mathbf{W}_{n-1})\mathbf{W}_n\right), \tag{7.19}$$

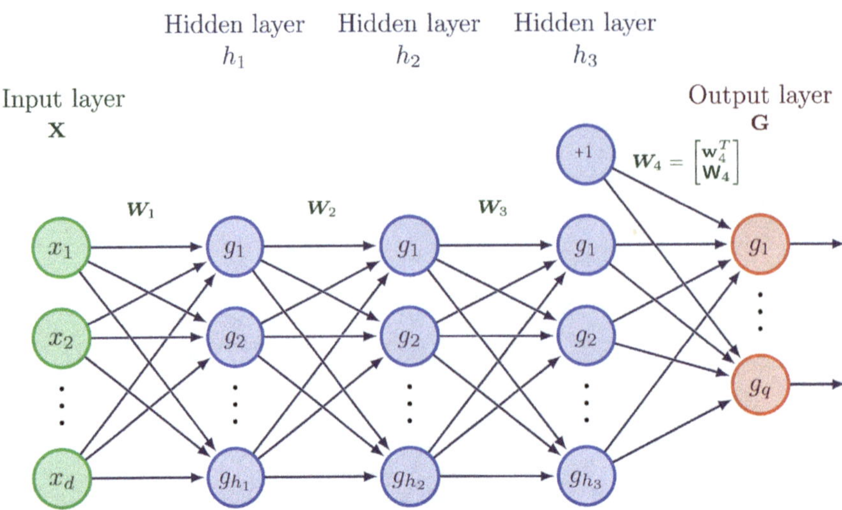

Fig. 7.3 A fully connected feedforward network with only bias at the output layer

where $\mathbf{X} = [\mathbf{1}, X] \in \mathbb{R}^{m \times (d+1)}$ is the augmented input data matrix (i.e., input with bias), $\mathbf{W}_1 \in \mathbb{R}^{(d+1) \times h_1}$, $\mathbf{W}_2 \in \mathbb{R}^{h_1 \times h_2}$, ..., $\mathbf{W}_{n-1} \in \mathbb{R}^{h_{n-2} \times h_{n-1}}$, and $\mathbf{W}_n \in \mathbb{R}^{h_{n-1} \times q}$ ($h_n = q$ is the output dimension) are the weight matrices without bias at each layer, and $\mathbf{G} \in \mathbb{R}^{m \times q}$ is the network output of q dimension.

Alternatively, the single bias or intercept node can be added only to the output layer as shown in Fig. 7.3. For this case, the input weight matrix can be represented as $\mathbf{X} = \in \mathbb{R}^{m \times d}$, while the hidden weights are $\mathbf{W}_2, \cdots, \mathbf{W}_{n-1}$ are in $\mathbb{R}^{h_1 \times h_2}, \cdots, \mathbb{R}^{h_{n-2} \times h_{n-1}}$ respectively. The output weight matrix is represented as $\mathbf{W}_n \in \mathbb{R}^{(h_{n-1}+1) \times q}$.

7.2.2 Receptive Field Network

Figure 7.4 shows a network with *receptive field*. This means that the network is sparsely connected with limited reception of outputs from the previous layer. In this figure, each of the neural node in the subsequent layer receives only 3 inputs which correspond to the outputs of the previous layer. This network is termed to have a receptive field of 3.

When the kth layer of the network, where $k = 1, \cdots, n$, has a receptive field of 3, each column of its weight matrix contains 3 non-zero entries. For instance, if d hidden nodes has been chosen for the 1st-layer, then with a receptive field of 3, the following matrix in circulant form can represent $\mathbf{W}_1$:

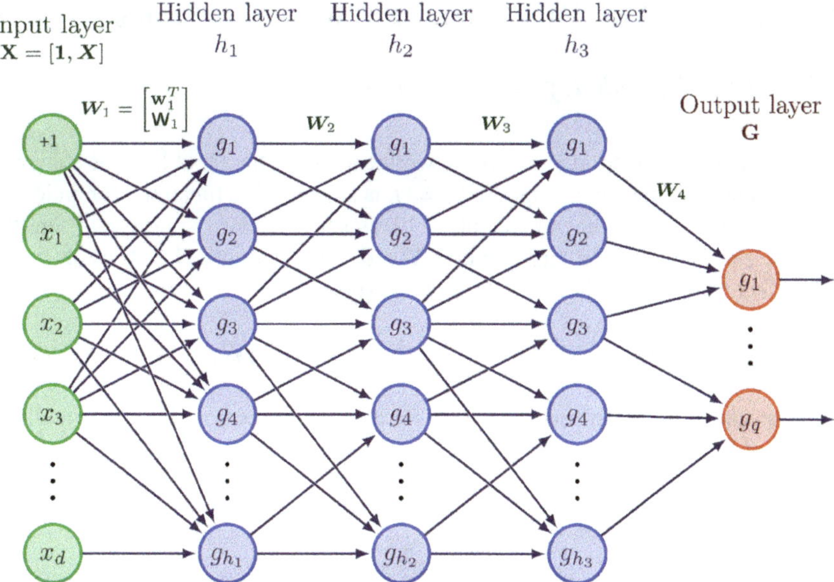

Fig. 7.4 A 4-layer feedforward network with a receptive field of 3

$$
\mathbf{W}_1 =
\begin{bmatrix}
w_{1,1} & 0 & 0 & \cdots & 0 \\
w_{2,1} & w_{2,2} & 0 & \cdots & 0 \\
w_{3,1} & w_{3,2} & w_{3,3} & \cdots & 0 \\
0 & w_{4,2} & w_{4,3} & \cdots & 0 \\
0 & 0 & w_{5,3} & \cdots & 0 \\
0 & 0 & 0 & \cdots & 0 \\
0 & 0 & 0 & \ddots & w_{d-1,d} \\
0 & 0 & 0 & \cdots & w_{d,d} \\
0 & 0 & 0 & \cdots & w_{d+1,d}
\end{bmatrix}
. \tag{7.20}
$$

Essentially, the sparse structure of the receptive field network corresponds to convolutional operation in local regions.

Our focus in this chapter shall be limited to deterministic learning solutions for these two network structures (dense/fully connected network and receptive field network).

7.3 Gradient-Free Fully Connected Network Learning

7.3.1 Interdependency of Network Weights

For simplicity, consider an unbiased network except for the input layer as shown in
Fig. 7.2. For the time being, suppose $f_k = f$ is an invertible function. The indexing
of f_k, $k = 1, \ldots, n$ for each layer of (7.19) is kept for clarity purpose. Then, the
network can learn a given target matrix $\mathbf{Y} \in \mathbb{R}^{m \times q}$ by putting $\mathbf{G} = \mathbf{Y}$ as shown in
(7.21) such that the weight matrices are to be solved.

$$\mathbf{Y} = f_n \left(f_{n-1}(f_{n-2}(\cdots f_2(f_1(\mathbf{X}\mathbf{W}_1)\mathbf{W}_2)\cdots)\mathbf{W}_{n-1})\mathbf{W}_n \right). \tag{7.21}$$

By taking the functional inverse of f_n to both sides of the equation, we have a modified
version of learning, i.e.,

$$f_n^{-1}(\mathbf{Y}) = f_{n-1}(f_{n-2}(\cdots f_2(f_1(\mathbf{X}\mathbf{W}_1)\mathbf{W}_2)\cdots)\mathbf{W}_{n-1})\mathbf{W}_n. \tag{7.22}$$

Due to the formulation in terms of the matrix products with nested nonlinear function f, the learning in (7.22) is *different* from that of (7.21). In other
words, minimizing the trace$\{(\mathbf{G} - \mathbf{Y})^T (\mathbf{G} - \mathbf{Y})\}$ is *different* from minimizing the
trace$\{(f_n^{-1}(\mathbf{G}) - f_n^{-1}(\mathbf{Y}))^T (f_n^{-1}(\mathbf{G}) - f_n^{-1}(\mathbf{Y}))\}$.

Suppose there exists a *transformation* function ϕ such that trace$\{(\phi(\mathbf{G}) -$
$\phi(\mathbf{Y}))^T (\phi(\mathbf{G}) - \phi(\mathbf{Y}))\} = $ trace$\{(\mathbf{G} - \mathbf{Y})^T (\mathbf{G} - \mathbf{Y})\}$. Then, based on the kernel and
range projection, a least squares approximation to solving the learning equation can
be obtained. This is achieved via solving for $\mathbf{A}$ in the system of linear equations
$\mathbf{AB} = \mathbf{Y}$ according to Lemma 7.2 by defining

$$\mathbf{A} \triangleq f_{n-1}(f_{n-2}(\cdots f_2(f_1(\mathbf{X}\mathbf{W}_1)\mathbf{W}_2)\cdots)\mathbf{W}_{n-1}), \text{ and } \mathbf{B} \triangleq \mathbf{W}_n,$$

where the equation can be manipulated as

$$\phi_n(\mathbf{Y})\mathbf{W}_n^\dagger = f_{n-1}(f_{n-2}(\cdots f_2(f_1(\mathbf{X}\mathbf{W}_1)\mathbf{W}_2)\cdots)\mathbf{W}_{n-1}). \tag{7.23}$$

Subsequently, such a *transformation* can again be taken for both sides of the equation
for the next layer of hidden nodes with kernel and range projection for least squares
error approximation. An iteration of this process gives rise to a series of equations
each solving the least squares error:

$$\phi_{n-1}(\phi_n(\mathbf{Y})\mathbf{W}_n^\dagger) = f_{n-2}(\cdots f_2(f_1(\mathbf{X}\mathbf{W}_1)\mathbf{W}_2)\cdots)\mathbf{W}_{n-1} \tag{7.24}$$

$$\phi_{n-1}(\phi_n(\mathbf{Y})\mathbf{W}_n^\dagger)\mathbf{W}_{n-1}^\dagger = f_{n-2}(\cdots f_2(f_1(\mathbf{X}\mathbf{W}_1)\mathbf{W}_2)\cdots) \tag{7.25}$$

$$\vdots$$

$$\phi_2(\cdots \phi_{n-1}(\phi_n(\mathbf{Y})\mathbf{W}_n^\dagger)\mathbf{W}_{n-1}^\dagger \cdots) = f_1(\mathbf{X}\mathbf{W}_1)\mathbf{W}_2 \tag{7.26}$$

$$\phi_1(\phi_2(\cdots \phi_{n-1}(\phi_n(\mathbf{Y})\mathbf{W}_n^\dagger)\mathbf{W}_{n-1}^\dagger \cdots)\mathbf{W}_2^\dagger) = \mathbf{X}\mathbf{W}_1. \tag{7.27}$$

By iterating the projections on (7.22) to (7.27), the matrices of weights can be expressed as

$$\mathbf{W}_1 = \mathbf{X}^\dagger \phi_1(\phi_2(\phi_3(\cdots \phi_{n-1}(\phi_n(\mathbf{Y})\mathbf{W}_n^\dagger)\mathbf{W}_{n-1}^\dagger \cdots)\mathbf{W}_3^\dagger)\mathbf{W}_2^\dagger) \tag{7.28}$$

$$\mathbf{W}_2 = [f_1(\mathbf{X}\mathbf{W}_1)]^\dagger \phi_2(\phi_3(\cdots \phi_{n-1}(\phi_n(\mathbf{Y})\mathbf{W}_n^\dagger)\mathbf{W}_{n-1}^\dagger \cdots)\mathbf{W}_3^\dagger) \tag{7.29}$$

$$\vdots$$

$$\mathbf{W}_{n-1} = [f_{n-2}(\cdots f_2(f_1(\mathbf{X}\mathbf{W}_1)\mathbf{W}_2)\cdots \mathbf{W}_{n-2})]^\dagger \phi_{n-1}(\phi_n(\mathbf{Y})\mathbf{W}_n^\dagger)$$
$$\tag{7.30}$$

$$\mathbf{W}_n = [f_{n-1}(f_{n-2}(\cdots f_2(f_1(\mathbf{X}\mathbf{W}_1)\mathbf{W}_2)\cdots)\mathbf{W}_{n-1})]^\dagger \phi_n(\mathbf{Y}). \tag{7.31}$$

The manipulation above demonstrates that even if the transformation function ϕ exists, learning of FFN is characterized by *cross-coupling*. In other words, the weights in each layer are dependent on those in every other layer. Fundamentally, the two main questions for global network learning are whether the transformation function ϕ exists and if the cross-coupling of network weights can be decoupled. The next section introduces an approach to analytic network learning that relies on initialization and stacking.

7.3.2 Learning Initialization and Stacking

The above discussion shows that even if the transformation ϕ exists, the resulting least squares solution to the network weights comes in a cross-coupling form where the weight matrices are interdependent. This poses difficulty toward an analytical learning. Fortunately, attributed to the high redundancy of the network weights that are constrained within an array structure, the solutions to network learning are vastly available (see Sect. 7.3.3 for an analysis).

Consider m data samples of d dimension which are packed as $\mathbf{X} \in \mathbb{R}^{m \times (d+1)}$ with augmentation. Given random matrices $\mathbf{R}_i, i = 2, \ldots, n$ where their dimensions matches those of $\mathbf{W}_j, j = 2, \ldots, n$ in (7.19) and suppose there exists a *transformation* function ϕ such that $\text{trace}\{(\phi(\mathbf{G}) - \phi(\mathbf{Y}))^T (\phi(\mathbf{G}) - \phi(\mathbf{Y}))\} = \text{trace}\{(\mathbf{G} - \mathbf{Y})^T (\mathbf{G} - \mathbf{Y})\}$. Then, learning of the network toward the target $\mathbf{Y} \in \mathbb{R}^{m \times q}$ admits the following form:

$$\mathbf{W}_1 = \mathbf{X}^\dagger \phi_1(\phi_2(\phi_3(\cdots \phi_{n-1}(\phi_n(\mathbf{Y})\mathbf{R}_n^\dagger)\mathbf{R}_{n-1}^\dagger \cdots)\mathbf{R}_3^\dagger)\mathbf{R}_2^\dagger) \tag{7.32}$$

$$\mathbf{W}_2 = [f_1(\mathbf{X}\mathbf{W}_1)]^\dagger \phi_2(\phi_3(\cdots \phi_{n-1}(\phi_n(\mathbf{Y})\mathbf{R}_n^\dagger)\mathbf{R}_{n-1}^\dagger \cdots)\mathbf{R}_3^\dagger) \tag{7.33}$$

$$\vdots$$

$$\mathbf{W}_{n-1} = [f_{n-2}(\cdots f_2(f_1(\mathbf{XW}_1)\mathbf{W}_2)\cdots)]^\dagger \phi_{n-1}(\phi_n(\mathbf{Y})\mathbf{R}_n^\dagger) \tag{7.34}$$

$$\mathbf{W}_n = [f_{n-1}(f_{n-2}(\cdots f_2(f_1(\mathbf{XW}_1)\mathbf{W}_2)\cdots)\mathbf{W}_{n-1})]^\dagger \phi_n(\mathbf{Y}). \tag{7.35}$$

Since the above equations involved application of the unknown transformation ϕ on random matrices, the unknown transformation can be absorbed into the random weights initialization. This leads to the following learning sequences:

$$\mathbf{W}_1 = \mathbf{R}_1, \qquad\qquad\qquad\qquad\qquad \mathbf{R}_1 \in \mathbb{R}^{(d+1)\times h_1} \tag{7.36}$$

$$\mathbf{W}_2 = [f_1(\mathbf{XW}_1)]^\dagger \mathbf{R}_2, \qquad\qquad\qquad \mathbf{R}_2 \in \mathbb{R}^{m\times h_2} \tag{7.37}$$

$$\vdots$$

$$\mathbf{W}_{n-1} = [f_{n-2}(\cdots f_2(f_1(\mathbf{XW}_1)\mathbf{W}_2)\cdots)]^\dagger \mathbf{R}_{n-1}, \qquad \mathbf{R}_{n-1} \in \mathbb{R}^{m\times h_{n-1}} \tag{7.38}$$

$$\mathbf{W}_n = [f_{n-1}(f_{n-2}(\cdots f_2(f_1(\mathbf{XW}_1)\mathbf{W}_2)\cdots)\mathbf{W}_{n-1})]^\dagger \phi_n(\mathbf{Y}). \tag{7.39}$$

Here, we observe that the remaining unknown is the transformation ϕ in (7.39). Suppose the network has a linear mapping at the output layer (which means that $\phi_n(\mathbf{Y})$ can be replaced by $\mathbf{Y}$ in (7.39)), then the above learning becomes feasible. This feasibility has been taken in view of the large number of network layers available so that little learning capacity is compromised.

Apart from the above weights setting based on random initialization, the following formulation shows an example of non-random initialization.

$$\mathbf{W}_1 = \mathbf{X}^\dagger \tag{7.40}$$

$$\mathbf{W}_2 = [f_1(\mathbf{XW}_1)]^\dagger \tag{7.41}$$

$$\vdots$$

$$\mathbf{W}_{n-1} = [f_{n-2}(\cdots f_2(f_1(\mathbf{XW}_1)\mathbf{W}_2)\cdots)]^\dagger \tag{7.42}$$

$$\mathbf{W}_n = [f_{n-1}(f_{n-2}(\cdots f_2(f_1(\mathbf{XW}_1)\mathbf{W}_2)\cdots)\mathbf{W}_{n-1})]^\dagger \phi_n(\mathbf{Y}). \tag{7.43}$$

This formulation leads to weights learning being hinged upon the output layer only. Moreover, the number of network hidden nodes has been fixed according to the training sample size. Hence, this formulation is limited to problems with sample size constrained by the operating memory.

In summary, the learning based on initialization and stacking leads to learning only the output weights where all hidden layers have been randomly initialized and stacked. Hence, this learning can only be used as network initialization, and fine-tuning based on gradient descent (such as backpropagation) is needed to adjust the random hidden weights eventually.

7.3.3 Network Solution Space

Consider a single-layer network $\mathbf{G} = f_1(\mathbf{X}\mathbf{W}_1)$ which is targeted to fit $\mathbf{Y}$. Suppose a transformation ϕ_1 exists such that $\text{trace}\{(\phi_1(\mathbf{G}) - \phi_1(\mathbf{Y}))^T(\phi_1(\mathbf{G}) - \phi_1(\mathbf{Y}))\} = \text{trace}\{(\mathbf{G} - \mathbf{Y})^T(\mathbf{G} - \mathbf{Y})\}$. Then, when the network can be expressed as $\phi_1(\mathbf{Y}) = \mathbf{X}\mathbf{W}_1$, the learning becomes a linear regression problem. Due to the uniqueness of the pseudoinversion, the least squares error approximate solution for the weight matrix $\mathbf{W}_1 = \mathbf{X}^\dagger \phi_1(\mathbf{Y})$ is also unique.

Next, a network with two layers can be trained by setting $\mathbf{G} = f_2(f_1(\mathbf{X}\mathbf{W}_1)\mathbf{W}_2)$ equal to $\mathbf{Y}$. This leads to a significant increase in the number of possible solutions. This is clear from the modified systems of linear equations represented by $\phi_2(\mathbf{Y}) = f_1(\mathbf{X}\mathbf{W}_1)\mathbf{W}_2$ and $\phi_1(\phi_2(\mathbf{Y})\mathbf{W}_2^\dagger) = \mathbf{X}\mathbf{W}_1$ where arbitrarily fixing one of $\mathbf{W}_1$ and $\mathbf{W}_2$ determines the other. Such numerous options for either weight matrix give rise to a multitude of viable solutions.

As for training the three-layer network, setting $\mathbf{G} = f_3(f_2(f_1(\mathbf{X}\mathbf{W}_1)\mathbf{W}_2)\mathbf{W}_3)$ equal to $\mathbf{Y}$ further expands the feasible solution space. This is evident from the three modified systems of linear equations $\phi_3(\mathbf{Y}) = f_2(f_1(\mathbf{X}\mathbf{W}_1)\mathbf{W}_2)\mathbf{W}_3$, $\phi_2(\phi_3(\mathbf{Y})\mathbf{W}_3^\dagger) = f_1(\mathbf{X}\mathbf{W}_1)\mathbf{W}_2$ and $\phi_1(\phi_2(\phi_3(\mathbf{Y})\mathbf{W}_3^\dagger)\mathbf{W}_2^\dagger) = \mathbf{X}\mathbf{W}_1$, where arbitrarily fixing two of the weight matrices determines the other. Due to the additional degrees of freedom in fixing two weight matrices, an exponential number of feasible solutions can be anticipated.

Proposition 7.1 *The number of feasible solutions for a feedforward network with two layers is infinite. This number increases exponentially for each increment of the network layer.*

Proof For a two-layer network, the number of feasible solutions is determined by the size of either $\mathbf{W}_1$ or $\mathbf{W}_2$ which is infinite in the real domain. Let us denote this number by N. Then, for a three-layer network, this number increases to $(N \times N) \times C_2^3 = N^2 \times 3$ since two arbitrary weights out of $\mathbf{W}_i$, $i = 1, 2, 3$ determine the remaining. For a n-layer network, we have $N^{n-1} \times C_{n-1}^n$ possibilities. Hence the proof. ∎

7.3.4 Network Representation

For simplicity, we shall use d as the dimension for generic matrices instead of the augmented dimension $d + 1$ when no ambiguity arises.

Definition 7.3 A matrix $\mathbf{A} \in \mathbb{R}^{m \times d}$ is said to be representative for $\mathbf{Y} \in \mathbb{R}^{m \times q}$ if $\mathbf{A}\mathbf{A}^\dagger\mathbf{Y} = \mathbf{Y}$ where $m, d, q \geq 1$.

Certainly, if $\mathbf{A}$ has full rank in the sense that $\mathbf{A}\mathbf{A}^T$ is invertible for $\mathbf{A}^\dagger = \mathbf{A}^T(\mathbf{A}\mathbf{A}^T)^{-1}$, then $\mathbf{A}\mathbf{A}^\dagger = \mathbf{I}$ for which $\mathbf{A}$ is representative for $\mathbf{Y}$. However, this definition of representation is weaker than the full rank requirement in $(\mathbf{A}^T\mathbf{A})^{-1}\mathbf{A}^T$ or $\mathbf{A}^T(\mathbf{A}\mathbf{A}^T)^{-1}$ because $\mathbf{A}$ needs not have full rank when $\mathbf{Y}$ falls within the range of $\mathbf{A}$.

Definition 7.4 A function f, which operates elementwise on its matrix domain, is said to be representative if $f(\mathbf{XW}_1)$ is representative for $\mathbf{Y} \in \mathbb{R}^{m \times q}$ on $\mathbf{X} \in \mathbb{R}^{m \times d}$ with distinct rows and $\mathbf{W}_1 \in \mathbb{R}^{d \times h_1}$ where $m, d, h_1, q \geq 1$, i.e., f is a function such that $f(\mathbf{XW}_1)f(\mathbf{XW}_1)^\dagger \mathbf{Y} = \mathbf{Y}$.

The definition above is applicable to linear activation such as $f(x) = x$, under certain conditions. For instance, when $\mathbf{XW}_1$ has a column rank similar to the row size of $\mathbf{Y}$, then $(\mathbf{XW}_1)(\mathbf{XW}_1)^\dagger \mathbf{Y} = \mathbf{Y}$. However, the linear activation function does not hold for representative target mapping when the data matrix does not have sufficient range rank. For activation functions of nonlinear nature, for instances $\log(1 + \exp(\cdot))$ and $\tan(\cdot)$, it is observed that the column rank of $\mathbf{XW}_1$ can increase beyond that utilizing linear activation, thereby improving representative target mapping.

Theorem 7.1 (Two-layer Network) *Given m distinct samples of input-output data pairs and suppose the activation functions are representative according to Definition 7.4. Then there exists a feedforward network of two layers with at least $m \times q$ adjustable weights that can represent these data samples of $q \geq 1$ output dimensions.*

Proof Consider a 2-layer network, with linear activation at the output layer, that takes an augmented set of inputs:

$$\mathbf{G} = f(\mathbf{XW}_1)\mathbf{W}_2, \tag{7.44}$$

where $\mathbf{X} \in \mathbb{R}^{m \times (d+1)}$, $\mathbf{W}_1 \in \mathbb{R}^{(d+1) \times h_1}$, $\mathbf{W}_2 \in \mathbb{R}^{h_1 \times q}$ and $\mathbf{G} \in \mathbb{R}^{m \times q}$ is the network output of $q \geq 1$ dimensions. Then, $\mathbf{G}$ (7.44) can be used to learn the target matrix $\mathbf{Y} \in \mathbb{R}^{m \times q}$ by putting

$$\mathbf{Y} = f(\mathbf{XW}_1)\mathbf{W}_2. \tag{7.45}$$

Since f is given to be a representative function according to Definition 7.4, we have $f(\mathbf{XW}_1)f(\mathbf{XW}_1)^\dagger \mathbf{Y} = \mathbf{Y}$ such that $\mathbf{Y}$ falls within the range space of $f(\mathbf{XW}_1)$. To ensure that $\mathbf{Y}$ falls within the range space of $f(\mathbf{XW}_1)$, it is sufficient that $\mathbf{W}_1$ having at least a column size of $h_1 = m$ in order to match with the number of samples in $\mathbf{Y}$. Suppose we have a non-trivial setting for all the weight elements in $\mathbf{W}_1$ such that not all of its elements are zeros. A good example for the non-trivial setting is to put $\mathbf{W}_1 = \mathbf{X}^\dagger$ (such as (7.40)). Then, due to the representativeness of f, the entire weight matrix $\mathbf{W}_2 \in \mathbb{R}^{m \times q}$ can be determined uniquely as $f(\mathbf{XW}_1)^\dagger \mathbf{Y}$ to represent the target $\mathbf{Y}$. In other words, $\mathbf{W}_1$ needs only distinct and arbitrary numbers in its m columns of entries to maintain the representation sufficiency of $f(\mathbf{XW}_1)$, and all those elements in $\mathbf{W}_2$ are the only adjustable parameters needed for the representation. Hence the proof. ∎

The above result shows that the two-layer network is a universal approximator for a finite set of data given sufficient hidden neurons at the output layer. Different from most existing results in literature (Funahashi, 1989; Hornik et al., 1989; Cybenko,

1989; Hecht-Nielsen, 1987; Kůrková, 1992; Leshno et al., 1993), the minimum number of adjustable hidden weights required here is dependent on the number of data samples and the output dimension. This result also stretches beyond that of Zhang et al. (2017) to include nonlinear activation functions.

Definition 7.5 A function f is said to be compositional representative if $f(\cdots f(\mathbf{X}\mathbf{W}_1)\cdots \mathbf{W}_n)$ is representative for $\mathbf{Y} \in \mathbb{R}^{m\times q}$ on $\mathbf{X} \in \mathbb{R}^{m\times d}$ with distinct rows and $\mathbf{W}_k \in \mathbb{R}^{h_{k-1}\times h_k}, k = 1,\ldots,n$ where $h_0 = d + 1, m, d, h_k, q \geq 1$, i.e., f is a function such that $f(\cdots f(\mathbf{X}\mathbf{W}_1)\cdots \mathbf{W}_n)f(\cdots f(\mathbf{X}\mathbf{W}_1)\cdots \mathbf{W}_n)^\dagger \mathbf{Y} = \mathbf{Y}$.

Theorem 7.2 (Multilayer Network) *Given m distinct samples of input-output data pairs and suppose the activation functions are representative according to Definition 7.5. Then there exists a feedforward network of n layers with at least $m \times q$ adjustable weights that can represent these data samples samples of $q \geq 1$ output dimensions.*

Proof Consider a feedforward network with linear activation at the output layer given by

$$\mathbf{G} = f_{n-1}(\mathbf{A}_{n-1})\mathbf{W}_n, \tag{7.46}$$

where $\mathbf{A}_{n-1} = f_{n-2}(\mathbf{A}_{n-2})\mathbf{W}_{n-1},\ldots, \mathbf{A}_1 = \mathbf{X}\mathbf{W}_1$. Learning toward $\mathbf{Y}$ using this network is analogous to using (7.44) in the proof of Theorem 7.1 where we require $m \times q$ number of adjustable weight elements in $\mathbf{W}_n$ here for the desired representation. The representation property of each $\mathbf{A}_i$, $i = 1,\ldots,n - 1$ shall not change if each $\mathbf{W}_i$, $i = 1,\ldots,n - 1$ is a representative matrix with sufficient column size (such as (7.40)–(7.42)) and if f is compositional representative according to Definition 7.5. Since $\mathbf{W}_i$, $i = 1,\ldots,n - 1$ can be arbitrarily prefixed, the elements of $\mathbf{W}_n \in \mathbb{R}^{m\times q}$ are the only adjustable parameters needed in learning the representation. This completes the proof. ∎

7.4 Gradient-Free Convolutional Neural Network Learning

7.4.1 Image Convolution

Consider an input feature map $\mathbf{X}_l \in \mathbb{R}^{C_l \times W_l \times H_l}$ at layer l having C_l channels with $W_l \times H_l$ pixels in each channel. Let $\mathbf{W}_l^{[j]} \in \mathbb{R}^{C_l \times K_l \times K_l}$ represents the jth filter weight matrix with kernel size $K_l \times K_l$ at layer l. Then, a convolution operation using filters $\mathbf{W}_l := \{\mathbf{W}_l^{[1]}, \mathbf{W}_l^{[2]}, \ldots, \mathbf{W}_l^{[J]}\}$ on $\mathbf{X}_l$ yields an output tensor given by

$$\mathbf{Z}_l = f_{\mathrm{conv}}(\mathbf{X}_l, \mathbf{W}_l), \tag{7.47}$$

where f_{conv} denotes an operator for image convolution. Suppose the convolution adopts P and S padding and stride operations respectively. Then, the filtered output would be $\mathbf{Z}_l \in \mathbb{R}^{J \times ([(W_l - K_l + 2P)S]+1) \times ([(H_l - K_l + 2P)S]+1)}$. For simplicity as well as for illustration purpose, a zero padding and a stride of one is adopted here and this gives $\mathbf{Z}_l \in \mathbb{R}^{J \times (W_l - K_l + 1) \times (H_l - K_l + 1)}$. The corresponding convolution operation can then be written as

$$\mathbf{Z}_l[j, w, h] = \sum_{c=1}^{C_l} \mathbf{X}_l[c, w : w + K_l, h : h + K_l] \circ \mathbf{W}_l^j[c, :, :], \tag{7.48}$$

where $\circ$ denotes an elementwise multiplication between two matrices with $1 \leq j \leq J$, $1 \leq w \leq (W_l - K_l + 1)$, and $1 \leq h \leq (H_l - K_l + 1)$.

7.4.2 Convolution via Matrix Product

The convolutional box filtering of a feature map $\mathbf{X}_l$ by a filter $\mathbf{W}_l^{[1]}$ at layer l can be written as

$$\mathbf{X}_l * \mathbf{W}_l^{[1]} = \mathbf{Z}_l$$

$$\begin{bmatrix} a & b & c & d \\ e & f & g & h \\ i & j & k & l \\ m & n & o & p \end{bmatrix} * \begin{bmatrix} A_1 & B_1 \\ C_1 & D_1 \end{bmatrix} = \begin{bmatrix} \alpha_1 & \beta_1 & \gamma_1 \\ \delta_1 & \epsilon_1 & \zeta_1 \\ \eta_1 & \theta_1 & \iota_1 \end{bmatrix} \tag{7.49}$$

where

$$\alpha_1 = aA_1 + bB_1 + eC_1 + fD_1$$
$$\beta_1 = bA_1 + cB_1 + fC_1 + gD_1$$
$$\vdots$$
$$\iota_1 = kA_1 + lB_1 + oC_1 + pD_1.$$

This operation can be written in matrix product form as

$$\begin{bmatrix} a & b & e & f \\ b & c & f & g \\ c & d & g & h \\ e & f & i & j \\ f & g & j & k \\ g & h & k & l \\ i & j & m & n \\ j & k & n & o \\ k & l & o & p \end{bmatrix} \begin{bmatrix} A_1 \\ B_1 \\ C_1 \\ D_1 \end{bmatrix} = \begin{bmatrix} \alpha_1 \\ \beta_1 \\ \gamma_1 \\ \delta_1 \\ \epsilon_1 \\ \zeta_1 \\ \eta_1 \\ \theta_1 \\ \iota_1 \end{bmatrix}. \tag{7.50}$$

By defining a *flattening operator* $\mathcal{F}$ which converts a N-dimensional tensor into an $(N-1)$-dimensional one and a corresponding transformation $\mathcal{T}$ of feature map $\mathbf{X}_l$, we can simplify the above equation into the notation given by $\mathcal{T}(\mathbf{X}_l)\mathcal{F}(\mathbf{W}_l^{[1]}) = \mathcal{F}(\mathbf{Z}_l)$. An inverse of the transformations, $\mathcal{F}^{-1}$ and $\mathcal{T}^{-1}$, shall reverse the operation to recover the original matrix or tensor.

For multiple channels (such as red, green blue image color channels), their processing can be merged, for instance adding a blue channel image to the red channel image, as follows

$$
\begin{bmatrix}
a & b & e & f & a & b & e & f \\
b & c & f & g & b & c & f & g \\
c & d & g & h & c & d & g & h \\
e & f & i & j & e & f & i & j \\
f & g & j & k & f & g & j & k \\
g & h & k & l & g & h & k & l \\
i & j & m & n & i & j & m & n \\
j & k & n & o & j & k & n & o \\
k & l & o & p & k & l & o & p
\end{bmatrix}
\begin{bmatrix}
A_1 \\ B_1 \\ C_1 \\ D_1 \\ A_1 \\ B_1 \\ C_1 \\ D_1
\end{bmatrix}
=
\begin{bmatrix}
\alpha_1 \\ \beta_1 \\ \gamma_1 \\ \delta_1 \\ \epsilon_1 \\ \zeta_1 \\ \eta_1 \\ \theta_1 \\ \iota_1
\end{bmatrix}.
\tag{7.51}
$$

Suppose we have a second filter $\mathbf{W}_l^{[2]}$ which is flattened as $\mathcal{F}(\mathbf{W}_l^{[2]})$. Then, the filtering can be packed together with $\mathcal{F}(\mathbf{W}_l^{[1]})$ giving

$$
\begin{bmatrix}
a & b & e & f & a & b & e & f \\
b & c & f & g & b & c & f & g \\
c & d & g & h & c & d & g & h \\
e & f & i & j & e & f & i & j \\
f & g & j & k & f & g & j & k \\
g & h & k & l & g & h & k & l \\
i & j & m & n & i & j & m & n \\
j & k & n & l & j & k & n & l \\
k & l & o & p & k & l & o & p
\end{bmatrix}
\begin{bmatrix}
A_1 & A_2 \\ B_1 & B_2 \\ C_1 & C_2 \\ D_1 & D_2 \\ A_1 & A_2 \\ B_1 & B_2 \\ C_1 & C_2 \\ D_1 & D_2
\end{bmatrix}
=
\begin{bmatrix}
\alpha_1 & \alpha_2 \\ \beta_1 & \beta_2 \\ \gamma_1 & \gamma_2 \\ \delta_1 & \delta_2 \\ \epsilon_1 & \epsilon_2 \\ \zeta_1 & \zeta_2 \\ \eta_1 & \eta_2 \\ \theta_1 & \theta_2 \\ \iota_1 & \iota_2
\end{bmatrix}.
\tag{7.52}
$$

In order to simplify the notations, we denote a concatenation of J flattened filters as

$$
\mathcal{W}_l = [\mathcal{F}(\mathbf{W}_l^{[1]}), \mathcal{F}(\mathbf{W}_l^{[2]}), \ldots, \mathcal{F}(\mathbf{W}_l^{[J]})],
\tag{7.53}
$$

where $\mathcal{W}_l \in \mathbb{R}^{(C_l K_l^2) \times J}$. For the feature map $\mathbf{X}_l$, the flattening is performed on each convolutional patch of the filtering given by $\mathbf{X}_l[c, w : w + K_l, h : h + K_l]$ of (7.48). These flattened patches are concatenated as a row vector as follows:

$$
\begin{aligned}
c_{l,w,h} = [&\mathcal{F}(\mathbf{X}_l[1, w : w + K_l, h : h + K_l])^T, \\
&\mathcal{F}(\mathbf{X}_l[2, w : w + K_l, h : h + K_l])^T, \ldots, \\
&\mathcal{F}(\mathbf{X}_l[C_l, w : w + K_l, h : h + K_l])^T].
\end{aligned}
\tag{7.54}
$$

Next, such rows are stacked to form a matrix as shown below:

$$
\mathcal{X}_l = \begin{bmatrix} \boldsymbol{c}_{l,1,1} \\ \boldsymbol{c}_{l,2,1} \\ \vdots \\ \boldsymbol{c}_{l,(W_l-K_l+1),1} \\ \boldsymbol{c}_{l,1,2} \\ \boldsymbol{c}_{l,2,2} \\ \vdots \\ \boldsymbol{c}_{l,(W_l-K_l+1),2} \\ \vdots \\ \boldsymbol{c}_{l,(W_l-K_l+1),(H_l-K_l+1)} \end{bmatrix}, \tag{7.55}
$$

where $\mathcal{X}_l \in \mathbb{R}^{(W_l-K_l+1)(H_l-K_l+1)\times C_l K_l^2}$ is a transformed two-dimensional matrix. This transformation has been denoted as

$$
\mathcal{X}_l = \mathcal{T}(\mathbf{X}_l). \tag{7.56}
$$

Conversely, a reverse mapping can be obtained as $\mathcal{T}^{-1}(\mathcal{T}(\mathbf{X}_l)) = \mathbf{X}_l$. The learning system can then be represented as

$$
\mathcal{X}_l \mathcal{W}_l = \mathcal{F}(\mathbf{Z}_l), \tag{7.57}
$$

where $\mathcal{F}(\mathbf{Z}_l) \in \mathbb{R}^{(W_l-K_l+1)(H_l-K_l+1)\times J}$ which is identical to that in (7.48) with some rearrangements. The transformation in (7.53)–(7.57) is similar to an operation called general matrix multiplication (or image-to-column transformation) implemented by various deep learning platforms (e.g., PyTorch) to accelerate the CNN operations.

7.4.3 Label Encoding

The target label $\mathbf{Z}_l$ at layer l is unknown in general. In order to facilitate learning, the target label $\mathbf{Y}$ can be projected into each hidden layer based on the network structure. Suppose there are N training samples with K categories, then we have $\mathbf{Y} \in \mathbb{R}^{N \times K}$. Its projection into the hidden layer can be generated using

$$
\bar{\mathbf{Z}}_l = \mathbf{Y}\mathbf{Q}_l, \tag{7.58}
$$

where $\mathbf{Q}_l$ is a label encoding matrix being generated according to the hidden learning structure. For example, if a hidden layer $\mathcal{F}(\mathbf{Z}_l)$ has $(W_l - K_l + 1)(H_l - K_l + 1)J$ entries, then $\mathbf{Q}_l \in \mathbb{R}^{K \times (W_l-K_l+1)(H_l-K_l+1)J}$. Since the target is projected from a lower dimension to a higher one, the label encoding does not lose information.

7.4.4 *Learning of the CNN Filter Weights*

Single-sample learning:

After the filtering transformation, the learning for a single data sample at each layer can be obtained by

$$\arg\min_{\boldsymbol{W}_l} \|\mathcal{F}^{-1}(\bar{\mathbf{Z}}_l) - \boldsymbol{X}_l\boldsymbol{W}_l\|_2^2 + \lambda\|\boldsymbol{W}_l\|_2^2, \tag{7.59}$$

where $\mathcal{F}^{-1}(\bar{\mathbf{Z}}_l) \in R^{(W_l-K_l+1)(H_l-K_l+1)\times J}$ and this gives

$$\hat{\boldsymbol{W}}_l = (\boldsymbol{X}_l^T\boldsymbol{X}_l + \lambda\mathbf{I})^{-1}\boldsymbol{X}_l^T\mathcal{F}^{-1}(\bar{\mathbf{Z}}_l). \tag{7.60}$$

Multiple-sample learning:

$$\arg\min_{\boldsymbol{W}_l} \sum_{n=1}^{N} \|\mathcal{F}^{-1}(\bar{\mathbf{Z}}_l)^{(n)} - \boldsymbol{X}_l^{(n)}\boldsymbol{W}_l\|_2^2 + \lambda\|\boldsymbol{W}_l\|_2^2, \tag{7.61}$$

with solution given by

$$\hat{\boldsymbol{W}}_l = \left(\sum_{n=1}^{N}\boldsymbol{X}_l^{(n)T}\boldsymbol{X}_l^{(n)} + \lambda\mathbf{I}\right)^{-1}\left(\sum_{n=1}^{N}\boldsymbol{X}_l^{(n)T}\mathcal{F}^{-1}(\bar{\mathbf{Z}}_l)^{(n)}\right). \tag{7.62}$$

Based on (7.62), the training of the CNN layers is conducted in a stacking manner. Starting from layer $l = 1$ nearest to the inputs, the estimated convolution weights and the input feature map are fed into the activation function $g_l(\cdot)$ to form the feature map of the next layer as follows:

$$\boldsymbol{X}_{l+1}^{(n)} = g_l(\boldsymbol{X}_l^{(n)}\hat{\boldsymbol{W}}_l), \quad n = 1,\ldots,N. \tag{7.63}$$

These transformed features can be reshaped back into the original feature map using

$$\mathbf{X}_{l+1}^{(n)} = \mathcal{T}^{-1}(\boldsymbol{X}_{l+1}^{(n)}). \tag{7.64}$$

Eventually, the MLP layer is trained based on the flattened features

$$\boldsymbol{X} = ([\mathcal{F}(\mathbf{X}_l^{(1)}), \mathcal{F}(\mathbf{X}_l^{(2)}), \ldots, \mathcal{F}(\mathbf{X}_l^{(N)})])^T, \tag{7.65}$$

utilizing the cost function

$$\arg\min_{\boldsymbol{W}_l} \|\mathbf{Y} - \boldsymbol{X}\boldsymbol{W}_l\|_2^2 + \|\boldsymbol{W}_l\|_2^2, \tag{7.66}$$

where the solution is

$$\hat{\mathcal{W}}_l = (\mathcal{X}^T \mathcal{X} + \lambda \mathbf{I})^{-1} \mathcal{X}^T \mathbf{Y}. \tag{7.67}$$

Algorithm 7.1 (Zhuang et al., 2022a)
Inputs: Training image samples $\{\mathbf{X}_1^{(1)}, \mathbf{X}_1^{(2)}\}$ with corresponding target label matrix $\mathbf{Y}$, a network of length L including an MLP as last layer with activation functions $g_1, \ldots, g_{L-1}$, label encoders $\mathbf{Q}_1, \ldots, \mathbf{Q}_{L-1}$, and regularization factor λ.
for $l \leftarrow 1$ **to** L **do**
    if $l < L$ (CNN layers) **then**
        LE process: obtain encoded label $\bar{\mathbf{Z}}$ using (7.58);
        Feature flattening: obtain the transformed features $\mathcal{X}_l^{(n)}$
        for $n = 1, \ldots, N$ using (7.54)–(7.55);
        LS solution: compute $\hat{\mathcal{W}}_l$ using (7.62);
        Compute the activated features $\mathbf{X}_{l+1}^{(n)}$ using (7.64);
    else
        Obtain flattened feature map matrix $\mathcal{X}$ using (7.65);
        Compute the MLP weights $\hat{\mathcal{W}}_l$ using (7.67);
    end
end
Outputs: $\hat{\mathcal{W}}_1, \ldots, \hat{\mathcal{W}}_L$.

7.4.5 CNN from the Regularization Perspective

Recall that the original input $\mathbf{X}_l \in \mathbb{R}^{C_l \times W_l \times H_l}$ at layer l corresponds to a single sample of the input image. After the transformation to facilitate matrix product in place of the convolution operation, the input becomes much larger in size as follows $\mathcal{X}_l \in \mathbb{R}^{(W_l - K_l + 1)(H_l - K_l + 1) \times C_l K_l^2}$. Due to the expansion of the matrix product according to the local arrangement of neighborhood cells, this transformation can be interpreted as an addition of linear constraints to the system of equations. When a small convolution kernel has been adopted, such an increase of constraints can be tremendous. Since an increase of constraints leads to a more restricted solution, the convolution operation, as evidenced by the equivalent transformation in matrix product form, can be interpreted as a regularized operation comparing with its Multilayer Perceptron counterpart.

7.5 Bibliographic Notes

Representer and Universal Approximator

The bibliographic notes of Chap. 3 traced the functional approximation of the polynomials to Weierstrass's approximation theorem back in 1885. For functional structures resembling those two-layer neural networks, the theoretical development of their universal approximation capability can be traced to the *Kolmogorov's theorem* in 1957. From the kernel perspective, the *representer theorem* (Kimeldorf & Wahba, 1971) provided a form for data representation for a class of regularized risks. In the late 1980s, the neural network was shown to be a universal approximator by several researchers (see, e.g., Hecht-Nielsen (1987), Funahashi (1989), Hornik et al. (1989), Cybenko (1989), Kůrková (1992), Leshno et al. (1993)). From the viewpoint of random projection, the *Johnson-Lindenstrauss's lemma* (Vempala, 2004) showed that a set of n points in $\mathbb{R}^d$ space can be projected down to that in $\mathbb{R}^k$ space such that all pairwise distances are preserved.

Analysis and Learning of Two-layer Networks

In terms of learning a system of equations or its alike under the framework of neural network, the following includes a non-exhaustive list of early works in the field.

An early attempt to analyze the network learning landscape of the least squares error was found in Baldi and Hornik (1989), Baldi (1988).[1] The learning landscape was analyzed based on a linear network of two layers in matrix form Baldi and Hornik (1989) and then extended to a linear network of multiple layers also in matrix form Baldi (1988). Apart from the analysis of decision landscapes, Baldi (1988) proposed a learning algorithm in iterative form. The study showed that the learning rate needs to be chosen to be lower than a certain factor which is inversely proportional to the eigenvalue of the covariance matrix for convergence. During the similar time, the relationship between *learning* in layered networks and *data fitting* with high-dimensional surfaces was discussed in Broomhead & Lowe (1988). Particularly, a specific class of *radial basis function* networks, which possessed the property that *learning* is equivalent to *the solution of a set of linear equations*, was identified. The multivariable functional interpolation property was illustrated based on the exclusive-OR and chaotic time series problems. In Lowe (1989), the necessity of the choice of the radial basis function *centers* based on the training points was removed. The *centers* were treated as a set of nonlinear parameters to be optimized based on the partial derivatives.

Subsequently in Barton (1991), a two-layer network with the exponential activation was studied in terms of learning the weights and biases of the output side of the network given any set of weights and biases for the input side of the hidden layer. Particularly, the solution of the output weights and biases was given in analytical form when the input weights and biases were known (see Eq. (2.16) in Barton (1991)). In Pao & Takefuji (1992); Pao et al. (1994), a functional-link net which supported different functionalities with the same system architecture was proposed. Essentially, this constitutes a two-layer network in our context where the hidden weights linking to the input can be generated randomly. The random initialization

---

[1] It was mentioned in Baldi (1988) that the results of Baldi (1988) was an extension of that in Baldi and Hornik (1989).

of the hidden weights of a two-layer network was also seen in the extreme learning machine (Huang et al., 2006), the random features for large-scale kernel machines (Rahimi & Recht, 2008), and the no-prop algorithm (Widrow et al., 2013). The no-prop algorithm in Widrow et al. (2013) extended the random initialization of hidden weights beyond a single hidden layer with only the weights of the output layer being trained by the least mean square (LMS) algorithm of Widrow and Hoff. In the non-random context of hidden weights initialization, a reduced set of polynomial terms was adopted in Toh and Yau (2002), Toh et al. (2004a) for deterministic parametric learning. The idea of polynomial basis expansion had been extended to optimize a counting-based objective function for pattern classification in Toh (2008). In Rahimi and Recht (2007), a two-layer neural network with random first-layer weights based on the kernel formulation was studied. The focus was on two particular random features namely, the random Fourier features and the random binning features. Belkin et al. (2018) pointed out that the concepts of interpolation were not contradicting to generalization. An asymptotic analysis of minimum ℓ_2-norm was subsequently carried out in Hastie et al. (2019) for high-dimensional linear regression. The study included a two-layer network (termed a single-layer network in their context) with nonlinear activation. The bias and variance of the risk were studied for both the under-parameterized and the over-parameterized cases.

Analysis of Networks Beyond Two Layers

In 2012, Baldi and Lu (2012) extended the analysis of convexity to a deep linear network and conjectured that every local minimum was a global minimum. Recently, Kawaguchi (2016) showed that the loss surface of deep linear networks was non-convex and non-concave, every local minimum was a global minimum, every critical point that was not a global minimum was a saddle point, and the rank of weight matrix at saddle points. These observations, represented in the form of Directed Acyclic Graph (DAG), were carried forward to the nonlinear networks with fewer unrealistic assumptions than that in Choromanska et al. (2015a), Choromanska et al. (2015b) which were inspired from the Hamiltonian of the spherical spin-glass model. Subsequently, Kawaguchi and Bengio (2019) showed that depth in linear networks created no bad local minima. More recently, Yun et al. (2018) provided sufficient conditions for global optimality in deep nonlinear networks when the activation functions satisfied certain invertibility, differentiability, and bounding conditions.

From the geometrical view of the loss function, Dinh et al. (2017) argued that an appropriate definition and use of minima in terms of their sharpness could help in understanding the generalization behavior. With the understanding that the difficulty of deep search was originated from the proliferation of saddle points and not local minima, Dauphin et al. (2014) attempted a saddle-free Newton's method to search for the minima in deep networks. In Li et al. (2018), a visualization technique based on filter normalization was introduced for studying the loss landscape. By comparing networks with and without skip connections, the study showed that the smoothness of loss surface depended highly on the network architecture. The landscape of the empirical risk for deep learning of convolutional neural networks was investigated in Liao and Poggio (2017). By extending the case of linear networks in Gunasekar et al. (2018) to the case of nonlinear networks, Poggio et al. (2018) showed that deep networks did not usually overfit the classification error for low-noise data sets and the solution corresponded to margin maximization.

From the view of having a theoretical bound for the estimation error, in 2002, Langford and Caruana (2002) investigated into generalization in terms of the true error rate bound of a distribution over a set of neural networks using the PAC-Bayes bound. This approach had sparked off several follow-ups (see, e.g., Dziugaite and Roy (2017), Neyshabur (2017), Neyshabur et al. (2018)). Recently, Bartlett et al. (2017) proposed a margin-based generalization bound which was spectrally-normalized. Apart from an empirical investigation of several capacity bounds based on the ℓ_2, ℓ_1-path, ℓ_2-path and the spectral norms (Neyshabur et al., 2017), in Neyshabur et al. (2018), the margin-based perturbation bound was combined with the PAC-Bayes analysis to derive the generalization bound. Using the gap between the expected risk and the empirical risk, Kawaguchi et al. (2023) provided generalization bounds for deep models which do not have an explicit dependency on the number of weights, the depth and the input dimensionality. In view of the lack of an analytic structure in learning, many of these analyses were hinged upon the Stochastic Gradient Descent (SGD, see e.g., Brunelli (1994), Patrick van der Smagt (1994), Barnard (1992)) search toward certain stationary points (e.g., Bartlett et al. (2017), Liao and Poggio (2017), Dziugaite and Roy (2017), Neyshabur (2017)). Moreover, many of the analyses of the linear network or the nonlinear one treated the network as a Directed Acyclic Graph (DAG) model (see, e.g., Gunasekar et al. (2018), Neyshabur (2017), Kawaguchi et al. (2023)) where it turned into a complex problem of topological sorting.

In light of the inherent least squares approximation carried by matrix inversion, Toh (2018c) proposed to solve the network learning problem deterministically by successive Moore-Penrose inverse operations of network matrices. This proposal has enabled an analytical approximate solution form for network learning beyond two layers. With full consideration of biases in all layers, the representation capability of the network to learn a finite set of data was shown (Toh et al., 2018a), taking into account non-invertible activation functions. It was subsequently shown (Toh, 2018a) that the bias term in one layer can be considered a scaled version in another layer, hence removing the need for bias in each layer except for the input layer. Apart from the corresponding representation capability, the vast solution space and the variance of network output were also shown. This analysis is related to the network generalization ability. In view of the heavy computational requirement of the Moore-Penrose inverse, a simplified approximate solution was proposed in Toh (2018b) by removing all the pseudoinverse operations except for one at the output layer.

Recursive and Sparse Network Learning

In view of the large learning overhead for bulk matrix formulation, efficient analytic learning beyond two layers was investigated in Zhuang et al. (2019) where a low-memory learning formulation was constructed by recursive means. To facilitate flexibility in memory size utilization, a blockwise recursive Moore-Penrose inverse was formulated for network learning (Zhuang et al., 2022a). The recursion strategy was subsequently extended to addressing class-incremental learning with absolute memorization and privacy protection (Zhuang et al., 2022b).

The convolution neural network can be considered a sparse construction with limited receptive field which can be represented in matrix product form based on the circulant matrix (Davis, 1970; Golub & Van Loan, 1996). Based on a rearrangement of the image blocks aligning with corresponding filter elements (Vasudevan et al., 2017) such as the implementation in deep learning platforms (Paszke et al., 2019), an analytic formulation of convolutional neural network learning was investigated in Zhuang et al. (2025). Moreover, an accumulated decoupled learning with gradient staleness mitigation was studied for Convolutional Neural Networks in Zhuang et al. (2021).

7.6   Exercises

7.1 Solving the parameters of the system of linear equations under the kernel and the range spaces result in the minimum ℓ_2-norm of the parameters. True or False?

7.2 Utilization of the ℓ_1 loss in network residual learning leads to sparseness of the network weights. True or False?

7.3 Select from below any ways that are effective to reduce overfitting in neural networks.

   (a) Including an L_2 regularization.
   (b) Increasing the number of hidden nodes.
   (c) Increasing the learning rate.
   (d) Augmenting the training data with synthetic samples.

7.4 Consider two networks using the same activation function in all their layers. Then, when trained appropriately, a neural network with 3 layers will not have a higher training error than that of a single-layer neural network.

7.5 Suppose you have trained a network for a huge number of iterations with high training and test errors. Which of the following, done in isolation, has a better chance of reducing both the training and test errors?

   (a) Add more hidden layer(s).
   (b) Add more nodes to hidden layers.
   (c) Add more training data.
   (d) Reduce connections of hidden layer(s).

(e) None of these.

7.6 Consider a three-layer neural network utilizing the sigmoid function given by $1/(1 + \exp(-\beta x))$ as the activation in all layers. The parameter β of the sigmoid controls the slope of the activation function. A training error of 89.35 and a testing error of 91.8 are observed. Which of the following is true?

(a) Increasing β will most likely increase the test error.
(b) Increasing β will most likely reduce the test error.
(c) Decreasing β will most likely reduce the test error.
(d) Decreasing β will most likely increase the test error.
(e) Not enough information is provided to determine how β should be changed.
(f) None of these.

7.7 A fully connected network of 3 layers can be written as

$$\mathbf{G} = f([\mathbf{1},\ f([\mathbf{1},\ f(\mathbf{XW}_1)]\mathbf{W}_2)]\mathbf{W}_3)$$

where

$$\mathbf{X} = \begin{bmatrix} 1 & 1 & 3.0 \\ 1 & 2 & 2.5 \end{bmatrix}, \quad \mathbf{W}_1 = \begin{bmatrix} -1 & 0 & 1 \\ 0 & -1 & 0 \\ 1 & 0 & 1 \end{bmatrix},$$

$$\mathbf{W}_2 = \mathbf{W}_3 = \begin{bmatrix} -1 & 0 & 1 \\ 0 & -1 & 0 \\ 1 & 0 & 1 \\ 1 & -1 & 1 \end{bmatrix}.$$

Suppose the Rectified Linear Unit (ReLU) has been used as the activation function (f) for all the nodes. Compute the network output matrix $\mathbf{G}$ based on the given network weights and data.

7.8 The following graph shows the structure of a simple neural network with a single hidden layer. The input layer consists of three dimensions $\boldsymbol{x} = [x_1, x_2, x_3]^T$. The hidden layer includes three units $\boldsymbol{y} = [y_1, y_2, y_3]^T$. The output layer includes one unit z. The bias terms are ignored for simplicity.

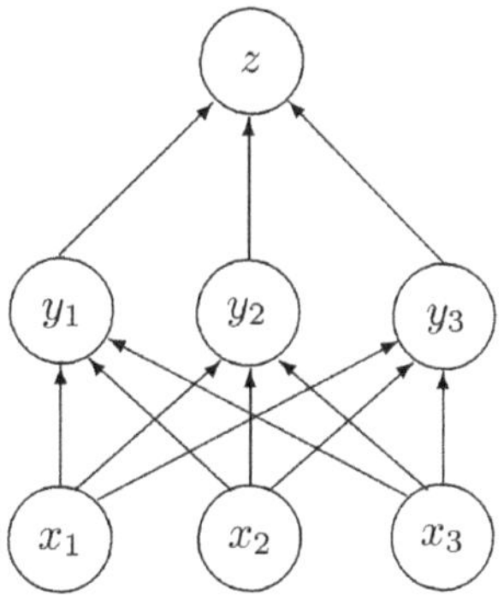

A linear rectified unit $\mathrm{ReLU}(a) = \max(0, a)$ is adopted as activation function for the hidden layer and the output layer. Given the loss function $L(z, t) = \frac{1}{2}(z - t)^2$ with t being the target value for learning. Denote by W_1 and W_2 weight matrices connecting input and hidden layer, and hidden layer and output respectively. They are initialized as follows:

$$W_1 = \begin{bmatrix} 1 & 0 & 1 \\ 0 & 1 & 0 \\ 1 & 0 & 1 \end{bmatrix}, \quad W_2 = \begin{bmatrix} 1 & -1 & 1 \end{bmatrix}.$$

Write out symbolically (no need to put any specific values of W_1 and W_2) the feedforward functional mapping $x \to z$ using ReLU, W_1, W_2. In other words, express y in terms of ReLU, W_1, W_2 and x.

7.9 Consider the network N below. The input layer of the network consists of three dimensions $x = [x_1, x_2, x_3]^T$ and the bias. The hidden layer includes three units $h = [h_1, h_2, h_3]^T$ and the bias. The output layer includes one unit y.

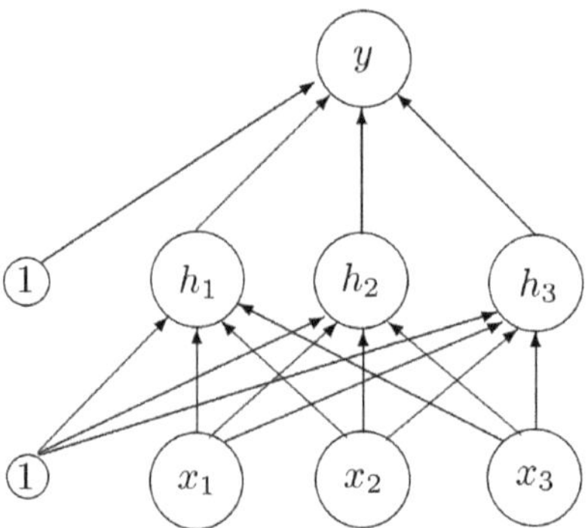

After training this network, the values of the weights (including the bias weights) are

$$W_1 = \begin{bmatrix} 1 & -1 & 0.5 & 1 \\ 0.5 & -1 & 0.2 & -1 \\ 1 & -0.5 & 1 & 1 \end{bmatrix}, \quad w_2 = \begin{bmatrix} 1 & 0.5 & 1 & -1 \end{bmatrix}^T.$$

(a) Suppose the same activation function σ has been adopted for all the hidden and output nodes. Write out symbolically (no need to plug in the specific values of W_1 and w_2 just yet) the feedforward functional mapping N : $x \mapsto y$ using σ, W_1, w_2. In other words, express y in terms of σ, W_1, w_2 and x.

(b) Write down the gradient of the loss function with respect to the weights. In other words, compute the following terms symbolically:

(i) The gradient with respect to w_2.

(ii) The gradient with respect to $\mathbf{W}_1$.

(c) Suppose you input a three-dimensional sample $x = [1, 3, 1]^T$ to the above network N. You tried four different types of activation functions $(\phi_1, \phi_2, \phi_3, \phi_4)$ (on all the hidden and output nodes) which respectively output the values $(y_1, y_2, y_3, y_4) = (0, \ 0.6818, \ -3.1500, \ -0.9963)$. What is a valid guess for the activation functions $(\phi_1, \phi_2, \phi_3, \phi_4)$?

7.10 Consider a binary classification task of classifying images as apple versus non-apple. You design a deep network with a single output neuron z for the learning. To make the binary decision, this output neuron goes through an activation given by $\hat{y} = \sigma(\text{ReLU}(z))$ where $\sigma(\cdot)$ denotes the sigmoid function and $\text{ReLU}(\cdot)$ denotes the linear rectified unit. Identify all correct setting(s) below to get full credit.

(a) We can set the target as $y \in \{-1, +1\}$ for apple and non-apple respectively and then classify the output as apple when $\hat{y} \geq 0$ and non-apple when $\hat{y} < 0$.

(b) We can set the target as $y \in \{0, 1\}$ for apple and non-apple respectively and then classify the output as apple when $\hat{y} \geq 0.5$ and non-apple when $\hat{y} < 0.5$.

(c) We can set the target as $y \in \{-1, +1\}$ for non-apple and apple respectively and then classify the output as non-apple when $\hat{y} \geq 0$ and apple when $\hat{y} < 0$.

(d) We can set the target as $y \in \{0, 1\}$ for non-apple and apple respectively and then classify the output as non-apple when $\hat{y} \geq 0.5$ and apple when $\hat{y} < 0.5$.

(e) None of these

Chapter 8
Ensemble Learning

8.1 Introduction

Ensemble learning stands as a pivotal methodology in machine learning, harnessing the collective power of multiple models to achieve superior predictive performance. This approach, rooted in the wisdom of crowds, posits that a group of models, when judiciously combined, can outperform any single model, no matter how sophisticated. Ensemble methods like *bagging* and *boosting* not only enhance prediction accuracy but also offer robust solutions to the perennial challenge of overfitting, making them indispensable in the toolkit of modern data scientists.

The following sections offer an in-depth exploration of ensemble learning methods, each focusing on distinct aspects. Section 8.2 introduces the basics and core concepts of ensemble learning, discussing how multiple models are combined in the ensemble framework. Section 8.3 explores the core principles of decision trees, establishing the essential framework that underpins the subsequent ensemble methods discussed throughout the chapter. Section 8.4 delves into Bagging (Bootstrap Aggregating), explaining its methodology and applications in reducing variance and avoiding overfitting. In Sect. 8.5, we explore Random Forest, discussing this popular modeling algorithm based on bagging. Section 8.6 covers Boosting, detailing its strategy for creating a powerful ensemble by focusing on errors of previous models. Section 8.7 examines AdaBoost, highlighting the insights behind this classical boosting algorithm. Section 8.8 presents Gradient-Boosted Trees, discussing their optimization strategies and improvements over Adaboost. Section 8.9 introduces XGBoost, emphasizing its novel approaches in optimizing individual decision trees. Section 8.10 discusses LightGBM and its innovative approaches to processing large data sets with high efficiency. Section 8.11 provides a comparative analysis of Bagging and Boosting, discussing their merits, differences, and suitable applications. Finally, Sect. 8.12 summarizes the various algorithms discussed in the chapter.

8.2 Fundamentals of Ensemble Learning

At its core, ensemble learning constructs a composite model from a set of base learners or models, often referred to as "weak learners" when individually their performance is limited, but collectively they form a "strong learner" capable of achieving superior accuracy. The fundamental premise is that by combining the predictions from multiple models, the ensemble can compensate for individual weaknesses, reduce the likelihood of overfitting, and handle the bias-variance trade-off more effectively than single models.

8.2.1 Framework for Combining Weak Learners

The process of combining weak learners in ensemble learning aims to construct a predictive model with enhanced accuracy and robustness compared to individual learners. We present a generalized mathematical framework that encapsulates the essence of both bagging and boosting techniques, offering a unified approach to aggregation for classification and regression.

General Aggregation Strategy

For $x \in \mathbb{R}^d$, the aggregation of weak learners, $f_m(x)$, where $m = 1, \ldots, M$, into a strong learner, $F(x)$, can be formalized as follows:

$$F(x) = \Phi \left(\sum_{m=1}^{M} \alpha_m f_m(x) \right), \tag{8.1}$$

where:

- Φ is an activation function that translates aggregated predictions into the final output, which can vary based on the task (classification or regression) and the specific ensemble method (boosting or bagging).
- α_m represents the weight of the m-th weak learner in the ensemble. In bagging, α_m is typically uniform, reflecting the equal contribution of each model, whereas in boosting, α_m is derived based on the performance or error of $f_m(x)$.

Classification

For **classification** tasks, Φ often embodies a decision rule:

- **Binary Classification**: $\Phi(z) = \mathrm{sign}(z)$, where the sign function determines the class based on the sign of the aggregated weighted sum.
- **Multi-Class Classification**: $\Phi(z) = \arg\max(z_k)$, where z_k represents the aggregated weighted votes or scores for class k, and the final prediction is the class with the highest score.

Regression

For **regression** tasks, Φ is typically the identity function, $\Phi(z) = z$, reflecting the direct use of the aggregated weighted sum as the prediction:

$$F(\boldsymbol{x}) = \sum_{m=1}^{M} \alpha_m f_m(\boldsymbol{x}),$$

emphasizing the linear combination of weak learners' predictions.

Determining Weights α_m

We will discuss two approaches to determine the weights:

- **Uniform**: Weights are uniform, $\alpha_m = \frac{1}{M}$. This is the standard approach used in *bagging*
- **Adaptive**: Weights are calculated based on the error reduction each weak learner contributes. For example, in the famous *AdaBoost* algorithm, $\alpha_m = \frac{1}{2} \ln\left(\frac{1-\epsilon_m}{\epsilon_m}\right)$, where ϵ_m is the error rate of the m-th model.

Side Note

A special case of the binary classification problem is outlined in *Condorcet's Jury Theorem*. It states that if each member of a jury has an independent probability p of making the correct decision, where $p > 0.5$, and if decisions are made by majority vote, then increasing the number of jurors increases the probability that the majority decision is correct. Moreover, as the jury size approaches infinity, the probability that the majority decision is correct approaches 1.

Choice of Weak Learners

In subsequent sections in this chapter, we will discuss the choice of weak learners in the context of bagging and boosting. In general, the choice of weak learners in ensemble methods hinges on a nuanced understanding of *bias* and *variance*, as these learners need to collectively overcome their individual limitations to form a robust model. For example, weak learners, typically simpler models with high bias and low variance, are foundational in boosting methods to iteratively correct errors, focusing on hard-to-predict samples. Mathematically, bias and variance are defined with precise statistical meanings, which we will elaborate on next.

8.2.2 The Bias-Variance Trade-Off

The bias-variance trade-off is a fundamental principle in statistical learning that provides insight into the behavior of predictive models. It decomposes the expected prediction error of a model into three distinct parts: bias, variance, and irreducible error, offering a framework to understand how different model complexities impact overall performance.

Bias and Variance: Definitions and Implications

Bias measures the error introduced by approximating the real-world problem, which may involve complex relationships, with a simpler model. High bias can lead to models that are too simplistic to capture the underlying structure of the data, a phenomenon known as underfitting. Mathematically, the bias of a model for a given input x is defined as

$$\text{Bias}[\hat{f}(x)] = \mathbb{E}[\hat{f}(x)] - f(x), \tag{8.2}$$

where $\hat{f}(x)$ is the model's prediction, $f(x)$ is the true function, and $\mathbb{E}[\hat{f}(x)]$ represents the expected prediction over different training sets. The squared bias, $(\text{Bias}[\hat{f}(x)])^2$, contributes to the overall error by quantifying how far the model's average prediction is from the true value.

Variance quantifies how much the predictions of a model vary for a given input x when trained on different subsets of the training data set. High variance can cause an algorithm to model random noise in the training data, leading to overfitting. The variance of a model is given by

$$\text{Variance}[\hat{f}(x)] = \mathbb{E}[(\hat{f}(x) - \mathbb{E}[\hat{f}(x)])^2], \tag{8.3}$$

indicating the spread of the model's predictions around their mean. Variance captures the model's sensitivity to the specific data it was trained on.

Prediction Error Decomposition

With bias and variance well-defined, we can now discuss the components of the general prediction error. For a model predictor $\hat{f}$, the expected prediction error at a point x can be decomposed as follows:

$$\text{Err}(x) = (\text{Bias}[\hat{f}(x)])^2 + \text{Variance}[\hat{f}(x)] + \sigma^2, \qquad (8.4)$$

where σ^2 represents the irreducible error, stemming from the noise in the data. This part of the error is not affected by the model's complexity. It is inherent in the problem itself due to noise or other factors outside the model.

The general learning goal is to minimize the reducible part of the error (bias and variance). Typically, as we increase the model complexity (like moving from linear models to high-degree polynomials), the bias decreases but the variance increases, and vice versa. This is known as the **Bias-Variance Trade-off**, which implies that a model cannot simultaneously have both low bias and low variance.

Thus, to achieve the lowest possible error, one must find an optimal balance between bias and variance. For this we will explore two pivotal ensemble strategies: Bagging and Boosting. These methods embody the ensemble learning philosophy by combining multiple weak learners to form a more accurate and robust model. Moreover, they offer nuanced approaches to navigating the bias-variance trade-off, enhancing model performance in distinct ways.

8.2.3 Bagging (Bootstrap Aggregating)

Bagging, short form for Bootstrap Aggregating, is an ensemble technique that aims to reduce variance and prevent overfitting. It involves creating multiple bootstrapped data sets from the original training data and training a separate model on each. The final prediction is obtained by averaging the predictions from all models (for regression) or by majority voting (for classification).

Mathematical Framework:

Given a training data set D of size N, bagging generates M bootstrapped data sets $\{D_1, D_2, \ldots, D_M\}$, each of size N, by sampling from D with replacement. A model f_m is trained on each data set D_m. The ensemble prediction for an input x is

- For regression:

$$\hat{f}_{\text{bag}}(x) = \frac{1}{M} \sum_{m=1}^{M} f_m(x)$$

- For classification:

$$\hat{f}_{\text{bag}}(x) = \text{sign}\left(\sum_{m=1}^{M} f_m(x)\right)$$

Bias-Variance Analysis:

Bagging primarily targets the reduction of variance. The aggregation of predictions from models trained on bootstrapped samples tends to cancel out the noise, as errors uncorrelated across models do not amplify when averaged. Mathematically, if the variance of each model's prediction is σ^2 and the predictions are uncorrelated, the variance of the bagged ensemble is $\frac{\sigma^2}{M}$, demonstrating a reduction in variance with an increasing number of models.

8.2.4 Boosting

Boosting is a sequential ensemble method designed to reduce both bias and variance. It iteratively trains models, with each model focusing on correcting the errors made by the previous models in the sequence. The final prediction is a weighted sum of all the models' predictions, where weights may reflect each model's accuracy or contribution to the ensemble's performance.

Mathematical Framework:

Boosting starts with assigning equal weights to all training instances. For each iteration m, a model f_m is trained, focusing on the instances that previous models misclassified. The ensemble prediction for an input x after M models have been trained is

$$\hat{f}_{\text{boost}}(x) = \sum_{m=1}^{M} \alpha_m f_m(x)$$

where α_m is the weight assigned to the m-th model, often determined by its performance (e.g., its error rate).

Bias-Variance Analysis:

Boosting can significantly reduce bias, as each successive model focuses on the hardest to predict instances, thereby correcting the ensemble's overall bias. Additionally, by carefully adjusting the weights α_m and focusing training on correcting previous errors, boosting can also manage variance, ensuring that the ensemble does not overfit the training data.

8.2.5 Section Summary

Bagging and Boosting exemplify the ensemble learning philosophy by demonstrating how multiple models, even if individually weak, can be combined to achieve superior predictive performance. Bagging reduces variance, making it particularly effective for high-variance models, while boosting addresses both bias and variance, offering a powerful tool for models suffering from either or both. In the following sections, we will discuss these two methodologies in greater detail and discuss several practical algorithms based on these methods.

8.3 Decision Trees

Before delving into the details of Bagging and Boosting, it is crucial to understand decision trees. All subsequent ensemble algorithms, Random Forest, Gradient-Boosted Trees, XGBoost, and LightGBM, all rely on decision trees as their underlying model, leveraging their simplicity and effectiveness to build more robust and accurate predictive systems.

8.3.1 Introduction to Decision Trees

Decision Trees are versatile machine learning algorithms used for both classification and regression tasks. They predict the value of a target variable by learning simple decision rules inferred from the data features.

Mathematical Formulation: A decision tree splits the data into subsets using a tree structure, where each internal node represents a "test" on an attribute, each branch represents the outcome of the test, and each leaf node represents a class label or a continuous outcome. The objective is to create a model that predicts the value of a target variable by learning simple decision rules from the data features.

The optimal decision tree is an NP-hard problem because finding the smallest tree that fits the data can be computationally expensive and infeasible for large data sets. Therefore, heuristic methods, such as the greedy algorithm, are used to approximate the solution by making locally optimal decisions at each step of the tree construction.

8.3.2 Building a Decision Tree: Split Criteria

The construction of a decision tree involves selecting the best attribute to split the data at each step, aiming to maximize the homogeneity of the resultant child nodes.

Split Criterion for Regression

For regression tasks in decision trees, the objective during tree construction is to minimize the total variance within the child nodes after a split. Variance in this context is defined as the average of the squared differences from the mean, calculated as:

$$\mathrm{Var}(\mathcal{N}) = \frac{1}{N} \sum_{i=1}^{N} (y_i - \bar{y})^2$$

where y_i represents the value of the target variable for each observation in the node $\mathcal{N}$, $\bar{y}$ is the mean of these values, and N is the total number of observations in the node. The variance reduction from a split is then calculated as follows:

$$\Delta\mathrm{Var} = \mathrm{Var}(\mathcal{N}) - \left(\frac{N_L}{N} \mathrm{Var}(\mathcal{N_L}) + \frac{N_R}{N} \mathrm{Var}(\mathcal{N_R}) \right),$$

where N, N_L, and N_R represent the number of observations in the parent node $\mathcal{N}$, left child node $\mathcal{N_L}$, and right child node $\mathcal{N_R}$, respectively. By choosing splits that maximize this variance reduction, the algorithm effectively creates subgroups that are more homogeneous in terms of the target variable, leading to better prediction accuracy in the resulting model.

Split Criterion for Classification

For classification tasks, two common measures are Gini impurity and entropy. The Gini impurity for a node $\mathcal{N}$ is defined as:

$$\mathrm{Gini}(\mathcal{N}) = 1 - \sum_{k} (p_k)^2,$$

where p_k is the proportion of samples in node $\mathcal{N}$ that belong to class k. The reduction in Gini impurity for a split is thus:

$$\Delta\mathrm{Gini} = \mathrm{Gini}(\mathcal{N}) - \left(\frac{N_L}{N} \mathrm{Gini}(\mathcal{N_L}) + \frac{N_R}{N} \mathrm{Gini}(\mathcal{N_R}) \right).$$

Entropy and the information gain for a node can be calculated as:

$$\mathrm{Entropy}(\mathcal{N}) = - \sum_{k} p_k \log_2(p_k),$$

$$\Delta\mathrm{Entropy} = \mathrm{Entropy}(\mathcal{N}) - \left(\frac{N_L}{N} \mathrm{Entropy}(\mathcal{N_L}) + \frac{N_R}{N} \mathrm{Entropy}(\mathcal{N_R}) \right).$$

8.3.3 Decision Tree Algorithm: Mathematical Representation

The decision tree algorithm builds the model in a hierarchical manner through recursive partitioning. It proceeds as follows:

1. **Function** BuildTree($D, depth$):

   - **If** D contains all samples of the same class or if $depth$ exceeds the maximum depth specified (stopping condition):

     - **Return** a leaf node with the majority class (in classification) or the average outcome (in regression), which represents the predicted value for that node.

   - **Else**:

     - **Find the Best Split**:

       · For each feature in D, evaluate potential splits.
       · Calculate the split criterion for each split. This could be Gini impurity, entropy for classification tasks, or variance reduction for regression tasks.
       · Choose the split that results in the highest gain, i.e., the greatest reduction in impurity or variance.

     - **Split the Data set** D into D_L and D_R based on the best split, where D_L contains all samples that meet the split condition and D_R contains the rest.
     - **Recursively Build the Tree**:

       · left_child $=$ BuildTree($D_L, depth + 1$)
       · right_child $=$ BuildTree($D_R, depth + 1$)

     - **Return** a node with the chosen split and branches to left_child and right_child.

   Additional Stopping Criteria: Apart from reaching a uniform class within a node or the maximum depth, other stopping criteria can include:

- **Minimum Samples Split**: The node will not be split if the number of samples in the node is less than the minimum specified.
- **Minimum Samples Leaf**: A split proposal will be rejected if it results in any leaf having fewer than the specified minimum number of samples.

This representation encapsulates the recursive and hierarchical nature of decision tree construction, emphasizing the importance of split criteria and stopping conditions to prevent overfitting and underfitting, balancing the complexity and depth of the tree for optimal predictive performance.

8.3.4 Advantages and Limitations

- **Advantages**:

  - **Interpretability**: Decision trees are easy to understand and interpret, making them useful in applications where decision logic needs to be explained.
  - **Flexibility**: They handle both numerical and categorical data and do not require normalization.
  - **Nonparametric**: There are no assumptions about the space distribution and the structure of the classifier.

- **Limitations**:

  - **Lack of Complexity**: At their core, decision trees are simple models and can struggle to capture complex behavioral patterns without extensive growth. This inherent simplicity often leads to models that cannot effectively represent all the nuances in complex data sets.
  - **Overfitting/High Variance**: If allowed to grow deep without constraints, decision trees can easily overfit the training data. This is because a fully grown tree can end up with leaves that represent very few samples, each tailored to the outliers or noise in the training data set, resulting in high variance.
  - **High Bias**: Conversely, when overly restricted (e.g., limited to shallow trees), decision trees exhibit high bias. This setup hinders their ability to model the underlying relationships accurately, failing to capture essential complexities in the data.

Bias-Variance Discussion Due to their structure, decision trees often do not inherently provide a good bias-variance tradeoff. Effective management of this tradeoff typically requires integrating decision trees into ensemble methods such as bagging and boosting, which leverage multiple trees to enhance performance and stability. These ensemble approaches help mitigate individual trees' limitations by averaging out biases, reducing variance, and improving the model's ability to generalize to unseen data.

8.3.5 Section Summary

Decision Trees are fundamental components used across various ensemble learning methods like Random Forest and Gradient-Boosted Trees. The methodology involves segmenting the data into branches to form predictions, which are simple to interpret and implement. These models operate by recursively selecting the best attribute to split the data, maximizing homogeneity in each resultant node. This binary splitting process continues until a stopping condition is met, such as achieving minimal node impurity or reaching a predefined tree depth.

While decision trees are highly interpretable and do not require data normalization, their simplicity can limit their ability to model complex patterns unless extensively tuned, often leading to overfitting or high variance when depth is not controlled. Conversely, overly simplified trees may not capture essential data relationships, exhibiting high bias. These limitations underscore the necessity of using ensemble methods to optimize the bias-variance tradeoff, enhancing predictive accuracy and model robustness by integrating multiple trees into a collective decision-making framework.

8.4 Bagging (Bootstrap Aggregating)

Bagging, or Bootstrap Aggregating, is a powerful ensemble learning technique designed to improve the stability and accuracy of machine learning algorithms by reducing variance and avoiding overfitting. The essence of bagging lies in its two-step process: bootstrap and aggregation, each underpinned by rigorous mathematical principles.

8.4.1 Bootstrap: The Foundation of Bagging

Bootstrap sampling is a statistical method for estimating quantities about a population by averaging estimates from multiple small data samples, ideally independent and identically distributed (i.i.d.), drawn with replacement from the original data set. This process creates "bootstrap samples", each serving as a distinct data set for training individual models within the ensemble.

Mathematical Framework

Given a data set D of size N, a bootstrap sample D_i is generated by randomly selecting N observations from D with replacement. After M bootstrap samples are generated, we have a collection $\{D_1, D_2, \ldots, D_M\}$.

The key property of bootstrap samples is their ability to mimic the distribution of the original data set, allowing for the estimation of true population parameters with high accuracy and precision. In particular, let $\hat{\theta}$ be an estimate of a true population parameter θ based on a bootstrap sample, then the variance of the estimator $\hat{\theta}$ can be expressed as

$$\mathrm{Var}(\hat{\theta}) = \frac{1}{M} \sum_{i=1}^{M} (\hat{\theta}_i - \hat{\theta}_{avg})^2,$$

where $\hat{\theta}_i$ is the estimate from the i-th bootstrap sample, and $\hat{\theta}_{avg}$ is the average of these bootstrap estimates. Thus, as we increase the number of bootstraps M, the variance decreases, which implies an increase in *precision*.

8.4.2 Aggregation: Combining Predictions Analytically

The aggregation stage in bagging involves synthesizing the predictions from models trained on bootstrap samples into a single prediction. This can be done through simple averaging or a linearly weighted combination, depending on the specific goals and the nature of the problem.

Variance Reduction Through Averaging

Consider an ensemble of M models $\{f_1(x), f_2(x), \ldots, f_B(x)\}$, each trained on a bootstrap sample of the original data set. Assuming these models are unbiased estimators of the true function $f(x)$, the expected prediction for any model $f_i(x)$ is equal to the true value, i.e., $\mathbb{E}[f_i(x)] = f(x)$. The variance of the aggregated prediction $\hat{f}_{\mathrm{bag}}(x)$, when models are equally weighted, is

$$\mathrm{Var}(\hat{f}_{\mathrm{bag}}(x)) = \mathrm{Var}\left(\frac{1}{M} \sum_{i=1}^{M} f_i(x)\right).$$

Expanding this, considering the independence of models, we get

$$\mathrm{Var}(\hat{f}_{\mathrm{bag}}(x)) = \frac{1}{M^2} \sum_{i=1}^{M} \mathrm{Var}(f_i(x)) = \frac{1}{M^2} \cdot M \cdot \sigma^2 = \frac{\sigma^2}{M}. \qquad (8.5)$$

This derivation shows that the variance of the bagged prediction decreases inversely with the number of models M. Hence, as more models are aggregated, the ensemble's prediction becomes more stable, effectively reducing the impact of variance.

Variance Reduction with Correlated Models

It is often difficult to achieve independence of models. Assuming that each model has a variance σ^2 and the correlation between any two different models $f_i(x)$ and $f_j(x)$ is ρ for $i \neq j$, the derivation of the variance of $\hat{f}_{\text{bag}}$ proceeds as follows: The variance of the average prediction can be expressed in terms of the variances and covariances of individual models:

$$\text{Var}(\hat{f}_{\text{bag}}(x)) = \frac{1}{M^2} \text{Var}\left(\sum_{i=1}^{M} f_i(x)\right)$$

The variance of the sum of correlated variables includes contributions from the variances of each term and the covariances between every pair of different terms:

$$\text{Var}\left(\sum_{i=1}^{M} f_i(x)\right) = \sum_{i=1}^{M} \text{Var}(f_i(x)) + \sum_{i \neq j} \text{Cov}(f_i(x), f_j(x))$$

Substituting $\text{Var}(f_i(x)) = \sigma^2$ and $\text{Cov}(f_i(x), f_j(x)) = \rho\sigma^2$, we find:

$$\text{Var}\left(\sum_{i=1}^{M} f_i(x)\right) = M\sigma^2 + M(M-1)\rho\sigma^2$$

Plugging the expression from above back into the formula:

$$\text{Var}(\hat{f}_{\text{bag}}(x)) = \frac{1}{M^2}\left(M\sigma^2 + M(M-1)\rho\sigma^2\right) = \frac{\sigma^2}{M} + \frac{M-1}{M}\rho\sigma^2 \tag{8.6}$$

This result demonstrates that while the variance component $\frac{\sigma^2}{M}$ decreases with an increasing number of models, the component $\frac{M-1}{M}\rho\sigma^2$ grows with the correlation ρ. This highlights a crucial aspect of bagging: the effectiveness of variance reduction is compromised as the correlation between the base learners increases. Therefore, it is essential for the individual models to maintain as much independence from each other as possible to maximize the benefits of variance reduction through bagging. Ensuring low correlation among models helps in achieving greater diversity in the ensemble, thereby enhancing the ensemble's overall performance and reliability. We will later see that one of the important features of the Random Forest Algorithm is specifically designed to reduce correlation between base models.

Impact on Bias

While the primary effect of aggregation in bagging is variance reduction, it's essential to consider its impact on bias. Since the models are assumed to be unbiased and the aggregation does not introduce any systematic shift away from the true function $f(x)$, the bias of the aggregated prediction remains unchanged. Mathematically, the bias of the bagged estimator is

$$\text{Bias}(\hat{f}_{\text{bag}}(x)) = \mathbb{E}[\hat{f}_{\text{bag}}(x)] - f(x) = \frac{1}{M}\sum_{i=1}^{M}(\mathbb{E}[f_i(x)] - f(x)) = 0$$

assuming $\mathbb{E}[f_i(x)] = f(x)$ for all i. Note that if each individual estimator has a bias, the above equation shows that the bias of the overall model is simply the average of the individual biases.

Out-of-Bag Error Estimation

In bagging, the Out-of-Bag (OOB) error serves as a convenient metric for model validation without the need for a dedicated test set. During the training of each model in the ensemble, a portion of instances is left out of the bootstrap sample these instances are termed "out-of-bag". The OOB error is calculated by making predictions on each instance using only those models for which the instance was out-of-bag.

To calculate the OOB error for the entire ensemble, we proceed as follows:

1. For each instance i, aggregate the predictions from all models where i was out-of-bag. For regression tasks, this involves calculating the average of the predictions. For classification tasks, this typically means selecting the class that was predicted most frequently.
2. Compare these aggregated predictions $\hat{y}_i^{\text{oob}}$ to the actual values y_i.
3. Compute the overall error rate as the proportion of instances where the aggregated out-of-bag prediction does not match the actual value:

$$\text{OOB Error} = \frac{1}{N}\sum_{i=1}^{N}\mathbf{1}(y_i \neq \hat{y}_i^{\text{oob}})$$

where $\mathbf{1}$ is an indicator function that outputs 1 when the predicted value does not match the actual value and 0 otherwise.

This approach leverages the inherent cross-validation mechanism of bagging, as each instance is typically left out of the training set for several models, allowing it to be used effectively as a test case. The OOB error thus provides a robust estimate of model performance, reflecting how well the ensemble can generalize to new data based on internal validation.

8.4.3 Diverse Techniques in Bootstrapping and Aggregation

In general, there are several the strategies for bootstrapping and aggregation. These techniques not only form the basis for model training and prediction but also significantly influence the ensemble's performance. We outline the various bootstrapping and aggregation methods in this section.

Representative Bootstrapping Techniques

1. **Simple Bootstrap**
   The most straightforward bootstrapping approach involves randomly selecting samples from the data set with replacement, ensuring each sample has an equal probability of being selected. Given a data set D of size N, a simple bootstrap sample D_i can be generated as follows:

   $$D_i = \{x_1^*, x_2^*, \ldots, x_N^*\},$$

   where each x_j^* is drawn randomly from D with replacement.
2. **Stratified Bootstrap**
   Stratified bootstrapping ensures that each bootstrap sample maintains the original distribution of groups or strata within the data set. This is particularly useful in classification tasks with imbalanced classes.
3. **Poisson Bootstrapping**
   Poisson bootstrapping uses a Poisson distribution to determine the inclusion of each observation in a bootstrap sample. For an observation x_j in the data set D, the number of times x_j appears in the bootstrap sample D_i is a Poisson random variable N_j with mean parameter $\lambda = 1$:

   $$N_j \sim \text{Poisson}(1)$$

   The bootstrap sample D_i is then composed of each x_j included N_j times:

   $$D_i = \bigcup_{j=1}^{N} \{x_j^{(N_j)}\}$$

4. **Bayesian Bootstrapping**
   Bayesian bootstrapping assigns random weights to each observation, drawing these weights from a specified distribution, typically a uniform or exponential distribution. For an observation x_j in the data set D, the weight w_j associated with x_j is

   $$w_j \sim \text{Exp}(1) \quad \text{or} \quad w_j \sim \text{Uniform}(0, 1)$$

The resampled data set D_i is constructed by sampling from D with replacement according to the probabilities proportional to these weights. Each $\boldsymbol{x}_j$ is included in the bootstrap sample with a probability proportional to w_j.

Representative Aggregation Methods

1. **Simple Averaging**
   For regression problems, simple averaging is often employed, where the ensemble prediction is the mean of all model predictions:

$$\hat{f}_{\text{ens}}(\boldsymbol{x}) = \frac{1}{M} \sum_{i=1}^{M} f_i(\boldsymbol{x}).$$

2. **Weighted Averaging**
   Weighted averaging assigns different weights to different models based on their performance, allowing for more flexibility in leveraging model strengths:

$$\hat{f}_{\text{ens}}(\boldsymbol{x}) = \sum_{i=1}^{M} \alpha_i f_i(\boldsymbol{x}),$$

   where α_i is the weight for the i-th model.
3. **Aggregating with Differential Weights**
   This method involves dynamically adjusting weights based on real-time performance metrics or the specific characteristics of the input data, optimizing the ensemble's adaptability:

$$\hat{f}_{\text{ens}}(\boldsymbol{x}) = \sum_{i=1}^{M} \alpha_i(\boldsymbol{x}) f_i(\boldsymbol{x}),$$

   where $\alpha_i(\boldsymbol{x})$ represents the weight for the i-th model, which can vary with the input x.
4. **Stacking (Stacked Generalization)**: Stacking extends aggregation by training a meta-model to combine the predictions of base models, offering a powerful method for prediction synthesis:

$$\hat{f}_{\text{ens}}(\boldsymbol{x}) = f_{\text{meta}}(f_1(\boldsymbol{x}), f_2(\boldsymbol{x}), \ldots, f_M(\boldsymbol{x})).$$

Section Summary

Bagging (Bootstrap Aggregating) is an ensemble learning technique designed to enhance the stability and accuracy of machine learning models by reducing variance and preventing overfitting. It operates on two main principles: bootstrapping and aggregation. Bootstrapping involves generating multiple independent and identically distributed samples from the original data set, each of which trains individual models within the ensemble. This method helps accurately estimate population parameters by reducing variance as the number of bootstrap samples increases.

During the aggregation phase, predictions from these models are combined, typically through averaging, to produce a single consensus prediction. This process effectively reduces variance while maintaining bias neutrality. However, the correlation among models can reduce the efficacy of variance reduction, emphasizing the importance of maintaining low model correlation to maximize bagging's benefits. Bagging thus balances diversity in model training with precision in predictive performance, making it valuable in contexts where robust and accurate predictions are crucial.

8.5 Random Forest

Random Forest is a hallmark of ensemble learning leveraging the bagging framework. It is renowned for its robustness and versatility across a wide array of machine learning tasks. By integrating multiple decision trees to form a "forest", Random Forest leverages the strengths of individual trees while mitigating their weaknesses through aggregation, thereby achieving superior prediction accuracy and generalization.

8.5.1 *Mathematical Formulation*

The core of the Random Forest algorithm lies in creating an ensemble of decision trees trained on bootstrapped subsets of the training data and using random subsets of features at each split. This process can be mathematically represented as follows:

1. **Bootstrap Sampling**: For each tree T_m, where $m = 1, \ldots, M$, a bootstrap sample Z_m^* is drawn from the original data set D. Given a data set D of size N, each bootstrap sample is generated by randomly selecting N observations from D with replacement, thus allowing some observations to appear more than once in the same bootstrap sample.

2. **Tree Growth**: Each decision tree T_m is grown using its respective bootstrap sample Z_m^*. The growth of a tree involves recursively partitioning the data based on feature splits that best separate the target variable. The process is detailed as follows:

- At each decision node, rather than examining all possible features, a random subset of k features is selected out of the total p features. This randomness helps in making the trees less correlated.
- The best split on these k features is determined by the criterion specific to the task (regression or classification). For regression, the split might aim to minimize the variance within the nodes. For classification, criteria like Gini impurity or information gain are often used.
- This process is repeated, splitting the node into two child nodes, until a stopping criterion is met. This could be a maximum depth of the tree, a minimum number of samples required to split a node, or a minimum number of samples required to be at a leaf node.

Random Forest Prediction

For a regression task, the Random Forest prediction for a new input x is the average of the predictions from all individual trees:

$$\hat{f}_{\text{RF}}(x) = \frac{1}{M} \sum_{m=1}^{M} T_m(x),$$

where $T_m(x)$ is the prediction of the m-th tree for input x.

For classification tasks, the final prediction is made by majority voting among the predictions of all trees.

8.5.2 Use of Decision Trees

Decision trees form the backbone of Random Forests due to their unique properties, which are complemented by the ensemble approach to mitigate their limitations. The key properties can be summarized as the following.

1. **Inherent Simplicity and Versatility**
   Decision trees partition the feature space into distinct regions through a series of decisions, making them simple yet versatile models. Mathematically, this process can be represented as a series of conditional splits that guide an input x through the tree from the root to a leaf:

$$x \in R_1 \rightarrow x \in R_2 \rightarrow \ldots \rightarrow x \in R_{\text{leaf}}$$

where each R_i denotes a region of the feature space. This model structure allows decision trees to handle both numerical and categorical data and model complex, nonlinear relationships.

2. **Reduction of Correlation Between Trees**

   An essential aspect of why decision trees are suitable for Random Forests is the method's ability to reduce the correlation between trees. In a standard bagging approach, if there are strong predictors in the data set, most trees will use these predictors near the root, making them correlated. Random Forest combats this by using a random subset of k features out of the total p features for splitting at each node, which mathematically can be expressed as

   $$\mathrm{Var}(\hat{f}_{\mathrm{RF}}) < \mathrm{Var}(\hat{f}_{\mathrm{tree}})$$

   because the correlation between any two trees T_i and T_j in the forest, $\rho(T_i, T_j)$ is less than 1, reducing the overall variance of the ensemble prediction.

3. **Complex Decision Boundaries**

   Decision trees naturally partition the feature space into distinct regions based on a series of conditional decisions, allowing them to capture nonlinear relationships and interactions between variables. In a Random Forest, each tree explores different aspects of the data due to the randomness introduced in feature selection and bootstrap sampling. This results in an ensemble where the integrated outputs of various trees can adapt to intricate patterns in the data more flexibly than other model types. By aggregating these diverse trees, Random Forest achieves a granularity in decision-making that is nuanced and tailored to complex data sets, significantly enhancing the ensemble's predictive accuracy and robustness.

4. **Scalability and Parallelizability**

   The computational efficiency of decision trees, combined with the independent nature of trees in a Random Forest, allows for parallel training processes. This feature makes Random Forest highly scalable and effective for large data sets and feature spaces.

5. **Handling High Variance Through Ensemble Methods**

   A key challenge with decision trees is their tendency to overfit, exhibiting high variance. We will discuss how Random Forest effectively reduces variance in the next subsection.

8.5.3 *Impact on Bias and Variance*

The beauty of Random Forest lies in its dual approach to reducing variance without a commensurate increase in bias.

Impact on Variance

Random Forests excel in reducing the variance component of the error, which reflects the model's sensitivity to the specific training data it is trained on. High variance models, such as individual decision trees, can overfit the training data, capturing noise as if it were signal, thus performing poorly on unseen data.

The reduction in variance in Random Forests is achieved through:

1. **Bagging** Training each tree on a bootstrap sample introduces diversity among the trees, as each tree sees a slightly different subset of the data. The model's final prediction is typically the average (regression) or majority vote (classification) of all trees. If σ^2 represents the variance of a single tree and ρ the correlation between any two trees' predictions, we have seen that the variance of the forest can be expressed as

$$\mathrm{Var}(\hat{f}_{\mathrm{RF}}) = \rho\sigma^2 + \frac{1-\rho}{M}\sigma^2$$

   where M is the number of trees. Increasing M reduces the overall variance due to the diminishing effect of the second term.
2. **Feature Subsampling**: Random selection of a subset of features for each split further ensures that trees are less correlated (reducing ρ), contributing to variance reduction.

Impact on Bias

Decision trees inherently exhibit low bias and high variance. This characteristic arises because they make minimal assumptions about the structure of the data distribution and can fit highly complex models.

In a Random Forest, each tree is developed by training on a random subset of the data, which might suggest a potential increase in bias due to less comprehensive data coverage per tree. However, the typically deep and fully grown trees within a Random Forest compensate for this effect, essentially preserving the low-bias nature of a single decision tree. Moreover, the introduction of randomness in selecting features and bootstrap samples does not significantly elevate the bias. Instead, it allows the ensemble to robustly capture complex patterns without a substantial bias increase. Thus, while the ensemble approach might slightly elevate the bias compared to a single tree, it remains relatively low, ensuring the model's ability to generalize well across diverse data scenarios.

8.5.4 *Practical Implementation*

Implementing Random Forest effectively requires careful consideration of several hyperparameters and practical aspects:

- **Number of Trees** (M): Increasing M generally improves model performance up to a point, beyond which gains diminish. A larger number of trees can provide a more stable and accurate prediction but at the cost of increased computational resources.
- **Number of Features at Each Split** (k): This parameter controls the degree of feature subsampling. A lower value for m increases tree diversity but can raise bias; a higher value reduces bias but can make trees more correlated.
- **Minimum Node Size** (n_{min}): Setting a minimum size for tree leaves can help prevent overfitting by stopping tree growth when a node reaches a specified number of instances.
- **Feature Importance**: Random Forest can assess the importance of each feature in prediction, offering insights into the data. Features contributing most to the model's predictive power can be identified, informing feature selection and engineering.
- **Handling Missing Values**: Random Forest can inherently handle missing values either by imputation or by using surrogate splits, making it robust to incomplete data.
- **Parallelization**: The independent nature of tree construction in Random Forest makes it highly amenable to parallel computation, significantly speeding up training on large data sets.

8.5.5 Section Summary

Random Forest is a robust ensemble learning technique that enhances the accuracy and generalization of decision trees through a bagging framework. It constructs a "forest" of decision trees, each trained on bootstrapped data subsets with random feature selection at each split, effectively reducing the high variance often associated with individual trees while maintaining their low bias.

The methodology involves growing each tree from a unique subset of data, using a random subset of features for determining optimal splits based on task-specific criteria minimizing variance in regression and optimizing Gini impurity or entropy in classification. This ensemble approach helps mitigate overfitting by averaging numerous tree predictions, which smooths out noise and reduces model sensitivity to variations in any single data sample. Highly scalable and adaptable, Random Forest performs well across complex data structures and large data sets, making it a versatile choice for both regression and classification tasks.

8.6 Boosting

Boosting is an ensemble learning method designed to improve the stability and accuracy of machine learning algorithms. It achieves this by sequentially applying a base learning algorithm to repeatedly modified versions of the data, and then combining these through a weighted aggregation to produce a final model. The base learners are typically weak, meaning that their performance is slightly better than random guessing.

8.6.1   General Concept

The general boosting approach can be summarized as below:

1. **Initialization**: Start with training data $D = \{(x_1, y_1), (x_2, y_2), \ldots, (x_N, y_N)\}$, where x_i are features and y_i are labels, and set an equal initial weight for each instance:
$$w_i^{(1)} = \frac{1}{N}, \quad \text{for } i = 1, 2, \ldots, N.$$

2. **Iterative Learning**:

    (a) **For each iteration** $m = 1$ **to** M:
    (b) **Fit Model**: Train a weak learner $h_m(x)$ using the weighted samples $(x_i, y_i, w_i^{(m)})$.
    (c) **Compute Error**: Calculate the error of h_m not in terms of prediction accuracy, but through a general loss function that captures the performance under the current distribution $w^{(m)}$:

$$\epsilon_m = \frac{\sum_{i=1}^{N} w_i^{(m)} \mathbf{1}(y_i \neq h_t(x_i))}{\sum_{i=1}^{N} w_i^{(m)}},$$

    where $\mathbf{1}(\cdot)$ is the indicator function.
    (d) **Update Weights**: Adjust weights for the next round, where the weight change mechanism is determined by the boosting algorithms strategy, generally increasing weights for misclassified instances to focus subsequent learning on them:
$$w_i^{(m+1)} = w_i^{(m)} \times \text{adjustment factor},$$

    with normalization to ensure they sum to one.

3. **Final Model Construction**:

    (a) Combine the weak learners into a final model, which aggregates their decisions to classify new instances. The combination is typically a weighted sum:
$$F(x) = \sum_{m=1}^{M} \alpha_m h_m(x),$$

    where α_m are calculated based on the individual learner's accuracy or another relevant metric.

Theoretical Insights

The theory behind boosting involves enhancing the focus on those instances that previous learners misclassified. Each iteration's goal is not just to correct mistakes but to find an optimal boundary that divides classes by effectively reweighting the training instances. The sequential application of this strategy tends to reduce bias and variance as we will discuss at the end of this section.

8.6.2 Gradient Boosting

Gradient Boosting is a widely used form of boosting that focuses on minimizing a differentiable loss function through the sequential addition of weak learners, commonly decision trees. Unlike other boosting methods that primarily adjust the distribution weights of training instances, gradient boosting directly targets the residuals of the previous learners, enhancing precision in both regression and classification. This method's effectiveness stems from its ability to perform optimization directly within the function space, making it a powerful tool for tackling complex predictive tasks.

Mathematical Formulation of Gradient Boosting

Gradient Boosting constructs an additive model in a forward stage-wise manner, systematically minimizing the residuals of previous models. This model development is not merely an optimization of parameters but an optimization within the function space itself. The general algorithm can be described as below.

Given a training data set $D = \{(x_1, y_1), (x_2, y_2), \ldots, (x_N, y_N)\}$ with x_i as input features and y_i as targets, the goal of Gradient Boosting is to approximate a function $F(x)$ that minimizes the expected loss $L(y, F(x))$:

1. **Initialization**: Initialize with a constant model that approximates the target values well across the data set:

$$F_0(x) = \arg\min_\gamma \sum_{i=1}^{N} L(y_i, \gamma).$$

2. **Iterative Model Enhancement**: For each iteration $m = 1$ to M:

   (a) **Compute Negative Gradients**: Calculate the negative gradient of the loss function at each point, which points to the steepest ascent in error reduction:

$$r_i^{(m)} = -\left[\frac{\partial L(y_i, F_{m-1}(x_i))}{\partial F_{m-1}(x_i)}\right]_{F(x)=F_{m-1}(x)}, \tag{8.7}$$

   these $\{r_i^{(m)}\}$ are called pseudo-residuals.

(b) **Fit a Weak Learner**: Train a weak learner $h_m(x)$ to model these pseudo-residuals.

(c) **Optimize Step Size**: Determine the optimal scale γ_t to apply to $h_m(x)$ to best reduce the overall loss:

$$\gamma_m = \arg\min_{\gamma} \sum_{i=1}^{N} L(y_i, F_{m-1}(x_i) + \gamma h_m(x_i)).$$

(d) **Update Model**: Update the model by adding the scaled weak learner to the existing model:

$$F_m(x) = F_{m-1}(x) + \gamma_m h_m(x).$$

3. **Composite Model Construction**: The final model is a composite of all weak learners adjusted and scaled optimally:

$$F_M(x) = F_0(x) + \sum_{m=1}^{M} \gamma_m h_m(x).$$

Regularization in Gradient Boosting

Regularization techniques are crucial in managing model complexity and preventing overfitting in gradient boosting. These techniques include the use of shrinkage (learning rate) and stochastic gradient boosting, both of which enhance the robustness and generalizability of the model.

- **Shrinkage (Learning Rate)**: Shrinkage, often represented by the learning rate v, is used to scale the contribution of each weak learner in the ensemble. By multiplying the output of weak learner by v, typically a small number between 0.01 and 0.3, the model's learning is slowed down. This deliberate pacing prevents the model from fitting too closely to the training data's idiosyncrasies, thus reducing the risk of overfitting. The mathematical representation is

$$F_m(x) = F_{m-1}(x) + v\gamma_m h_m(x), \tag{8.8}$$

where $F_m(x)$ is the model at iteration m, $h_m(x)$ is the weak learner added at iteration m, γ_m is the optimization step size calculated for each weak learner, and v controls the overall step size in the direction suggested by $h_m(x)$.
- **Stochastic Gradient Boosting**: This variation of gradient boosting introduces a further element of randomness into the training process, akin to the subsampling seen in bagging techniques. In stochastic gradient boosting, each tree is trained on a random subset of the training data, rather than the full data set. Typically, about 50% to 80% of the training data is sampled (without replacement) to train

each tree. This approach not only speeds up the training process by reducing the amount of data processed at each step but also helps in reducing overfitting by increasing the diversity within the ensemble. The randomness helps in ensuring that the boosting algorithm does not concentrate too heavily on any single aspect or pattern in the training data, promoting a more generalized model.

Optimization in Function Spaces Explained

Gradient Boosting's iterative updates can be interpreted as gradient descent within a space of functions. Each function added is selected not just to reduce the loss, but to do so in the most effective direction available in the function space at that step.

- **Functional Gradient Descent**: The concept of updating $F_{m-1}(x)$ to $F_m(x)$ by adding $\gamma_m h_m(x)$ mimics the gradient descent step in function spaces. Here, $h_m(x)$ represents the direction of the steepest descent in the loss landscape, and γ_m is the step size akin to a learning rate in parameter optimization.
- **Minimization Process**: The minimization process involves finding a function $h_m(x)$ at each iteration that best addresses the residual errors of $F_{m-1}(x)$, guided by the loss function's gradient. The function space here consists of potential weak learners that could be added to the model. Selecting $h_m(x)$ and scaling it by γ_m is analogous to selecting the optimal direction and size of a step in a multidimensional space.
- **Loss Function and Space Mapping**: Each addition of $h_m(x)$ effectively transforms the space in which the function $F(x)$ operates, aiming to mold the cumulative model $F_M(x)$ into the contour of the loss function's minimum. This transformational approach allows for a dynamic fitting process that adapts to the complexities of the data and the specific characteristics of the loss function.

Applications to Regression and Classification

Gradient Boosting tailors its approach to both regression and classification through the choice of loss function and model updates.

- **Regression**: For regression tasks, Gradient Boosting typically uses the squared error loss function.

$$L(y, F(x)) = (y - F(x))^2.$$

- **Classification**: In classification, Gradient Boosting often utilizes the logistic loss, which is effective for binary outcomes.

$$L(y, F(x)) = \log(1 + \exp(-y F(x))).$$

Impact on Bias and Variance

Boosting sequentially focuses on reducing the errors of weak learners by adjusting weights, thereby impacting both bias and variance in specific ways:

- **Bias Reduction**: Each iteration of boosting adjusts the weights of incorrectly predicted instances to focus subsequent learners on these harder cases. This iterative correction reduces bias by enabling the ensemble to better approximate complex patterns in the data. Mathematically, let $F(x)$ be the final model and $h_m(x)$ the weak learner at iteration t. If $F_{m-1}(x)$ is the model before adding $h_m(x)$, then the model update equation is

$$F_m(x) = F_{m-1}(x) + \gamma_m h_m(x),$$

  where γ_m optimizes the contribution of $h_m(x)$ to minimize the ensemble's overall error. This focused correction enables the model to capture the data dynamics more accurately, thereby systematically reducing bias across iterations.
- **Variance Management**: The increase in complexity with each added learner could potentially raise the model's variance. However, boosting controls this by applying a learning rate v alongside the optimized step size γ_m for each tree, both of which act as forms of regularization. The learning rate v slows the overall learning process, while γ_m is optimized to ensure each tree's contribution is just right to reduce the loss without unnecessarily increasing variance. The cumulative impact of the learners is moderated, maintaining a balance in variance. The variance of the boosting model can be expressed as

$$\mathrm{Var}(F_m(x)) \approx \mathrm{Var}(F_{m-1}(x)) + (v\gamma_m)^2 \mathrm{Var}(h_m(x)),$$

  indicating that each additional learner contributes to the variance controlled by the square of the product of the learning rate and the optimal step size, thereby limiting the overall increase in variance.

Through these mechanisms, boosting efficiently decreases bias while managing variance, enhancing model accuracy and generalizability.

8.6.3 Section Summary

Boosting employs multiple weak learners in a sequential manner to focus on instances that previous learners misclassified. This method assigns and adjusts weights iteratively, improving the ensembles ability to minimize errors and enhance accuracy across iterations. The process not only refines predictions but also targets reductions in both bias and variance effectively.

Gradient Boosting builds on this by using a differentiable loss function for more precise optimization. It iteratively fits models to the steepest gradient of the loss function, allowing each addition to directly address the most significant errors left by the preceding models. This approach is especially effective in handling complex data patterns in both regression and classification tasks, optimizing performance through strategic updates in the function space.

8.7 AdaBoost

AdaBoost, short for Adaptive Boosting, is a seminal algorithm in the history of machine learning, demonstrating the power of boosting methods for creating robust classifiers from simple base learners. Developed in 1995 by Yoav Freund and Robert Schapire, it has significantly influenced the development of more sophisticated boosting algorithms such as the aforementioned Gradient Boosting.

8.7.1 Historical Context and Theoretical Framework

AdaBoost was one of the first algorithms to effectively use boosting to improve learning accuracy. It focuses on adaptively adjusting the weights of incorrectly classified instances, allowing subsequent learners to focus more on challenging cases. This adaptive mechanism is framed within an exponential loss function:

$$L(y, F(x)) = e^{-yF(x)}, \tag{8.9}$$

where $y \in \{-1, +1\}$ are class labels, and $F(x)$ is the model's prediction. This exponential loss function particularly emphasizes the classification errors by exponentially increasing the penalty for misclassifications. As $yF(x)$ becomes negative, indicating a disagreement between the actual and predicted labels, the loss exponentially increases. This sharp increase in penalty ensures that AdaBoost aggressively focuses on instances that previous iterations have misclassified, driving the algorithm to prioritize correcting these errors in subsequent iterations.

AdaBoost can be seen as a specialized case of Gradient Boosting configured specifically to minimize this exponential loss, which simplifies the learning problem to updating weights and combining weak learners.

8.7.2 AdaBoost Algorithm

The AdaBoost algorithm iteratively adjusts the weights of training instances to focus on those most difficult to classify in previous rounds. The details are as follows.

1. **Initialization**: Assign initial equal weights to all training instances, $w_{i,1} = 1/N$ for each $i = 1, 2, \ldots, N$, where N is the total number of training samples.
2. **Iterative Training**: For each iteration $m = 1$ to M:

   (a) **Weak Learner Training**: Fit a weak learner $h_m(x)$ using the weighted data $(x_i, y_i, w_{i,m})$.
   (b) **Error and Learner Weight Calculation**: Compute the weighted error of $h_m(x)$ and determine its influence in the final model:

   $$\epsilon_m = \sum_{\substack{i=1 \\ h_m(x_i) \neq y_i}}^{n} w_{i,m}, \quad \alpha_m = \frac{1}{2} \log\left(\frac{1 - \epsilon_m}{\epsilon_m}\right).$$

   (c) **Weight Update**: Adjust the weights for the next iteration to focus on misclassified instances:

   $$w_{i,m+1} = w_{i,m} \cdot e^{-y_i \alpha_m h_m(x_i))},$$

   and renormalize them so that the total weight of all instances remains 1.

3. **Final Model Construction**: Aggregate the contributions of all weak learners to form the final prediction model:

$$F(x) = \text{sign}\left(\sum_{m=1}^{M} \alpha_m h_m(x)\right).$$

This structured approach ensures that each weak learner focuses more on the instances that were harder to classify in the previous round, thereby making the overall ensemble progressively better at classifying the training data.

8.7.3 Applications in Multi-class Classification and Regression

- **Classification**: As shown above, the original AdaBoost algorithm was designed for binary classification, but there are variants that extend it to handle multiple classes. One of the most well-known methods for this is *AdaBoost.M1*. While AdaBoost.M1 is a straightforward extension for multi-class classification, it might not perform as well as more modern techniques like Gradient-Boosted Trees or Random Forests, which inherently handle multi-class problems more naturally and often with better generalization. We will not discuss the details of AdaBoost.M1.
- **Regression**: AdaBoost was originally designed for classification problems, particularly binary classification. However, it can be adapted for use in regression tasks;

this adaptation is commonly referred to as *AdaBoost.R2*. Although AdaBoost.R2 is an extension of AdaBoost for regression, it is less commonly used than other boosting methods tailored for regression, such as Gradient-Boosted Trees. We will not discuss the details of AdaBoost.R2.

8.7.4 Utilizing Tree Stumps as Weak Learners in AdaBoost

Tree stumps, which are decision trees with a depth of only one split, serve as the fundamental building blocks in the AdaBoost algorithm. These simple models, although limited in their complexity, play a pivotal role in AdaBoost's approach to ensemble learning.

Introduction to Tree Stumps: A tree stump makes a single decision based on one feature and its threshold, effectively dividing the data set into two groups. Due to their simplicity, tree stumps inherently have high bias but very low variance, as they consider only one feature and make broad generalizations about the data.

Why Use Tree Stumps? AdaBoost utilizes tree stumps for several strategic reasons:

- **Focus on Error Correction**: The core mechanism of AdaBoost involves sequentially correcting the errors of previous learners. Tree stumps, by focusing on only one feature at a time, allow the algorithm to hone in on specific errors without introducing significant complexity.
- **Control Overfitting**: With their simplicity, tree stumps reduce the risk of overfitting a common challenge in more complex models. This makes AdaBoost particularly robust in diverse data sets, as the aggregation of many simple models can achieve a balance of bias and variance without fitting excessively to noise in the data.
- **Boosting Model Diversity**: Each stump can focus on different features or different splits of the same feature across iterations. This diversity in the weak learners contributes to the overall strength and accuracy of the ensemble.

Impact on Variance and Bias: By aggregating multiple tree stumps, AdaBoost effectively reduces the high bias associated with individual stumps while keeping variance low. This aggregation transforms the ensemble into a more complex model capable of capturing intricate patterns in the data, which individual stumps cannot. Consequently, the AdaBoost ensemble, through its focus on sequential error correction and its robust aggregation strategy, can achieve a more optimal balance of bias and variance.

Thus, the use of tree stumps in AdaBoost is not merely a matter of computational convenience but a strategic choice that aligns with the algorithm's goals of creating a powerful, adaptive, and generalizable model.

8.7.5 Practical Implementation Considerations

Implementing AdaBoost effectively involves several strategic considerations to optimize model performance and efficiency. After selecting tree stumps as weak learners, further decisions around hyperparameter tuning, handling feature interactions, and leveraging AdaBoosts unique functionalities remain critical for maximizing performance and scalability.

- **Hyperparameter Tuning**:

  - **Number of Weak Learners** (M): Controls the ensemble's complexity and its ability to learn from the data. An optimal number of learners balances learning effectiveness with the risk of overfitting.
  - **Learning Rate**: Adjusts the contribution of each weak learner, where a smaller rate may require more learners but can lead to more stable and generalizable models.

- **Advanced Modifications**:

  - **Subsampling**: Implementing stochastic boosting by training each learner on a random subset of the data can prevent overfitting and enhance model generalization.
  - **Learning Rate**: AdaBoost can include a learning rate that scales the contribution of each weak learner. By reducing the weight of each classifier's contribution, the learning rate can help in improving the generalization by making the boosting process more conservative, thereby reducing the risk of overfitting.

8.7.6 Section Summary

AdaBoost revolutionized machine learning by demonstrating the efficacy of simple weak learners in forming a powerful classifier. This algorithm, by using tree stumps, focuses learning precisely on challenging cases by adaptively adjusting the weights of training instances, thereby effectively enhancing model accuracy. These stumps, being decision trees with only one split, simplify the learning process, reducing variance while allowing targeted improvements in bias. Through iterative updates that concentrate on misclassified instances, AdaBoost minimizes an exponential loss function, establishing itself as a precursor to more complex boosting methods like Gradient Boosting. Initially designed for binary classification, AdaBoost has been adapted for multi-class classification and regression, although these variants are less common compared to other specialized boosting methods. Practical implementation of AdaBoost necessitates careful consideration of the number of weak learners, the number of iterations, and maintaining a balance of bias and variance to optimize performance.

8.8 Gradient-Boosted Trees

Gradient-Boosted Trees (GBT) is an advanced ensemble learning technique that improves predictive accuracy by building a series of decision trees, each one correcting errors made by its predecessors. Developed from the concept of Gradient Boosting, GBT specifically utilizes trees as base learners and is highly effective in both regression and classification tasks.

8.8.1 Historical Context and Theoretical Framework

Gradient-Boosted Trees extend the Gradient Boosting framework by using decision trees as the weak learners. This approach was popularized by Jerome H. Friedman in the late 1990s through his papers on greedy function approximation: a gradient boosting machine. GBT combines the advantages of tree-based models with the optimization capabilities of boosting, making it one of the most powerful and widely used predictive modeling techniques today.

The core of GBT lies in its use of a differentiable loss function to optimize the model:

$$J(y, F(x)) = \sum_{i=1}^{N} L(y_i, F(x_i)), \tag{8.10}$$

where y_i are the actual labels, $F(x_i)$ is the model's prediction, and L is a loss function that measures the difference between predicted and actual values.

8.8.2 Gradient-Boosted Trees Algorithm

1. **Initialization**: Start with a base model, typically a constant value in regression or the log odds in classification, which minimizes the initial loss:

$$F_0(x) = \arg\min_{\gamma} \sum_{i=1}^{N} L(y_i, \gamma).$$

2. **Iterative Enhancement**: For each iteration $m = 1$ to M:

   (a) **Compute Pseudo-Residuals**: Calculate the negative gradient of the loss function, which serves as pseudo-residuals:

$$r_{i,m} = -\left[\frac{\partial L(y_i, F_{m-1}(x_i))}{\partial F_{m-1}(x_i)}\right].$$

(b) **Fit a Tree**: Train a decision tree $h_m(x)$ to fit these pseudo-residuals.

(c) **Determine Optimal Step Size**: Find the best step size γ_m that minimizes the loss when adding $h_m(x)$:

$$\gamma_m = \arg\min_\gamma \sum_{i=1}^{N} L(y_i, F_{m-1}(x_i) + \gamma h_m(x_i)).$$

(d) **Update the Model**: Update the ensemble model by adding the optimized tree:

$$F_m(x) = F_{m-1}(x) + \gamma_m h_m(x).$$

3. **Final Model**: The final predictive model is an aggregation of all sequentially enhanced trees:

$$F(x) = F_0(x) + \sum_{m=1}^{M} \gamma_m h_m(x).$$

Enhancements to Gradient-Boosted Trees Algorithm

- **Introduction of Learning Rate**: As discussed in the gradient boosting section, we can introduce an overall learning rate to the boosting steps as below:

$$F_m(x) = F_{m-1}(x) + \nu \gamma_m h_m(x),$$

where $\nu \in (0, 1]$ is the learning rate. This modification ensures that each tree's influence on the final model is tempered, allowing for more gradual improvements and typically leading to a more robust model.

- **Region-specific Step Size Optimization**: Since we are using decision trees as our weak learners, instead of a single step size γ_m for the entire tree, the step size can be made specific to different regions within each tree. This approach acknowledges that different subsets of data might benefit from different levels of adjustments. Let's denote the number of regions within each tree as R, where typically $4 \le R \le 10$. Each region r in the tree $h_m(x)$ will have its own step size $\gamma_{m,r}$, leading to a more finely tuned model. The update rule for the model then becomes:

$$F_m(x) = F_{m-1}(x) + \sum_{r=1}^{R} \gamma_{m,r} \mathbf{1}_{\{x \in \text{region } r\}} h_m(x),$$

where $\mathbf{1}_{\{x \in \text{region } r\}}$ is an indicator function that is 1 if x falls into region r and 0 otherwise. The optimal $\gamma_{m,r}$ for each region is computed by minimizing the loss specific to the region:

$$\gamma_{m,r} = \arg\min_{\gamma} \sum_{i \in \text{region } r} L(y_i, F_{m-1}(\boldsymbol{x}_i) + \gamma h_m(\boldsymbol{x}_i)).$$

This enhancement allows the model to adapt more precisely to different characteristics exhibited by the data in different regions of the input space.

8.8.3 Applications in Regression and Classification

Gradient-Boosted Trees are versatile in handling both regression and classification tasks:

- **Regression**: GBT is commonly used with squared error loss, focusing on minimizing prediction errors in continuous variables.

$$L(y, F(\boldsymbol{x})) = (y - F(\boldsymbol{x}))^2,$$

  where y is the actual value and $F(\boldsymbol{x})$ is the predicted value.
- **Classification**: For binary classification, logistic loss is typically used to model the probability of class membership, enhancing the decision boundary's accuracy.

$$L(y, F(\boldsymbol{x})) = \log(1 + \exp(-yF(\boldsymbol{x}))),$$

  where $y \in \{-1, 1\}$ represents the class labels, and $F(\boldsymbol{x})$ is the model's prediction of the log odds of the positive class.

8.8.4 Practical Implementation Considerations

Implementing Gradient-Boosted Trees (GBT) involves nuanced considerations around decision tree characteristics to optimize performance and mitigate risks such as overfitting:

- **Choice of Weak Learners**: Decision trees excel in GBT for their ability to capture nonlinear patterns and interact seamlessly with both categorical and continuous features. Setting the tree parameters correctly such as depth and minimum samples per leaf is vital for controlling model complexity and preventing bias or variance from skewing results.
- **Number of Trees** (M): More trees generally enhance accuracy but increase the risk of overfitting and elevate computational demands. It is advisable to use validation techniques to determine an optimal count that ensures accuracy without excessive complexity.

- **Tree Depth**: The depth of trees influences how well the model can learn detailed information. Deeper trees might capture more complex patterns but can also learn noise, leading to overfitting. Limiting tree depth or setting minimum node size requirements can help balance detail with generalization.
- **Learning Rate**: A lower learning rate can make the model more robust by slowing down the learning process and requiring more trees to achieve model convergence. This setting affects how quickly the model adapts to the complex structure of the data.
- **Subsampling**: Implementing subsampling or stochastic boosting by training each tree on a random data subset prevents overfitting and promotes better generalization, enhancing the ensemble's predictive reliability.
- **Feature Importance and Model Interpretability**: The natural feature selection ability of decision trees within GBT aids in understanding which predictors most significantly impact the target, providing valuable insights into the model's decision-making process.

These considerations detail the strategic advantages of using trees in GBT and underscore the importance of parameter tuning and model design in achieving effective and generalizable results.

8.8.5 Comparison of AdaBoost and Gradient-Boosted Trees

While both AdaBoost and Gradient-Boosted Trees are fundamental boosting techniques that build strong predictors from a series of weak learners, they differ significantly in their approach and application:

- **Loss Function Optimization**: AdaBoost minimizes an exponential loss function, which inherently focuses on classification errors and enhances the weight of misclassified instances in subsequent models. In contrast, GBT can utilize a variety of loss functions, making it adaptable to both regression and classification tasks. This flexibility allows GBT to tailor the loss function to specific data characteristics and prediction goals.
- **Model Complexity and Overfitting**: AdaBoost typically uses simple models such as decision stumps as weak learners, which limits the complexity of the individual models but can also restrict the depth of interactions it can capture. GBT, however, often employs more complex decision trees, which can model deeper interactions but also carry a higher risk of overfitting. GBT's use of gradient descent to minimize the loss function provides a more controlled way to manage model complexity through learning rate and tree depth adjustments.
- **Versatility in Handling Different Types of Data**: The ability of GBT to work with various loss functions makes it broadly applicable across different types of data, including skewed and multi-class distributions. AdaBoost, while extremely effective in binary classification scenarios, requires modifications, like AdaBoost.M1

for multi-class classification, which may not perform as well as GBT in these situations.

- **Performance and Tuning**: GBT generally offers more parameters for tuning, such as the number of trees, tree depth, and learning rate, which can lead to better performance on a wider range of tasks if tuned correctly. AdaBoost's performance is largely influenced by the number of iterations and the learning algorithm used for the weak learners, with less flexibility in adjusting the learning dynamics during model training.

8.8.6 Section Summary

Gradient-Boosted Trees (GBT) represent an advanced ensemble learning technique that builds upon the principles of boosting to improve model accuracy and robustness. By iteratively fitting decision trees to the negative gradients of a differentiable loss function, GBT fine-tunes model predictions and systematically minimizes errors. The flexibility in choosing loss functions makes GBT highly adaptable, enabling effective handling of both regression and classification tasks across a diverse range of data types.

Key considerations for implementing GBT include managing tree complexity, adjusting learning rates, and employing subsampling to enhance model generalization and prevent overfitting. The ability to adjust these parameters allows for precise control over model training and optimization, contributing to GBT's widespread adoption in solving complex predictive problems. Compared with AdaBoost, GBT offers enhanced flexibility and performance, particularly in scenarios that require modeling of complex patterns and interactions within data.

Overall, GBT stands out for its powerful predictive capabilities and versatility, making it a preferred choice for many machine learning applications that require robust and accurate modeling.

8.9 XGBoost

XGBoost (eXtreme Gradient Boosting) represents a sophisticated extension of traditional gradient boosting, incorporating advanced features designed to enhance model performance and computational efficiency. It is particularly recognized for its ability to handle large data sets, its robustness against overfitting, and its flexibility in addressing both regression and classification problems.

8.9.1 Mathematical Formulation of XGBoost

XGBoost optimizes a comprehensive objective function that integrates both predictive accuracy and regularization mechanisms to control model complexity and avoid overfitting. This formulation not only aims at improving predictive performance but also ensures that the model remains generalizable across various data sets.

Objective Function

The objective function in XGBoost, designed to be minimized, is composed of two main components: the loss function and the regularization term. The function is expressed as

$$J = \sum_{i=1}^{N} L(y_i, \hat{y}_i) + \sum_{m=1}^{M} R(f_m), \tag{8.11}$$

where[1]:

- $L(y_i, \hat{y}_i)$ is the differentiable convex loss function that measures the prediction error for each instance in the data set, ensuring that the model accurately predicts the target variable. This loss can be customized to suit specific problems such as logistic regression for classification tasks or squared error for regression.
- f_m is the m-th decision tree learner.[2]
- $R(f_m)$ is the regularization term that adds a penalty for the complexity of each tree in the ensemble. This term is crucial for preventing overfitting and is defined as

$$R(f_m) = \gamma T_m + \frac{1}{2}\lambda \sum_{j=1}^{T_m} w_j^2, \tag{8.12}$$

where T_m is the number of leaves in the m-th tree, w_j is the weight associated with the j-th leaf, and γ and λ are regularization parameters. γ penalizes the number of leaves in the tree, discouraging overly complex trees, and λ penalizes the magnitude of the leaf weights, reducing model variance.

Regularization Details

The regularization component $R(f_m)$ in XGBoost is a key feature that differentiates it from other boosting algorithms. It directly controls the complexity of the model:

---

[1] In the original paper, the loss function is denoted by l, the regularization term is denoted by Ω. We use L and R here to be consistent with the earlier definitions in this book.

[2] We have been using h_m to denote the m learner previously, but for this and the next section, we switch to f_m and reserve h_i to denote components of Hessian.

- The term γT_m helps in trimming unnecessary branches from the trees during the learning process, effectively pruning the trees post-training to remove splits that provide minimal gain in reducing the loss.
- The term $\frac{1}{2}\lambda \sum_{j=1}^{T_m} w_j^2$ smoothes the output of the trees, stabilizing the learning process and helping to maintain consistency in prediction performance across varied data sets.

This dual approach to regularization controlling both the architectural complexity and the output values of the trees ensures that XGBoost models can handle a variety of overfitting scenarios, making it robust and effective for a wide range of applications. By integrating these elements directly into the core algorithm, XGBoost provides an efficient and scalable solution for advanced predictive modeling.

8.9.2 Specific Optimizations in XGBoost

XGBoost implements several advanced features to enhance model performance and computational efficiency. These optimizations not only improve accuracy but also make XGBoost highly scalable and effective in handling complex and large data sets:

- **Regularization**: XGBoost integrates both L_1 (Lasso) and L_2 (Ridge) regularization directly into the learning algorithm. This helps to reduce overfitting by penalizing more complex models, encouraging simpler models that perform well on new, unseen data. Regularization is particularly vital when dealing with high-dimensional data, where the risk of overfitting is more pronounced.
- **Utilization of Second-Order Optimization**: A distinctive feature of XGBoost is its use of the second-order partial derivatives of the loss function, which provides a more accurate curvature estimation than first-order gradient methods used in other boosting algorithms. This allows XGBoost to make more informed, optimal steps in the model's learning path, significantly speeding up convergence and improving the final model's accuracy and performance.
- **Tree Pruning**: In contrast to traditional gradient boosting methods that halt tree growth when no further gains are observed, XGBoost allows trees to grow to their maximum depth and then prunes them using a depth-first approach. This pruning removes branches that contribute minimally to the overall model's predictive power, based on a calculated gain that incorporates the regularization term:

$$\text{Gain} = \text{Loss}_{\text{before split}} - (\text{Loss}_{\text{left child}} + \text{Loss}_{\text{right child}} + \gamma),$$

where γ serves to control the complexity by penalizing the addition of new branches, thus optimizing the model's generalization capabilities.
- **Handling Missing Values**: XGBoost addresses missing data effectively through its sparsity-aware splitting algorithms. Rather than requiring preprocessing or imputation, it automatically determines the best way to handle missing values during

the training process. For each split in a tree, XGBoost learns whether to send data points with missing values to the left or right child, enhancing its ability to manage data sparsity. This capability not only simplifies data preparation but also preserves the integrity and distribution of the data set, optimizing performance.

- **Cache Awareness**: XGBoost is designed to make optimal use of hardware resources. It organizes data into blocks that are stored in a compressed sparse row format, making operations cache-efficient and minimizing computing overhead. This structure significantly speeds up the computations required for model training and predictions.
- **Block Structure for Parallel Learning**: XGBoost utilizes a block structure to support parallel tree learning. It sorts and bins the data into blocks that allow the algorithm to reuse the structure in subsequent iterations rather than recomputing it. This block structure is not only efficient for single-machine environments but also scales effectively in distributed systems, dramatically reducing the time required for training complex models over large data sets.

These optimizations ensure that XGBoost is not only precise in its predictions but also efficient in terms of computation, making it suitable for a wide range of practical applications from standard regression tasks to complex classification challenges in large-scale machine learning environments.

8.9.3 Building Trees in XGBoost

XGBoost enhances decision tree performance through a structured and optimized tree building process. This includes initial tree setup, optimized split finding, gain calculation, and effective tree pruning to ensure high model accuracy and robustness against overfitting.

Tree Initialization

Trees in XGBoost are initialized with a simple model that predicts a constant value, typically the value that minimizes the loss function over the initial data set.

Split Finding and Optimization

The process of finding the best splits involves calculating potential gains for various splits and choosing the one that maximizes the overall gain. The detailed steps are as follows:

- **Gradient and Hessian Computation**[3]: For each data point i, calculate the gradient (g_i) and Hessian (h_i), which are the first and second derivatives of the loss function:

$$g_i = \frac{\partial L(y_i, \hat{y}_i)}{\partial \hat{y}_i}, \quad h_i = \frac{\partial^2 L(y_i, \hat{y}_i)}{\partial \hat{y}_i^2} \tag{8.13}$$

These derivatives provide the necessary direction and curvature information about the loss function at each point.

- **Objective Function for Tree Learning**: The objective function for adding a new tree f_m to the ensemble is approximated using a second-order Taylor expansion around the predictions from the current ensemble:

$$J(f_m) \approx \sum_{i=1}^{N} \left[L(y_i, \hat{y}_i^{(m-1)}) + g_i f_m(x_i) + \frac{1}{2} h_i f_m(x_i)^2 \right] + R(f_m)$$

where $\hat{y}_i^{(m-1)}$ is the prediction from the previous ensemble, and $R(f_m)$ is the regularization term for the new tree.

Gain Calculation and Optimal Value Derivation

Ignoring constants, the objective for tree f_m simplifies to:

$$J(f_m) = \sum_{i=1}^{N} \left[g_i f_m(x_i) + \frac{1}{2} h_i f_m(x_i)^2 \right] + R(f_m)$$

The regularization term for the tree is given by

$$R(f_m) = \gamma T + \frac{1}{2} \lambda \sum_{j=1}^{T} w_j^2,$$

where T is the number of leaves, w_j are the leaf weights, γ and λ are parameters that control the complexity. For each leaf j, minimize the objective by setting the derivative with respect to w_j to zero, solving for w_j:

$$w_j^* = -\frac{\sum_{i \in I_j} g_i}{\sum_{i \in I_j} h_i + \lambda}$$

---

[3] We are using the terms Hessian loosely here. Technically Hessian refers to the matrix of second derivatives. Here we are simply computing the diagonal components of that matrix. In the boosting literature, this is often loosely referred to as computing the Hessian and denoted by these h_i.

Substituting w_j^* back into the objective function, derive the reduction in loss (gain) achieved by adding this tree:

$$\text{Gain} = \frac{1}{2}\left[\sum_{j=1}^{T} \frac{\left(\sum_{i \in I_j} g_i\right)^2}{\sum_{i \in I_j} h_i + \lambda}\right] - \gamma T$$

Tree Pruning

After a tree is fully constructed, XGBoost evaluates the gain from each potential split during the tree's construction:

$$\text{Gain} = \frac{1}{2}\left[\frac{\left(\sum_{i \in I_L} g_i\right)^2}{\sum_{i \in I_L} h_i + \lambda} + \frac{\left(\sum_{i \in I_R} g_i\right)^2}{\sum_{i \in I_R} h_i + \lambda} - \frac{\left(\sum_{i \in I} g_i\right)^2}{\sum_{i \in I} h_i + \lambda}\right] - \gamma \qquad (8.14)$$

If the gain from a potential split is negative or does not surpass a predefined threshold (often set as the complexity parameter γ), the split is not made. This pruning step happens retroactively; after the tree grows to its maximum depth, the algorithm prunes back branches that do not contribute positively to the model's predictive power, effectively reducing complexity and enhancing generalizability.

8.9.4 XGBoost Algorithm

XGBoost optimizes both model training speed and predictive accuracy by constructing an ensemble of trees in a sequential manner. Each tree is built to effectively improve upon the residuals of all the trees that preceded it.

1. **Initialization**: The model starts with a base prediction which might be the average target value (for regression) or another suitable initializer:

$$F_0(x) = \arg\min_{\gamma} \sum_{i=1}^{N} L(y_i, \gamma),$$

   where $F_0(x)$ represents the initial model output.

2. **Iterative Enhancement**: Each subsequent tree is added to the ensemble to refine the predictions, aiming to correct the residuals left by the previous ensemble of trees.

   (a) **Gradient and Hessian Computation**: For each training instance i, compute the gradients and Hessians of the loss function, which are essential for determining the direction and magnitude of the model updates:

$$g_{i,m} = \frac{\partial L(y_i, F_{m-1}(x_i))}{\partial F_{m-1}(x_i)}, \quad h_{i,m} = \frac{\partial^2 L(y_i, F_{m-1}(x_i))}{\partial^2 F_{m-1}(x_i)},$$

(b) **Tree Building**: Using the gradients and Hessians, a new tree $f_m(x)$ is fitted. The specifics of tree growth, including how splits are chosen and trees are pruned, are covered in detail in the previous subsection.

(c) **Update Model with Shrinkage**: The new tree is scaled by the learning rate η before being added to the model, controlling the speed of learning and helping to prevent overfitting:

$$F_m(x) = F_{m-1}(x) + \eta \cdot f_m(x),$$

where η is a constant, typically set between 0.01 and 0.3.

3. **Final Model**: After completing a predefined number of iterations or trees, the final model is a combination of the initial predictions and the contributions from all trees:

$$F(x) = F_0(x) + \eta \sum_{m=1}^{M} f_m(x),$$

where M is the total number of trees built.

Contrast with Classical Gradient Boosting: Unlike XGBoost, classical gradient boosting often calculates an optimal step size γ_m for each tree individually, potentially allowing for faster but less controlled convergence. XGBoost does not use a separate γ_m for each tree because it integrates optimization directly within its tree construction and pruning algorithm. By optimizing each trees structure to best reduce the residual loss and using a fixed learning rate η, XGBoost simplifies the model tuning process and enhances regularization. This method avoids the need for individual tree scaling, which classical gradient boosting handles with γ_m. The consistent use of η in XGBoost not only simplifies parameter tuning but also ensures that each tree's contribution is systematically moderated to enhance the model's overall robustness and prevent overfitting.

8.9.5 *Practical Implementation Considerations*

Implementing XGBoost effectively involves various considerations and hyperparameter tuning to optimize model performance and efficiency. Some of the most important hyperparameters include:

- **Learning Rate** (η): The learning rate controls the contribution of each tree to the ensemble. A lower learning rate requires more trees to achieve similar performance but can enhance model stability and convergence.

- **Number of Trees** (M): Determining the optimal number of trees in the ensemble involves balancing model complexity and computational resources. Increasing the number of trees can improve predictive performance, but excessive trees may lead to overfitting.
- **Max Depth**: The maximum depth of each tree influences the complexity of the learned patterns. Deeper trees can capture more intricate relationships in the data but may also increase the risk of overfitting.
- **Subsample**: This parameter controls the fraction of instances used to train each tree. Subsampling can help mitigate overfitting by introducing randomness into the training process.
- **Column Sample by Tree/Level/Node**: These parameters determine the subset of features considered when constructing each tree, offering additional control over model complexity and generalization.
- **Min Child Weight**: Specifies the minimum sum of instance weights required to make further splits in a node. Adjusting this parameter influences the tree structure and can help prevent overfitting.

8.9.6 Section Summary

In summary, XGBoost is a highly efficient algorithm that extends traditional gradient boosting with advanced features tailored for enhanced performance and efficiency. Its mathematical formulation integrates predictive accuracy and regularization mechanisms to control model complexity and prevent overfitting. The regularization terms, including γ for penalizing additional tree complexity and λ for controlling leaf weights, play a crucial role in optimizing the model's generalization capabilities. The use of the Hessian matrix for second-order optimization allows XGBoost to more accurately determine the most effective direction and size of steps in learning, markedly improving convergence speed and model performance. Specific optimizations such as tree pruning, handling missing values, cache awareness, and parallelization further contribute to XGBoost's effectiveness in handling large data sets and complex predictive tasks.

The XGBoost algorithm iteratively enhances the model through gradient-based updates and regularization, adding new trees to the ensemble optimized to minimize the overall loss function while considering model complexity. Practical implementation considerations include tuning hyperparameters such as the learning rate, number of trees, tree depth, and subsampling rate to achieve optimal performance and avoid overfitting. By balancing predictive accuracy and model complexity, XGBoost stands as a versatile and efficient tool for regression and classification tasks, making it a popular choice in machine learning applications.

8.10 LightGBM

LightGBM (Light Gradient Boosting Machine) is a gradient boosting framework designed for speed and efficiency. It introduces several novel technologies, including Gradient-based One-Side Sampling (GOSS), Exclusive Feature Bundling (EFB), and leaf-wise growth of trees. These features make LightGBM particularly effective for large data sets and high-dimensional features.

8.10.1 Mathematical Formulation of LightGBM

Similar to XGBoost, LightGBM optimizes an objective function that combines predictive accuracy with model complexity control through regularization techniques. The formulation is given by

$$J = \sum_{i=1}^{N} L(y_i, \hat{y}_i) + \sum_{m=1}^{M} R(f_m),$$

where N denotes the number of data points, $L(y_i, \hat{y}_i)$ represents the loss function, f_m denotes the m-th decision tree, and $R(f_m)$ is the regularization term. This term, defined as

$$R(f_m) = \gamma T_m + \frac{1}{2}\lambda \sum_{j=1}^{T_m} w_j^2,$$

includes penalties γT_m for the number of leaves and $\frac{1}{2}\lambda \sum_{j=1}^{T_m} w_j^2$ for the magnitude of leaf scores, echoing the regularization approach in XGBoost.

8.10.2 Specific Optimizations in LightGBM

Differing from XGBoost, LightGBM employs Gradient-based One-Side Sampling and Exclusive Feature Bundling to enhance processing efficiency and model scalability. Additionally, it adopts a leaf-wise tree growth strategy, which can lead to faster learning and greater model complexity compared to the level-wise approach of XGBoost.

Gradient-Based One-Side Sampling (GOSS)

Gradient-based One-Side Sampling (GOSS) retains a subset of data that has the largest gradients, thus maintaining accuracy during training while improving efficiency. It amplifies the gradients of randomly sampled smaller gradient data to compensate for their smaller presence in the sample.

GOSS Mechanism: GOSS retains a subset of the data with the largest gradients, which are considered more informative and therefore crucial for computing information gain. It performs random sampling on the remaining data, ensuring that training focuses on difficult cases without losing general gradient distribution essential for accurate learning.

Procedure and Mathematical Details:

1. **Sorting and Partitioning**: Data points are sorted by the absolute values of their gradients. Larger gradients indicate larger errors and a higher priority for learning.
2. **Data Retention**: Retain a significant percentage (e.g., 20%) of the data with the highest gradients.
3. **Random Sampling**: Randomly sample a smaller proportion (e.g., 30%) from the remaining data to ensure diversity and robustness.
4. **Gradient Adjustment**: Adjust the gradients of the randomly sampled data to maintain accurate gradient distribution across the training set. The adjustment factor is

$$g_i' = g_i \times \frac{1 - |A|/N}{|B|/N}, \tag{8.15}$$

where g_i is the original gradient, $|A|$ and $|B|$ are the sizes of the high gradient and randomly sampled subsets, respectively.

Exclusive Feature Bundling (EFB)

EFB reduces the number of features in high-dimensional data sets by grouping mutually exclusive features, those that rarely take non-zero values simultaneously.

Compatibility Graph Construction: A compatibility graph identifies which features can be bundled together without significant information loss.

Greedy Bundling Algorithm:

- Start with an empty set of bundles.
- Iteratively add features to existing bundles if no significant overlap exists, or create new bundles if necessary.

Mathematical Representation: Define binary compatibility $c(f_i, f_j) = 1$ if features f_i and f_j can coexist without loss. Minimize the number of bundles B such that:

$$\sum_{f_i, f_j \in b, i \neq j} (1 - c(f_i, f_j)) = 0.$$

Leaf-Wise Tree Growth

Unlike conventional gradient boosting methods that grow trees level by level, Light-GBM employs a leaf-wise (best-first) tree growth algorithm, which offers a more efficient way to reduce loss compared to the traditional level-wise approach. This method selects the leaf that maximizes loss reduction for growth at each step, optimizing the model's performance by focusing on the most promising splits.

Process of Leaf-wise Growth:

- For each leaf in the tree, calculate the possible splits across all features and identify the split that offers the maximum gain.
- Using the same derivation as XGBoost, the gain from splitting a leaf is calculated using the formula:

$$\text{Gain} = \frac{1}{2}\left(\frac{G_L^2}{H_L + \lambda} + \frac{G_R^2}{H_R + \lambda} - \frac{(G_L + G_R)^2}{H_L + H_R + \lambda}\right) - \gamma,$$

  where G_L and G_R are the sums of gradients of the loss function for the left and right splits, respectively; H_L and H_R are the sums of the second derivative of the loss function (Hessians) for the left and right splits, respectively; λ is the regularization parameter, and γ is the complexity cost of adding a new leaf.
- Choose the split that results in the highest gain. If the gain is less than a predefined threshold or another stopping criterion is met (e.g., maximum depth, minimum data in leaf), stop growing the tree.

8.10.3 LightGBM Algorithm

LightGBM employs a sophisticated gradient boosting framework that enhances predictive performance while maintaining computational efficiency. The algorithm is iterative, with each iteration consisting of the following steps:

1. **Initialization**: Start with an initial approximation for the output values across the data set:

$$F_0(x) = \arg\min_\gamma \sum_{i=1}^{N} L(y_i, \gamma).$$

2. **Iterative Model Enhancement**: Enhance the model iteratively by adding a new decision tree at each step to minimize the overall loss:

   (a) **Compute Gradients and Hessians**: Calculate the gradients and Hessians for each instance based on the current model predictions:

$$g_{i,m} = \frac{\partial l(y_i, F_{m-1}(x_i))}{\partial F_{m-1}(x_i)}, \quad h_{i,m} = \frac{\partial^2 l(y_i, F_{m-1}(x_i))}{\partial^2 F_{m-1}(x_i)},$$

where $g_{i,m}$ and $h_{i,m}$ are the gradient and Hessian for the i-th instance, respectively, and m denotes the iteration index.

(b) **Apply GOSS and EFB**: Apply Gradient-based One-Side Sampling to focus training on the most informative instances and use Exclusive Feature Bundling to reduce dimensionality.

(c) **Leaf-wise Tree Growth**: Employ leaf-wise growth strategy to choose the splits that yield the highest gain, optimizing tree structure and depth to enhance model performance.

3. **Update the Model**: Update the model by adding the output of the new tree scaled by the learning rate η:

$$F_m(x) = F_{m-1}(x) + \eta \cdot f_m(x),$$

where $f_m(x)$ represents the contribution from the m-th tree.

4. **Final Model**: After M iterations, the final model is

$$F(x) = F_0(x) + \eta \sum_{m=1}^{M} f_m(x),$$

representing the accumulated contributions from all iterations to form the predictive model.

8.10.4 *Practical Implementation Considerations*

Effective implementation of LightGBM requires careful consideration of several parameters and settings to optimize both model performance and computational efficiency. Key areas of focus include:

- **Tree-Specific Parameters**: LightGBM offers parameters that directly influence the construction and complexity of trees:

  - **Number of Leaves**: Controls the maximum number of leaves in each tree. This parameter is crucial for managing model complexity and can significantly affect both performance and overfitting.
  - **Min Data in Leaf**: Sets the minimum number of data points allowed in one leaf. This parameter helps prevent overfitting, especially in cases where the data is highly imbalanced.
  - **Max Depth**: Limits the depth of each tree. While deeper trees can model more complex patterns, they are also more likely to overfit, making this parameter important for controlling model behavior.

- **Learning Rate and Number of Estimators**: The learning rate, also known as shrinkage, affects how quickly the model learns. Smaller learning rates require more boosting rounds (trees) but can provide more robust performance on unseen data:

  - **Learning Rate** (η): Determines the impact of each individual tree on the final outcome. A lower rate requires more trees to achieve model convergence but generally yields more robust models.
  - **Number of Estimators** (M): Specifies the number of trees in the model. More trees can increase the model's ability to fit the data but also raise the risk of overfitting.

- **Regularization Techniques**: Regularization is critical for preventing overfitting in LightGBM, involving parameters that control both the tree-specific and overall model regularization:

  - **Lambda** (λ, L_2 **regularization**): Adds a penalty on the sum of squares of model weights, which helps reduce model complexity and enhance prediction robustness.
  - **Gamma** (γ, L_1 **regularization**): Penalizes the absolute value of the weights, encouraging sparsity in the model's features.

- **Handling Categorical Features**: LightGBM can handle categorical features natively, which is essential for maintaining efficiency and model accuracy:

  - **Categorical Feature Support**: LightGBM can directly use categorical features without the need for manual encoding, optimizing the splits based on categorical variables.

- **Advanced Features for Performance Optimization**: Use advanced features to fine-tune model performance and prevent overfitting:

  - **Feature Fraction**: Randomly selects a subset of features on each iteration (tree), which can help prevent overfitting and reduce training time, particularly in data sets with a large number of noisy features.
  - **Early Stopping**: Stops training if the validation score does not improve for a specified number of rounds, saving computational resources and preventing overfitting.

8.10.5 Comparison of LightGBM and XGBoost

LightGBM and XGBoost are both prominent gradient boosting frameworks, yet they exhibit distinct differences in terms of implementation and performance optimizations. Here's a high-level comparison:

Algorithmic Differences

LightGBM utilizes a leaf-wise tree growth algorithm, which tends to achieve better performance and efficiency on large data sets by reducing loss more rapidly. XGBoost, however, employs a level-wise approach that optimizes space complexity and can be more stable for smaller data sets.

Efficiency and Scalability

LightGBM is particularly optimized for speed and efficiency. It uses Gradient-based One-Side Sampling (GOSS) to maintain the most informative gradients and Exclusive Feature Bundling (EFB) to reduce the number of features, which enhances its ability to handle large-scale data more efficiently than XGBoost. XGBoost, while providing robust scalability and parallel processing capabilities, generally requires more resources for comparable tasks due to its less aggressive feature reduction techniques.

Handling of Sparse Data

LightGBM's EFB effectively bundles mutually exclusive features, significantly reducing the number of features and improving execution times on sparse data sets. XGBoost, while capable of handling sparse data, does not include a native mechanism to compress or bundle features, which can lead to increased computational times in high-dimensional spaces.

Model Performance and Overfitting

Both frameworks offer strong performance across a variety of tasks, but their different approaches to tree construction can lead to variations in model accuracy and overfitting. LightGBM's aggressive leaf-wise growth can lead to overfitting on smaller data sets, whereas XGBoost's level-wise growth generally provides more consistent performance across different data sizes.

In summary, the choice between LightGBM and XGBoost often comes down to specific project requirements and data set characteristics. LightGBM is typically preferred for large data sets or scenarios where computational efficiency is crucial, while XGBoost might be favored for scenarios where model stability and consistency are more critical.

8.10.6 Section Summary

LightGBM (Light Gradient Boosting Machine) excels in efficiency, particularly suitable for large data sets and high-dimensional features due to its unique methods like Gradient-based One-Side Sampling (GOSS), Exclusive Feature Bundling (EFB), and leaf-wise tree growth. These innovations allow LightGBM to enhance processing speed and model accuracy, optimizing an objective function that finely balances predictive accuracy and model complexity. This balance ensures robust generalization across diverse data sets, making LightGBM a preferred choice for applications where both speed and accuracy are paramount.

The algorithmic implementation of LightGBM refines the model through iterative enhancements, leveraging specialized techniques to optimize computational efficiency and performance. Features such as GOSS reduce data redundancy, and EFB minimizes feature dimensionality, which, alongside the leaf-wise growth strategy, accelerates training and improves the model's ability to manage complex data patterns more effectively than conventional models. These strategies necessitate precise hyperparameter adjustments and an understanding of LightGBMs categorical feature handling to fully exploit its capabilities in real-world scenarios, solidifying its status as a highly effective tool for advanced predictive modeling tasks.

8.11 Bagging Versus Boosting: A Brief Summary

Bagging (Bootstrap Aggregating) and Boosting are transformative ensemble techniques designed to enhance the accuracy and stability of machine learning models. While both methods aggregate the insights of multiple models to surpass the limitations of individual learners, they apply fundamentally different strategies to optimize performance and address data set complexities.

8.11.1 Shared Philosophical Underpinnings

Both bagging and boosting fundamentally aim to improve prediction performance by leveraging the collective power of multiple models, drawing from the wisdom of crowds theory which suggests that collective decisions are often superior to those made by individuals.

Error Reduction

Bagging primarily reduces variance without necessarily reducing bias, making it ideal for high-variance models. It operates by generating multiple independent copies of the original model, each trained on a random subset of the data, and averaging their predictions:

$$F_{\text{bag}}(x) = \frac{1}{M} \sum_{m=1}^{M} f_m(x),$$

where M denotes the number of models, and $f_m(x)$ is the prediction of the m-th model.

Conversely, Boosting focuses on reducing both bias and variance through a sequential correction of errors, where each successive model incrementally improves upon the previous by focusing more on the instances misclassified by the ensemble thus far:

$$F_{\text{boost}}(x) = \sum_{m=1}^{M} \alpha_m h_m(x),$$

with M representing the number of models, α_m the weight of the m-th model, and $h_m(x)$ its prediction.

8.11.2   Mechanisms for Generating Model Diversity

Data Sampling Versus Data Weighting

Bagging introduces diversity through bootstrap sampling, providing each model with a unique subset of the data set, thus ensuring reduced variance in the aggregated predictions.

Boosting introduces diversity through the sequential adjustment of instance weights, enhancing focus on those portions of the data where previous models performed poorly, thereby refining the accuracy and robustness of the overall model.

Model Independence Versus Interaction

In Bagging, models operate independently, allowing for parallel computation. This independence enables effective variance reduction through simple averaging of model outputs.

In Boosting, models are interdependent, building sequentially from one another, which allows for detailed error correction but also increases the risk of overfitting if not properly managed.

Adaptation to Data Changes

Bagging is robust against data shifts due to its averaging approach, which naturally mitigates outlier impacts and general noise.

Boosting, being highly adaptive to the training data through iterative learning, can swiftly adjust to changes in data distribution but may also amplify noise and overfit, necessitating careful tuning.

8.11.3 Strategic Considerations

The choice between bagging and boosting should consider expected noise in data, model complexity, and computational resources. Bagging is favored for its robustness in variable conditions and is particularly effective in domains requiring stability across diverse data sets. Boosting, known for its precision, is suited for tasks where error reduction is crucial, such as in predictive analytics where small improvements in accuracy can be significant.

8.11.4 Section Summary

Bagging and Boosting, while both powerful, are best used under different circumstances due to their inherent characteristics. Bagging is optimal for reducing variance and enhancing model stability, especially suitable for complex models prone to overfitting. Boosting, adept at minimizing both bias and variance, requires careful calibration to maximize its efficacy without incurring overfitting, making it ideal for applications where detailed error correction is beneficial to model performance.

8.12 Comparative Analysis of Ensemble Learning Algorithms

In this section, we explore a comparative analysis of five leading ensemble learning algorithms: Random Forest, AdaBoost, Gradient-Boosted Trees, XGBoost, and LightGBM.

Random Forest employs a combination of decision tree predictors where each tree depends on the values of a random vector sampled independently, and with the same distribution for all trees in the forest. This method excels in handling complex data sets with embedded nonlinearities and interactions, without the need for feature scaling. It is highly resilient to overfitting, although not entirely immune.

AdaBoost (Adaptive Boosting) refines the concept of boosting by adjusting the weights of incorrectly classified instances so that subsequent classifiers focus more on difficult cases. While extremely effective in reducing bias, AdaBoost can be sensitive to noisy data and outliers, which may lead to a decrease in performance if the noise is substantial.

Gradient-Boosted Trees build on the principles of boosting to form strong predictive models. These algorithms typically employ mechanisms like learning rate adjustments and constraints on tree complexity to control overfitting. These methods are particularly adept at managing structured data environments where relationships between features can be leveraged to improve prediction accuracy.

XGBoost (eXtreme Gradient Boosting) adds an advanced twist to gradient boosting, introducing regularization techniques to penalize complex models and prevent overfitting. It also features column sampling and advanced tree pruning methods, which enhance scalability and efficiency, making XGBoost a popular choice in competitive machine learning scenarios.

LightGBM (Light Gradient Boosting Machine) streamlines the boosting process even further by implementing novel techniques such as Gradient-based One-Side Sampling (GOSS) and Exclusive Feature Bundling (EFB). These innovations reduce the amount of data needed for training without compromising on model accuracy, facilitating faster training times and lower memory consumption. LightGBM is especially effective in scenarios where the handling of large-scale data and categorical features is paramount.

Each of these algorithms presents unique strengths and potential drawbacks. Random Forest is ideal for applications requiring robust performance across a variety of settings, particularly where model interpret ability is important. AdaBoost and Gradient-Boosted Trees offer powerful options for scenarios where precision is critical, though they require careful tuning to guard against overfitting. XGBoost and LightGBM, with their high efficiency and scalability, are suited for high-performance needs in competitive data science, although they might require more careful handling of parameter tuning and data preprocessing to achieve optimal results.

In summary, the choice among these ensemble methods should be guided by the specific requirements of the application, including the nature of the data set, the computational resources available, and the need for model accuracy versus training speed. Understanding the nuances of each algorithm allows practitioners to better harness their capabilities, tailoring their approach to meet diverse analytical needs effectively.

The table below presents a detailed comparison of these algorithms across several key metrics such as type, overfitting control, performance, scalability, feature handling, hyperparameters, model complexity, and their implementation (Table 8.1).

Table 8.1 Comparison of ensemble learning algorithms

Comparison metric	Random forest	AdaBoost	Gradient boosted trees	XGBoost	LightGBM
Type	Bagging	Boosting	Boosting	Boosting	Boosting
Overfitting control	Uses bootstrapping and random feature selection to reduce variance	Combines weak learners; sensitive to noise and outliers	Shrinkage via learning rate and tree complexity limits overfitting	Regularization techniques and column sampling enhance robustness	Leaf-wise growth with constraints helps prevent overfitting
Performance	High accuracy, robust across diverse data sets	Effective with clear margins and less noise	High performance on structured data when properly tuned	Often delivers top results in competitive scenarios; efficient on large data sets	Comparable or superior to XGBoost; excels with large data sets and categorical data
Scalability	Good; training of trees can be parallelized	Moderate; sequential model building limits scalability	Moderate; some implementations support parallel processing	Excellent; supports distributed computing and GPU utilization	Outstanding; designed for speed and distributed computing
Feature handling	Handles categorical and continuous data well; no input scaling required	Requires careful handling of outliers; benefits from feature preprocessing	Sensitive to feature scaling; requires well-prepared data	Effectively manages missing values; often requires feature scaling	Efficiently handles categorical features; robust to different data scales
Hyperparameters	Number of trees, maximum depth, features per split	Number of learners, learning rate	Number of trees, depth, learning rate, subsample size	Learning rate, number of estimators, maximum depth, subsample rates	Learning rate, number of leaves, minimum data in leaf, maximum depth
Model complexity	Moderate; ensemble of potentially deep trees	Low to moderate; dependent on the complexity of weak learners	High; multiple tunable parameters	High; extensive tuning may be required	High; complex due to leaf-wise growth
Implementation	Widely supported in major data science libraries	Commonly available in libraries like scikit-learn	Available in major libraries, functionalities extended by XGBoost	Dedicated package available, supports advanced configurations	Available in major languages, part of Microsoft's ML suite
Key Advantages	Robust to outliers, minimal preprocessing required, excellent for complex data sets	Efficient with binary classification, performs well with adequate data	Versatile across various tasks, highly customizable	Scalable, efficient, delivers top-notch performance in many settings	Extremely fast, minimal memory usage, performs well with large and categorical data

8.13 Bibliographical Notes

Ensemble learning methods have evolved as a fundamental technique in the field of machine learning, contributing significantly to improvements in predictive accuracy and robustness. The core idea behind ensemble learning, which involves combining multiple models to enhance overall model performance, has been explored extensively in various research papers and articles.

Bagging, introduced by Breiman (1996), represents one of the earliest and most influential ensemble techniques. Bagging, or Bootstrap Aggregating, aims to reduce variance and prevent overfitting by creating multiple versions of a predictor and using their average to make decisions.

Boosting, another cornerstone ensemble technique, was explored extensively by Freund and Schapire (1995), Freund and Schapire (1996). Their work on AdaBoost in laid the groundwork for many subsequent boosting algorithms, emphasizing iterative correction of errors made by previous models to improve performance.

Random Forests, as a development of the bagging idea applied specifically to decision trees, was also popularized by Breiman (2001). This method combines the simplicity of decision trees with the robustness of ensemble techniques to handle various types of data effectively.

The introduction of **Gradient Boosting Machines** by Friedman (2001) brought a new dimension to ensemble techniques, focusing on minimizing a differentiable loss function through the addition of new models that address the residuals of previous ones. This approach has been foundational for later algorithms like XGBoost and LightGBM.

XGBoost, developed by Chen and Guestrin (2016), enhanced traditional gradient boosting through system optimizations and advanced regularization, which have proven effective in various machine learning competitions.

LightGBM by Ke et al. (2017) built on the principles of gradient boosting but optimized further for computational efficiency and handling large-scale data through techniques such as Gradient-based One-Side Sampling and Exclusive Feature Bundling.

For a comprehensive understanding of the theoretical underpinnings and practical applications of these methods, readers are encouraged to refer to seminal works by Hastie et al. (2009) in their book "The Elements of Statistical Learning" which provides a deep dive into statistical modeling techniques including ensemble methods.

8.14 Exercises

8.1 Explain how the correlation between base learners in a bagging ensemble affects the overall variance of the ensemble prediction.

8.2 Explain the concept of "Irreducible Error" and how it impacts the performance of ensemble learning methods like bagging and boosting.

8.3 Discuss the importance of "Feature Importance" in the context of Random Forest and how it is calculated.

8.4 Describe the effect of "Learning Rate" in boosting algorithms like AdaBoost and XGBoost and how it influences model convergence.

8.5 Explain the role of "Loss Functions" in the formulation of boosting algorithms, specifically referring to Gradient Boosting and XGBoost.

8.6 Write down the mathematical expression for the general aggregation strategy used in ensemble learning as discussed in the chapter. Explain how this strategy works in the context of both Bagging and Boosting.

8.7 Provide a derivation of how bagging helps reduce variance without significantly affecting bias (assuming independence of individual models). Refer to the mathematical formulation of a bagged estimator and discuss its implications on a model's prediction error.

8.8 Derive the weight update rule used in the AdaBoost algorithm as mentioned in the chapter. Explain how this formula helps to focus learning on more difficult training examples.

8.9 Demonstrate mathematically how boosting can achieve lower bias with an increasing number of weak learners.

8.10 Implement a basic Random Forest classifier using Python's scikit-learn library on the Iris data set. Demonstrate the classifier's training process, prediction, and evaluate its accuracy on the test set.

Chapter 9
Performance Evaluation and Statistical Inference

According to Fig. 1.4 in Sect. 1.6, the main components of learning are learning model, learning loss function, learning cost function, and optimization/learning procedure. Main topics covered thus far include linear regression, penalized learning, network learning and ensemble learning. In this chapter, several existing methods for performance evaluation and statistical inference are presented. Performance evaluation provides a measure upon how good can a learned predictor or classifier perform and statistical inference provides the confidence of such performance on unseen data.

9.1 Empirical Evaluation

Apart from their use in classifier learning, the performance measures discussed in Chap. 4 can be adopted in this section for classifier performance evaluation. Essentially, the evaluation has to take into consideration how well can the predictor or classifier perform for unseen data. Without knowing the unseen data, the evaluation can be performed by holding out some data from the given samples for testing. The testing performance shall make use of those measures discussed in Chap. 4 for scoring purpose. Two popular approaches to partition the given data are presented below.

9.1.1 Hold-Out Test

Consider a database consisting of M samples (x_i, y_i), $i = 1, 2, \ldots, M$, where $x_i \in \mathbb{R}^d$ denotes the ith feature sample, and $y_i \in \{0, 1\}$ denotes the corresponding category indicator or target label.

A *hold-out* evaluation splits the database into two sets namely a training set which consists of m samples and a test set which consists of n samples (i.e., $m + n = M$). The training set is used for classifier learning and the test set is used for evaluation of the learned classifier's performance.

If $m \approx n$, then a two-fold test can be performed. This means that after the first round of training and testing, the training set and the test set can be swapped and the training and testing processes can be repeated. The two test results are then averaged to produce an overall test result.

9.1.2 Cross-Validation Test

It is argued that the hold-out test does not make use of all given data for classifier evaluation. Although the two-fold test serves the purpose of testing all data, it may not be statistically informative to describe the classifier performance particularly for small data sets.

Suppose we have a small set of M samples for training and testing. Without re-sampling, the largest possible number of test cases without under-training the system is given by the leave-one-out cross-validation (LOOCV) method. In this method, one piece of data is kept for testing while the remaining data is used for training. This gives rise to M test results, which are eventually reported in averaged form.

However, if M is huge, then the leave-one-out test may become impractical since it will take up huge computing effort. A practical way to produce the desired statistics without looping the training process for full M times is to perform an *N-fold cross-validation*. Here, the entire data set is partitioned into N equal portions where the positive and negative classes are evenly distributed among these partitions. Typical values for N include 2, 5 and 10. The N test results are reported in average form.

9.2 Null Hypothesis Significance Test

Statistical inference through hypothesis testing is a method of confirmatory analysis based on data. It is often used to decide whether experimental results contain enough information to challenge conventional wisdom. In other words, hypothesis testing in statistical inference assesses the confidence in the occurrence of an experimental observation using statistical methods.

In empirical studies, it is often desired to conduct experiments to compare between two or among several algorithms on one or more data sets. The number of data sets to be used depends much on the data availability in the field of research. In the following, common scenarios of comparison such as comparing two classifiers on a single data set, comparing two classifiers over multiple data sets, and comparing multiple classifiers over multiple data sets are covered.

Table 9.1 Contingency table for McNemar's test

(Classifier A)	(Classifier B)	
	'Yes'	'No'
'Yes'	a	b
'No'	c	d

9.2.1 Comparing Two Classifiers on a Single Data Set

McNemar's test: This is a nonparametric method applied to nominal[1] data. It is available in R and other statistical packages. The test uses a 2×2 contingency table (Table 9.1) based on the outcomes of two tests on a sample of n subjects. Under the null hypothesis (H_0) that the probability of occurrences $p_b = p_c$, the two classifiers have no significant difference in their marginal proportions. The numbers b and c are called *discordances* since they refer to the "No" occurrence for one classifier, but "Yes" occurrence for the other classifier. A large value of the test statistic gives a small p value and this supports H_1 to reject the null hypothesis H_0.

The main assumptions made in McNemar's test include: (i) The classifier pair (A, B) is mutually independent, and (ii) Each classifier A and classifier B can be assigned to one of two possible categories.

9.2.2 Comparing Two Classifiers over Multiple Data Sets

Averaging over data sets: A straightforward approach to compare the difference between two classifiers over multiple data sets is to compute their average classification accuracies across the tested data sets. For example, a comparison of average accuracies of different classifiers on a set of related problems such as medical databases for a certain diagnosis from different institutions could be useful.

However, it is argued that averaging accuracies across domains could be meaningless if the results on different data sets are not comparable. Moreover, averages are susceptible to outliers since one excellent performance on one data set can prevail the overall poor performance, or vice versa.

t-**Test**: Another way to compare the difference between two classifiers over several data sets is to compute a *t-test* when the data is normally distributed. It is available in MATLAB (statistics toolbox) and in Python (scipy.stats). The testing includes two

---

[1] In the theory of scales of measurement, though not universally accepted, all measurements in science can be categorized into four scales, namely, *nominal, ordinal, interval* and *ratio* (Ref: https://www.britannica.com/topic/measurement-scale). Data based on nominal scale refers to categorical type which uses labels and data based on ordinal scale is based on rank-ordering. Data based on interval scale refers to quantitative attributes made within an interval and data based on ration scale refers to measurements with a non-arbitrary zero value.

cases of considerations namely, *unpaired* and *paired* t-tests. An *unpaired* (or independent samples, also known as Student's) t-test is a two-sample location test of the null hypothesis which takes the means of two normally distributed populations being equal. For example, the effect of a treatment has been measured on one population before treatment and on another population after treatment. A *paired* (or dependent samples, or repeated measures) t-test (see, e.g., Demšar (2006)) however, assumes that the two responses measured are on the same statistical unit. For example, in cancer treatment, an evaluation of effectiveness is to measure the size of tumor before and after the treatment for the same population of patients.

When comparing the means of two independent samples, the following assumptions should be met: (i) Normal distribution of compared populations, (ii) Same variance for the two compared populations being compared (as though the distribution for one group was merely a shifted of the other group, without changing shape), and (iii) Data being sampled independently for the two compared populations.

Wilcoxon Signed-Ranks Test: The Wilcoxon signed-ranks test serves as a non-parametric alternative to the paired t-test (Demšar, 2006). Consider N data sets, and let d_i represent the difference in accuracies between two classifiers for each set $i = 1, \ldots, N$. These differences are ranked in ascending order, with the better-performed one having a smaller rank number. When two classifiers have the same accuracy, an average rank is assigned. We denote R^+ as the sum of ranks for data sets where the second algorithm outperformed the first, and R^- as the sum of ranks for the opposite case. In the event of ties (i.e., when $d_i = 0$), the ranks are evenly distributed among the sums. If there is an odd number of ties, one rank is disregarded. Mathematically, R^+ and R^- can be represented (Demšar, 2006) as

$$R^+ = \sum_{d_i > 0} \text{rank}(d_i) + \frac{1}{2} \sum_{d_i = 0} \text{rank}(d_i) \tag{9.1}$$

$$R^- = \sum_{d_i < 0} \text{rank}(d_i) + \frac{1}{2} \sum_{d_i = 0} \text{rank}(d_i). \tag{9.2}$$

When $\min(R^+, R^-)$ is smaller than or equal to the critical value in the statistical table, the null hypothesis is rejected.

The procedure can be illustrated by the following example which compares between TER-based learning and LSE-based learning on a projection predictor. As shown in Table 9.2, the average percentage accuracies obtained from tenfold cross-validations for the two classifiers are tabulated for $N = 16$ data sets from the UCI repository with a binary class attribute.

The sum of ranks for positive differences between the performances of TER-based learning and LSE-based learning is $R^+ = (7 + 15 + 16 + 13.5 + 5.5 + 8.5 + 5.5 + 13.5 + 12 + 10) + (1.5) = 108$. On the other hand, $R^- = (3.5 + 11 + 3.5 + 8.5) + (1.5) = 28$. At a confidence level of $\alpha = 0.05$ for 16 data sets, the critical value from the Wilcoxon signed-ranks test table is 29. Since we have 28 (which is equal or less

Table 9.2 Classification accuracies of LSE-based and TER-based learning

No.	Data	LSE-based Ave($x\%$ accuracy)	TER-based Ave($y\%$ accuracy)	Diff (%) ($d = y - x$)	Rank
1	Shuttle-l-contr	99.03	99.06	+0.03	7
2	BUPA-liver	72.17	72.16	−0.01	3.5
3	Monk-1	89.90	90.55	+0.65	15
4	Monk-2	66.16	68.27	+2.11	16
5	Monk-3	90.39	91.02	+0.62	13.5
6	Pima-diabetes	77.73	77.73	0.00	1.5
7	Tic-tac-toe	98.22	98.24	+0.02	5.5
8	Breast-cancer-W	97.26	97.37	+0.11	8.5
9	StatLog-heart	83.25	83.07	−0.19	11
10	Credit-app	86.65	86.64	−0.01	3.5
11	Votes	94.24	94.27	+0.02	5.5
12	Mushroom	100.00	100.00	0.00	1.5
13	Wdbc	96.18	96.80	+0.62	13.5
14	Wpbc	79.60	79.49	−0.11	8.5
15	Ionosphere	89.18	89.63	+0.45	12
16	Sonar	81.78	81.96	+0.17	10

than the critical value) as the smaller of the two sums, we reject the null hypothesis that the two classifiers are equivalent at 0.05 confidence level.

The Wilcoxon signed-ranks test is argued to be safer than the paired t-test since it does not assume a normal distribution. However, when the assumptions of the paired t-test are met, the Wilcoxon signed-ranks test is less preferred over the paired t-test.

Sign Test: Another popular choice, known as sign test, to compare the overall performances of classifiers is to count the number of data sets on which an algorithm is the overall winner (Demšar, 2006). When each of the two compared algorithms wins approximately half of the data sets (i.e., $N_{win} \approx N/2$), they are assumed to be equivalent under the null hypothesis. The number of wins follows a binomial distribution, and the critical number of wins ($N_{critical}$) can be determined from statistical tables. A classifier is considered significantly better than another if it outperforms it on at least $N_{critical}$ data sets.

For a larger number of data sets, the number of wins under the null hypothesis follows a distribution with mean $N/2$ and standard deviation $\sqrt{N}/2$. In this context, a z-test can be conducted, and a classifier is considered significantly better if the number of wins exceeds $N/2 + 1.96\sqrt{N}/2$ at a confidence level of 0.05. When ties occur, the counts are evenly divided between the two compared classifiers.

When we apply the counting-based sign test on the above example where the TER-based learning was better on 11 out of 16 data sets (counting also one of the two data sets on which the two classifiers were tied), the sign test table shows that null hypothesis cannot be rejected at confidence level of 0.05.

The sign test neither requires commensurability of scores nor assumes normal distributions. It is applicable to any data, as long as the data sets are independent. However, it is less powerful than the Wilcoxon signed-ranks test. The sign test will only reject the null hypothesis if one classifier consistently outperforms the other, as indicated by the sign test table.

For the same example above, consider another classifier where its winning count with respect to LSE-based learning is 12 (critical value) out of the 16 data sets but with all $d_i < 0.01\%$. Although the null hypothesis that the two classifiers are equivalent is rejected, and the conclusion that the two classifiers are significantly different remains questionable. Hence much care must be taken when testing the hypothesis for small number of data sets.

9.2.3 Comparing Multiple Classifiers over Multiple Data Sets

The above-paired tests could be applied to the case of multiple classifiers by enumerating all combinatorial pairs. An argument against such an application is due to the possibility of rejecting the null hypothesis by random chance since so many tests have been made. Under this situation, either the Analysis of Variance (ANOVA, which assumes a normal distribution of data) or its nonparametric counterpart, the Friedman test can be adopted. An outline of the Friedman test is presented as follows.

Friedman Test (see, e.g., Demšar (2006)): The Friedman test serves as a nonparametric alternative to the repeated-measures ANOVA. It is available in MATLAB (statistics toolbox) and in Python (scipy.stats). The procedure to conduct a Friedman test is as follows. The compared algorithms or classifiers are ranked separately for each data set. A rank of 1 is assigned to the best-performing classifier, rank 2 to the second best, and so on. In the case of ties, an average rank is assigned. A sample ranking tabulation is shown in Table 9.3 for four classifiers (RP-FLD, RP-LSE, RP-TER, and RP-AUC).

Consider K algorithms running on N data sets. Let $r_{k,n}$ be the rank of k-th ($k \in \{1, \ldots, K\}$) algorithm on the n-th ($n \in \{1, \ldots, N\}$) data set. The average rank of each algorithm is computed as

$$R_k = \frac{1}{N} \sum_n r_{k,n}. \tag{9.3}$$

Under the null hypothesis, all the algorithms are equivalent and so their average ranks are equal. Based on the number of data sets N and the number of classifiers $K = 4$, the degree of freedom can be computed for locating the critical value.

Table 9.3 Average accuracy and ranks

		Accuracy(Rank)			
	Database name	RP-FLD	RP-LSE	RP-TER	RP-AUC
1	Shuttle-l-contr	0.9811(3.0)	0.9861(2.0)	0.9909(1.0)	0.9574(4.0)
2	BUPA-liver	0.7038(2.0)	0.6995(3.0)	0.7217(1.0)	0.6513(4.0)
3	Monks-1	0.9171(1.0)	0.8959(4.0)	0.9016(3.0)	0.9161(2.0)
4	Monks-2	0.6570(1.0)	0.6174(3.0)	0.6157(4.0)	0.6471(2.0)
5	Monks-3	0.9123(1.0)	0.8977(3.0)	0.9041(2.0)	0.8745(4.0)
6	Pima-diabetes	0.7591(2.0)	0.7073(4.0)	0.7776(1.0)	0.7549(3.0)
7	Tic-tac-toe	0.9827(1.0)	0.9814(3.0)	0.9822(2.0)	0.8660(4.0)
8	Breast-cancer-W	0.9720(3.0)	0.9695(4.0)	0.9740(1.0)	0.9725(2.0)
9	Statlog-heart	0.8324(1.5)	0.8255(4.0)	0.8318(3.0)	0.8324(1.5)
10	Credit-app	0.8671(1.0)	0.8592(4.0)	0.8667(2.0)	0.8632(3.0)
11	Votes	0.9414(3.0)	0.9408(4.0)	0.9424(2.0)	0.9439(1.0)
12	Mushroom	1.0000(2.0)	1.0000(2.0)	1.0000(2.0)	0.9945(4.0)
13	Wdbc	0.9654(2.0)	0.9445(4.0)	0.9650(3.0)	0.9716(1.0)
14	Wpbc	0.6722(2.0)	0.6218(4.0)	0.7961(1.0)	0.6591(3.0)
15	Ionosphere	0.8869(1.0)	0.8372(4.0)	0.8695(2.0)	0.8663(3.0)
16	Sonar	0.8162(2.0)	0.8071(4.0)	0.8152(3.0)	0.8238(1.0)
	Average	0.8667(1.781)	0.8494(3.500)	0.8722(2.063)	0.8497(2.656)

Post-hoc Test (see, e.g., Demšar (2006)): If the null hypothesis is rejected, a post-hoc Nemenyi test can be conducted to analyze the difference among the classifiers. If the average ranks of two classifiers differ by at least the *critical difference* value (9.4)

$$CD = q_\alpha \sqrt{\frac{K(K+1)}{6N}}, \tag{9.4}$$

then the performance of the two classifiers is significantly different.

Graphical Presentation of Results: The results of the post-hoc test can be represented graphically as shown in Fig. 9.1.

Example: Consider the four classifiers in Table 9.3. Given $N = 16$, $K = 4$ and based on the averaged ranks listed at the bottom of the table, the statistics should be obtained from the tables for the Friedman rank test (which is 7.80 for the upper critical values). For illustration purpose (assuming $K > 5$), we follow the statistics in Demšar (2006) to compute the following:

$$\chi_F^2 = \frac{12 \times 16}{4 \times 5} \left[(1.781^2 + 3.5^2 + 2.063^2 + 2.656^2) - \frac{4 \times 5^2}{4} \right] = 16.63,$$

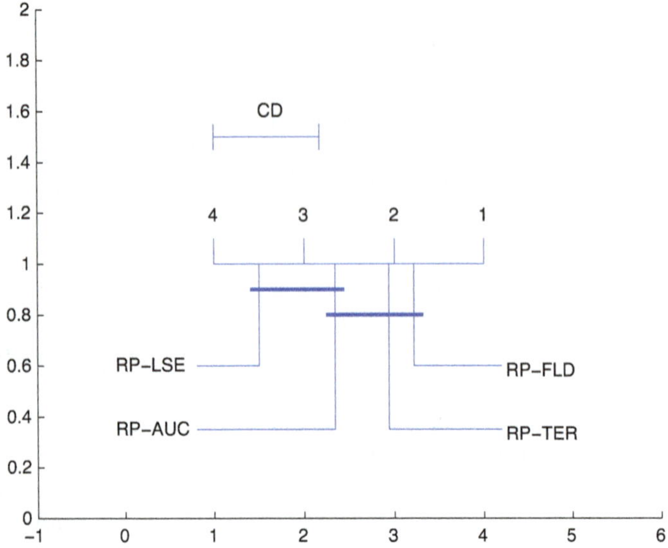

Fig. 9.1 Comparison of classifiers against each other with the Nemenyi test where groups of classifiers that are not significantly different are connected

$$F_F = \frac{(16 - 1) \times 16.63}{16 \times (4 - 1) - 16.63} = 7.95.$$

Since there are $K = 4$ algorithms and $N = 16$ data sets, we have $(4 - 1) = 3$ and $(4 - 1) \times (16 - 1) = 45$ degrees of freedom. The critical value of $F(3, 45)$ for $\alpha = 0.05$ is 2.8115. Since $F_F = 7.95$ is larger than the critical value, we reject the null hypothesis that the average ranks are of no significant difference from the mean rank $R_j = 2.5$.

Next, we can use the Nemenyi test for pairwise comparison if no classifier is singled out. The critical value for $\alpha = 0.05$ is 2.569 and this gives $CD = 2.569\sqrt{(4 \times 5)/(6 \times 16)} = 1.17$. From the graphical plot in Fig. 9.1, we see two distinct groups of performance: (RP-LSE) versus (RP-FLD and RP-TER). However, we cannot tell which group does the classifier RP-AUC fall within since it is seen to belong to two groups simultaneously. We can say that the experimental data is not sufficient to reach any conclusion regarding the performance of RP-AUC with respect to other compared classifiers.

9.3 Bayesian Analysis

Having the tools outlined for statistical inferences based on the Null Hypothesis Significance Test (NHST) in the previous section, we note here that the use of the p-value needs extra care. According to Ioannidis (2005), Amrhein et al. (2019), Nuzzo (2014), the NHST has not been utilized appropriately in many instances. This might have caused the high rate of non-replication of research discoveries. In a nutshell, the two fundamental issues pointed out in these references are, namely, (i) the bias and the extent of repeated independent testing, and (ii) the use of the p-value in a dichotomous way. The bias might come as a result of small sizes of studies, effects, and test relationships, flexibility in design, and relation to financial interests. Apart from the dichotomous "black-and-white thinking" where two methods with no statistically significant difference are not necessarily equivalent, Benavoli et al. (2017) pointed out several pitfalls of NHST such as $p = p(t(x) > \tau | H_0)$ (where $t(x)$ is the statistic computed from the data x, and τ is the critical value for responding to the test) being different from $p(H_0|x)$, practical falsehood of point-wise null hypothesis, omission of effect and sample sizes, magnitude and uncertainty, lack of information about the null hypothesis outcomes, lack of principled way to decide the confidence level, and lack of the link to sampling intention.

Since the point-wise null hypothesis is not preferred, Benavoli et al. (2017) recommended the approach based on Bayesian estimation, known as *Bayesian Analysis* which consists of three steps namely, (i) establishment of a descriptive mathematical model of the data, (ii) establishment of the credibility for each value of the parameter(s) before observing data and establishment of the prior distribution, and (iii) Utilization of the Bayes' rule to compute the posterior distribution based on the likelihood and prior distributions.

9.3.1 Comparing Two Classifiers on a Single Data Set

Consider two classifiers namely A and B, which have respectively the following accuracies in percentage for 10 folds:

$$
A : \begin{bmatrix} 95.8707 \\ 94.7935 \\ 93.5368 \\ 96.4093 \\ 95.6912 \\ 94.2549 \\ 94.6140 \\ 95.5117 \\ 96.2298 \\ 95.1526 \end{bmatrix} \quad B : \begin{bmatrix} 93.5368 \\ 92.8187 \\ 91.5619 \\ 93.8959 \\ 93.1777 \\ 93.1777 \\ 94.7935 \\ 91.7415 \\ 94.6140 \\ 94.0754 \end{bmatrix} \tag{9.5}
$$

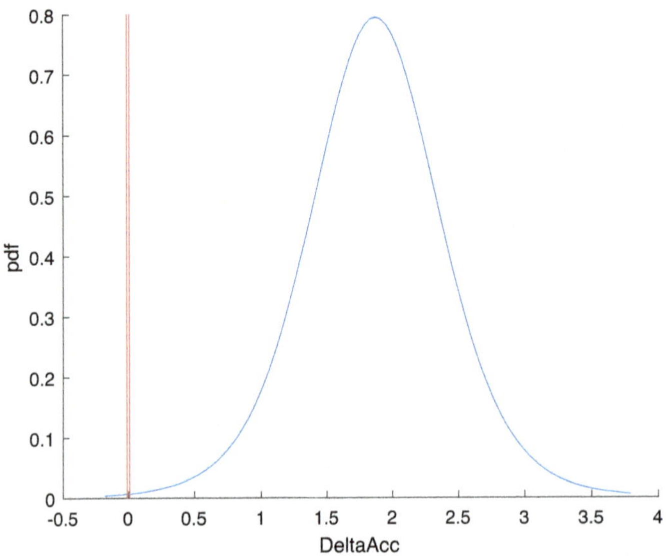

Fig. 9.2 Posterior for *A* versus *B* in one data set

Then, running of Bayesian Analysis based on

```
[probs,plot_data] = two_classifiers_one_data set(x,
rope, cvruns)
```

(https://github.com/LucyKuncheva/Bayesian-Analysis-for-Comparing-Classifiers) gives the posterior results: `probs = [0.0001, 0.0549, 0.9450]` where probs := $[A < B, \ A = B(rope), \ A > B]$. The distribution of `plot_data` is plotted in Fig. 9.2 where the *robe* has been defined as ± 0.01.

9.3.2 Comparing Two Classifiers over Multiple Data Sets

Consider the 16 data sets of Table 9.2 where we want to compare between the TER method and the LSE method in terms of their classification performances. Running of the following codes

```
[samples, probs] = signed_rank_test_diff(x,rope,prior,T)
```

(https://github.com/LucyKuncheva/Bayesian-Analysis-for-Comparing-Classifiers) gives the plot in Fig. 9.3. Due to the small differences in the accuracy percentage, the analysis shows a very narrow distribution with $p(\text{Rope}) = 99.834\%$, $p(\text{TER}) = 0.166\%$ and $p(\text{LSE}) = 0.000\%$.

As another example, if the differences in the accuracy percentage have been enlarged by 10 times, then the analysis results in Fig. 9.4 are observed. This modified

Fig. 9.3 Bayesian analysis comparing two classifiers (TER and LSE) on the 16 data sets of Table 9.2

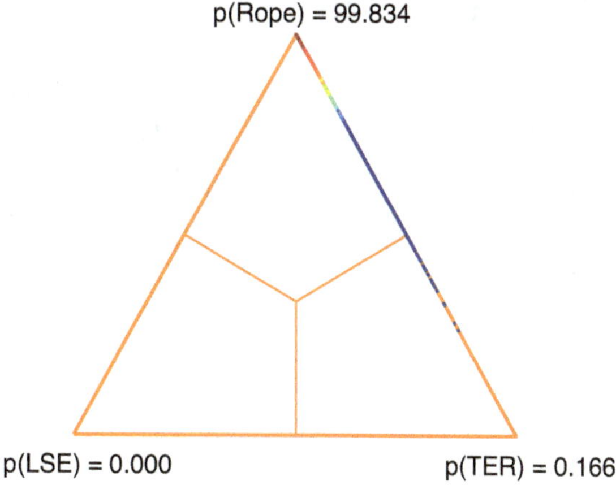

Fig. 9.4 Bayesian analysis comparing two classifiers (TER and LSE) on the 16 data sets of Table 9.2 with a 10× magnification of the difference values of the classifiers' accuracies

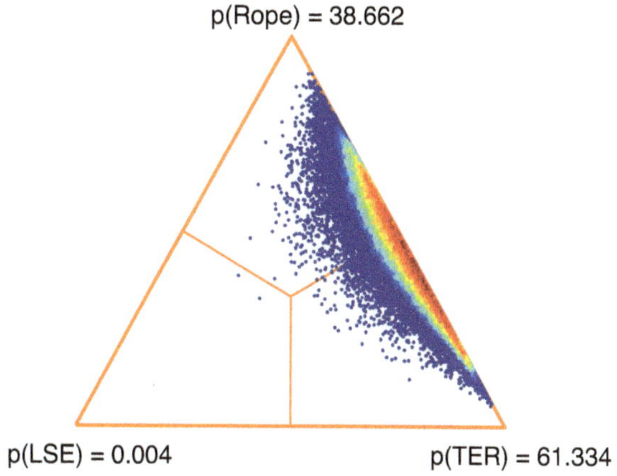

example demonstrates sensitiveness of the Bayesian Analysis to the effect size of differences in the accuracy percentage.

9.3.3 Comparing Multiple Classifiers over Multiple Data Sets

According to Benavoli et al. (2017) who quoted (Gelman et al. 2012) that "in Bayesian analysis we usually do not have to worry about multiple comparisons. The reason is that we do not worry about Type-I error, because the null hypothesis is hardly believable to be true". They suggested using multilevel analysis or a hierarchical Bayesian model on multiple classifiers. In other words, the various combinations of classifier pairs are to be evaluated.

9.4 Summary Notes

While acknowledging the tendency toward a "black-and-white" nature in null hypothesis significance testing (NHST), it is crucial to exercise caution, especially when presenting inference results and drawing conclusions. On the other hand, the argument in favor of Bayesian analysis (BA) that "point-wise null hypotheses are practically always false" (based on the assumption of the existence of H_0) can be intricate. This complexity arises because the null hypothesis may not apply equally to two groups with diverse observations drawn from the same classifier.

Given the current available tools, we recommend using BA for comparing two classifiers on a single data set or across multiple data sets. Additionally, consider reserving the post-hoc Nemenyi test for visualizing the ranking order of multiple classifiers. Most importantly, avoiding absolute "black-and-white" claims can prevent misrepresentation.

9.5 Bibliographic Notes

The t-statistic was introduced by William Sealy Gosset (1876–1937) in 1908 (Student, 1908). Gosset used the pen name "Student" for his publications due to the trade secret nature of his statistical work in Guinness Brewery in Dublin, Ireland. Gosset who was among the best graduates from Oxford, was hired to apply biochemistry and statistics to Guinness's industrial processes. A t-test refers to any statistical hypothesis test in which the test statistic follows a Student's t distribution. Gosset devised the test as a way to monitor the quality of stout.[2] It was Sir Ronald Aylmer Fisher (1890–1962) who appreciated the importance of Gosset's small-sample work and introduced the current t-form which gained wide recognition after.

The McNemar's test is named after Quinn McNemar who introduced it in 1947 (McNemar, 1947). Quinn McNemar, a US psychologist and statistician, was also known for his work on Intelligence Quotient (IQ) test revision and for his book on Psychological Statistics (1949). His primary motivation for devising the well-known test statistic, was that under various scenarios, the sampling variance of the difference between two proportions (or percentages) must consider that these proportions are not based on independent samples. For instance, this arises when comparing responses before and after an experience, assessing item difficulty, testing group opinions, or evaluating differences between paired proportions.

In terms of statistical tests for comparisons of more algorithms on multiple data sets, Demšar (2006) reviewed the then-current practice and examined several relevant tests. He then recommended the Wilcoxon signed-ranks test (Wilcoxon, 1945) for comparison of two classifiers and the Friedman test (Friedman, 1937) with the corresponding post-hoc tests (Nemenyi, 1963) for comparison of more classifiers

---

[2] A kind of dark beer made using roasted malt or barley.

over multiple data sets. To facilitate interpretation of results, CD (critical difference) diagrams were further introduced for the Friedman test.

In the recent years, there has been a call for retirement of the use of p-values (Ioannidis, 2005; Nuzzo, 2014; Amrhein et al., 2019) in view of the abuse of its conclusion in black-and-white sense. In response, the ASA came out with a statement on p-values and statistical significance (Wasserstein & Lazar, 2016). Subsequently, Benavoli et al. (2017) proposed a change in comparing classifiers through Bayesian Analysis.

9.6 Exercises

9.1 Suppose you are given a data set of 800 samples in total. Which of the following is/are among the best for the given options for evaluating an algorithm based on this data set?

(a) Use all samples as training and test sets.
(b) Based on single hold-out test, i.e., use M samples for testing and use $800 - M$ samples for training the algorithm.
(c) Based on multiple hold-out tests, i.e., perform hold-out test multiple times using randomly partitioned training and test sets.
(d) Based on N-fold cross-validation test.
(e) Based on leave-one-out test, i.e., use 1 sample for testing and use 799 samples for training the algorithm, and repeat this procedure until all samples have been tested. The final results are averaged.

9.2 How can the hold-out test be performed using Python?
9.3 What is the difference between regression error and classification error? Provide examples of their score measurements.
9.4 How to do a hyperparameter tuning based on grid search in Python?
9.5 Under what situation cross-validation should be avoided?
9.6 Is it possible to check if a classifier has underfit or overfit the data using only the training set? If yes, how? If no why?
9.7 A classifier that attains 100% accuracy on the training set and 70% accuracy on test set is better than another classifier that attains 80% accuracy on the training set and 75% accuracy on test set. True or False?
9.8 Suppose the hyperparameter η has been selected for a model using tenfold cross-validation. Among the following, the best way to decide a final model to use and estimate its error is to

(a) pick any of the 10 models from the 10 folds as the final model; use the average CV error for the 10 models as its error estimate.
(b) pick any of the 10 models from the 10 folds as the final model; use its error estimate on the held-out data.

   (c) train a new model on the full data set using the hyperparameter η obtained; use the average CV error as its error estimate.

   (d) average all of the 10 models from the 10 folds; use the average CV error as its error estimate.

   (e) average all of the 10 models from the 10 folds; use the error of the combined model on the full training set.

9.9 The _______ the p-value, the stronger the evidence against the null hypothesis provided by the data.

   (a) larger

   (b) smaller

9.10 One-sample t-test: To test the hypothesis that preparation study for more than 3 h for three days right before the test helps to score over 70 marks. Here are the results of a random sample of 12 students who had done the preparation study before the test: 86, 81, 71, 90, 69, 64, 76, 85, 77, 71, 80, 62. Test using the following hypotheses, report the test statistic with the p-value at 95% confidence level, then summarize your conclusion.

Chapter 10
Applications

This chapter consists of three main sections, namely, synthetic examples, benchmark examples, and biometric authentication examples. The section on synthetic examples provides low dimensional data constructions to aid understanding of the underlying fundamental concepts. Without needing to dwell into the often specialized data pre-processing and extraction processes, the readers can readily observe the effect of classification algorithms on representative one- and two-dimensional data with intuitive interpretation. The section on benchmark examples provides a slightly elevated view on classification of physical data of dimensions higher than two. Finally, in the section of biometric authentication examples, the full process of data preprocessing, feature extraction, and classification are showcased.

10.1 Synthetic Examples

10.1.1 Single-Dimensional Problems

(i) The non-overlapping case

Figure 10.1 shows two cases of an over-determined system based on different polynomial model orders. In Fig. 10.1a, a third-order polynomial model is adopted using three training methods such as the Least Squares Error (LSE), the Total Error Rate (TER), and the Area Under the ROC (AUC) methods. Since this is single-dimensional data, the total number of parameters including the intercept is four, i.e., $\boldsymbol{w} = [w_0, w_1, w_2, w_3]^T$ and $\boldsymbol{p}(x) = [1, x, x^2, x^3]^T$. The training data points are given as (x_i, y_i): $(-5.5, -1), (-4.5, -1), (-3.5, -1), (-2.5, -1), (-1.5, -1),$ $(-0.5, -1), (0.5, 1), (1.5, 1), (2.5, 1), (3.5, 1), (4.5, 1), (5.5, 1)$, where their polynomial transformed features $\boldsymbol{p}(x_i)$, $i = 1, \ldots, 12$ can be stacked as $\mathbf{P} \in \mathbb{R}^{12 \times 4}$.

© The Author(s), under exclusive license to Springer Nature Singapore Pte Ltd. 2025

K.-A. Toh et al., *Analytic Learning Methods for Pattern Recognition*,

https://doi.org/10.1007/978-981-96-2151-4_10

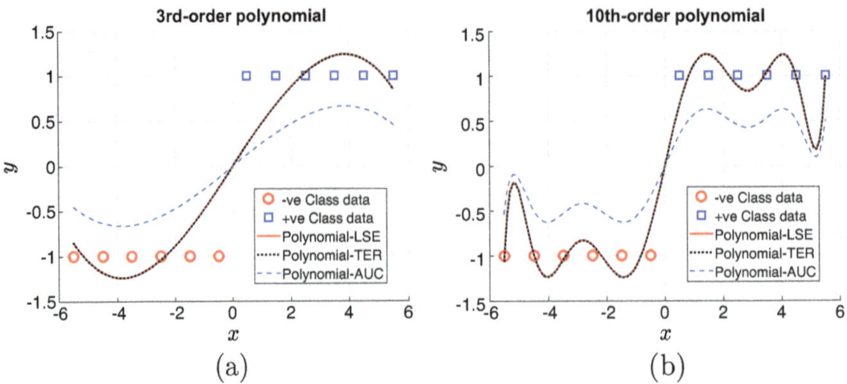

Fig. 10.1 Learning of twelve single-dimensional data points where the two categories are separable: **a** using a third-order polynomial model, **b** using a tenth-order polynomial model

The respective training methods adopted are LSE according to (5.6), TER according to (5.75) and AUC according to (5.90). No regularization has been adopted for LSE and TER because their covariance matrix is non-singular. Since the number of positively labeled data is equal to the number of negatively labeled data, the TER and LSE produce the same decision output. For the AUC method, a regularization of $b = 10^{-100}$ was adopted. Due to the ranking nature of AUC training, the decision output does not fit as well as that of LSE and TER. Nevertheless, all three training methods result in perfect classification when a decision threshold is set at $\tau = 0$. Figure 10.1b shows a tenth-order polynomial model being adopted to learn the same set of data. This example shows the complexity of the model with respect to the order of polynomials.

Figure 10.2 shows two cases of an under-determined system trained using six training data points (x_i, y_i): $(-2.5, -1), (-1.5, -1), (-0.5, -1), (0.5, 1), (1.5, 1), (2.5, 1), i = 1, \ldots, 6$. The learning is under-determined because the number of polynomial parameters (9 and 11 parameters for the eighth-order and the tenth-order polynomial models respectively) is larger than the number of training samples. The adopted learning methods are LSE according to (5.25) and TER according to (5.83) (regularization factor $b = 0$) using a polynomial model. The AUC according to (5.90) is not shown since the covariance matrix of the difference vectors does not have sufficient rank for inversion. For both Fig. 10.2a and b, the learning results show that the LSE and TER methods pass through all six data points with similar output values. This is because the system is under-determined and with balanced weighting between the two categories where TER learns the same weights as that of LSE. In terms of classification, both methods give the same classification error rate as they have similar output cross-over points at zero threshold value for decision.

Although the above examples (Figs. 10.1 and 10.2) constitute the trivial case for classification, they nevertheless serve the purpose of observing the fitting behavior of learning algorithms. We shall observe the non-trivial case with overlapping categorial observations in the next example.

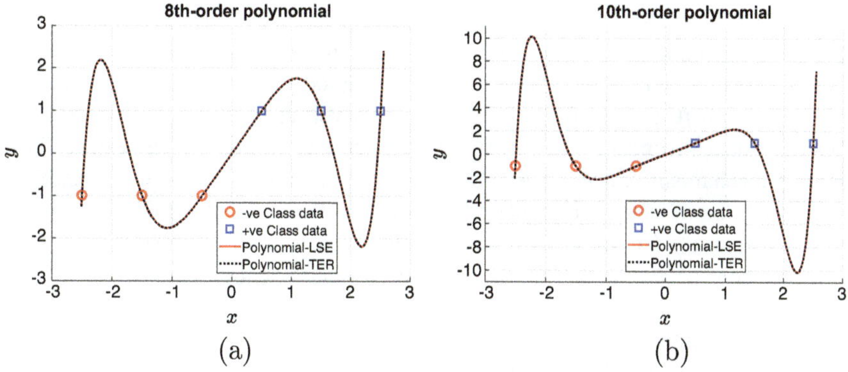

Fig. 10.2 Learning of six single-dimensional data points where the two categories are separable: **a** using a eighth-order polynomial model, **b** using a tenth-order polynomial model

(ii) The overlapping case

Figure 10.3 shows the training results based on eleven class overlapping data points given by (x_i, y_i): $(-4, -1)$, $(-3, -1)$, $(-2, -1)$, $(-2.5, 1)$, $(-1.5, 1)$, $(-0.5, 1)$, $(0.5, 1)$, $(1.5, 1)$, $(2.5, 1)$, $(3.5, 1)$, $(4.5, 1)$, $i = 1, \ldots, 11$. The learning system is over-determined for both the third-order polynomial (with 4 parameters) and the seventh-order polynomial (with 8 parameters) models since there are 11 learning samples. No regularization has been applied for both LSE and TER learning as the covariance matrix is non-singular. The AUC used a regularization of $b = 10^{-10}$ to deal with near singularity of the covariance matrix. For LSE (5.6) and TER (5.75) methods, the decision threshold is set at 0 since the target values are in $[-1, +1]$. Based on this decision threshold, the resultant classification error counts are respec-

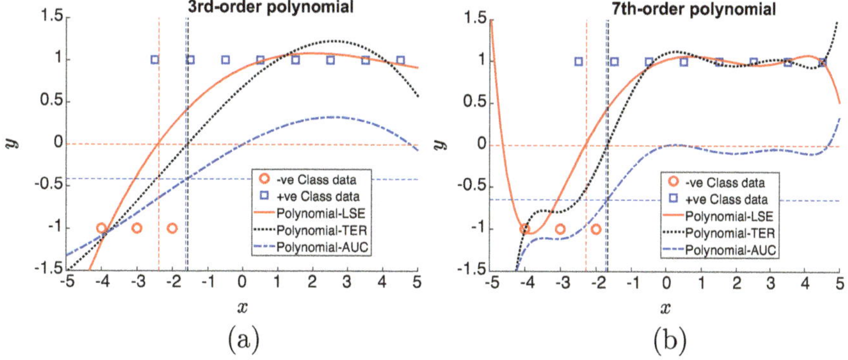

Fig. 10.3 Learning of eleven single-dimensional data points where the two categories are non-separable: **a** using a third-order polynomial model, **b** using a seventh-order polynomial model

tively 2 and 1 for LSE and TER as shown in Fig. 10.3a. For the LSE method, one blue square and one red circle are misclassified (with respect to the thin red-dashed vertical decision boundary line). For the TER method, only the blue square is misclassified (with respect to the thin black-dashed vertical decision boundary line). This illustrates the impact of error counting with respect to each category in the TER learning mechanism. Since the AUC does not use an explicit threshold in learning, the learned output is not zero centered. This means that a threshold of zero value is not applicable to AUC for classification decision. In view of this issue, an optimal TER threshold according to (5.91) is adopted for classification decision. The classification result for AUC based on this threshold is 1 error count similar to that of TER (see the thin blue-dashed decision boundary line).

In Fig. 10.3b, a seventh-order polynomial model is adopted to learn the eleven data samples using the same regularization setting as that in above. The system remains over-determined since there are only 8 model parameters including the intercept. The results show respectively 2, 1, and 1 error counts for LSE, TER, and AUC methods. The AUC method has the same error count with that of TER because the threshold decision has been based on the optimal total error rate according to (5.91).

Figure 10.4a shows an equal-determined system with 11 model parameters (based on a $10th$-order polynomial with intercept) and 11 training samples. A regularization setting of $b = 0.1$ has been utilized for both LSE and TER. For TER, such addition of regularization appears to weaken the effect of the class-specific weighting factors. The results show that the outputs of LSE have more closely fit training samples than that of TER. Figure 10.4b shows an under-determined system with 17 learning parameters (based on a $16th$-order polynomial with intercept) on the 11 training samples. The LSE shows a close fit of all data samples due to the high model complexity. For the TER method, the learning does not appear to fit some data samples. These two learning cases distinguish the weighted learning of TER from that of LSE.

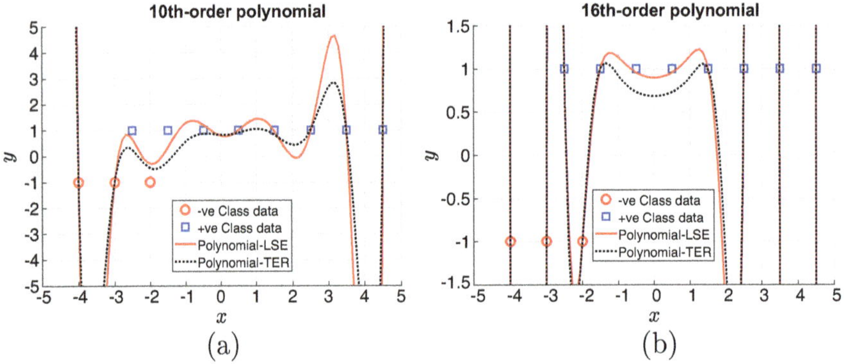

Fig. 10.4 Learning of eleven single-dimensional data points where the two categories are nonseparable and imbalance: **a** using a tenth-order polynomial model, **b** using a sixteenth-order polynomial model

10.1.2 Two-Dimensional Problems

(i) An extended XOR problem

This is an extension to the four points XOR problem with two additional points. The data samples are linearly non-separable. Figure 10.5 shows the x_1 and x_2 features distribution with corresponding label y, i.e., (x_1, x_2, y): $(0.4, 0.4, -1)$, $(0.8, 0.8, -1)$, $(1.2, 0.4, -1)$, $(0.4, 0.8, +1)$, $(0.8, 0.4, +1)$, $(1.2, 0.8, +1)$. A sixth-order bivariate polynomial is adopted to learn the 6 data points. Since there are 28 weight coefficients including the bias term for the polynomial model, the learning system is under-determined. Figures 10.5a, b and 10.6a and b shows respectively the decision boundaries for the p-bridge regression when $k = 2.0, k = 1.8, k = 1.5$ and $k = 1.1$. Their corresponding estimated polynomial coefficients are shown in the (c) and (d) plots in each figure. These plots show the difference in order of importance (or values) of the coefficients for different k values. When k approaches 2, the estimation is observed to be relatively "unbiased". When k approaches 1, many of the coefficients are suppressed.

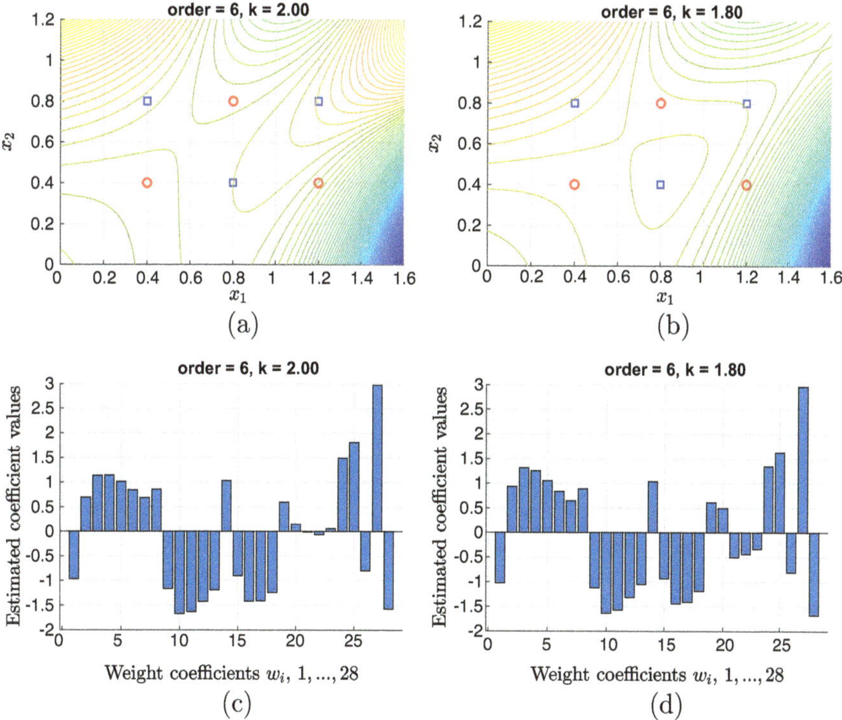

Fig. 10.5 Learning of 2D synthetic problem using a sixth-order polynomial model on p-bridge. **a–b** Decision contours for $k = 2.0$ and $k = 1.8$, **c–d** Estimated polynomial coefficients

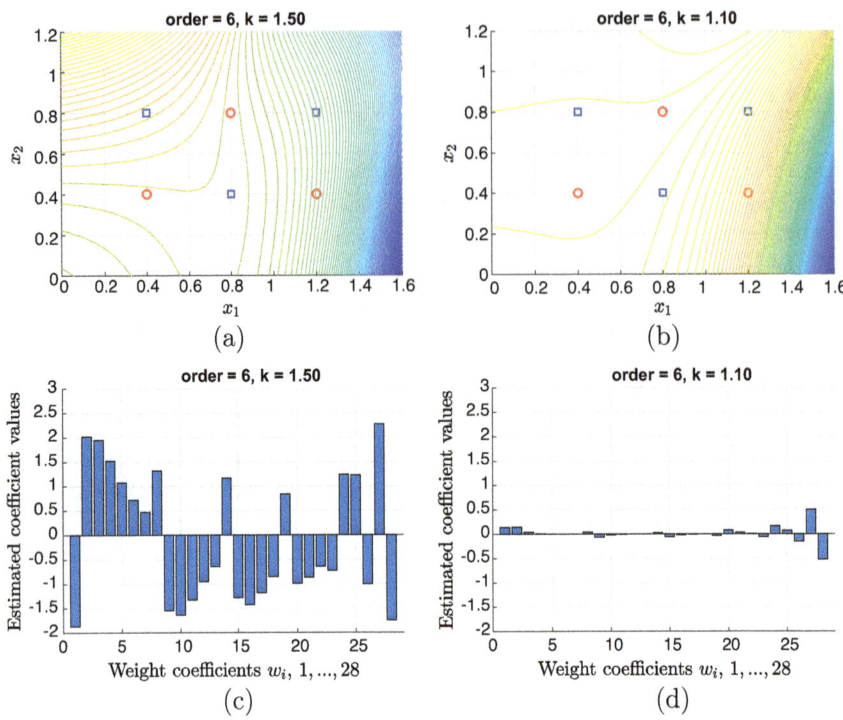

Fig. 10.6 Learning of 2D synthetic problem using a sixth-order polynomial model on p-bridge.
a–b Decision contours for $k = 1.5$ and $k = 1.1$, **c–d** Estimated polynomial coefficients

(ii) Two-Dimensional Single Gaussian per Category

In this example, a set of two-category data points is generated by using a single
Gaussian distribution for each class. The two categories are labeled as Class-0 and
Class-1 which respectively have 300 and 200 samples for training. Another 1000
samples are generated for each category for testing. The total training and test samples are thus respectively 500 and 2000. The learning target for Class-0 data is set
at 0 and the learning target for Class-1 data is set at 1. Figure 10.7 shows the training and test samples marked by heavy and light markers for each category. Apart
from the linear regression model (3.28), a full bivariate polynomial model (3.42) is
adopted to learn the training data. There are 6 and 21 polynomial coefficients for the
adopted second-order and fifth-order polynomial models respectively. Two learning
methods namely the least squares error (5.6) and the total error rate (5.75) methods
are adopted for fitting the data. We abbreviate the Linear Regression, the polynomial model adopting Least Squares Error, and the polynomial model adopting Total
Error Rate as LR, Polynomial-LSE, and Polynomial-TER respectively. Figure 10.7a
shows the decision boundaries of LR, second-order Polynomial-LSE, and second-
order Polynomial-TER with respective test errors of 2.5, 3.45, and 2.4%. Figure 10.7b

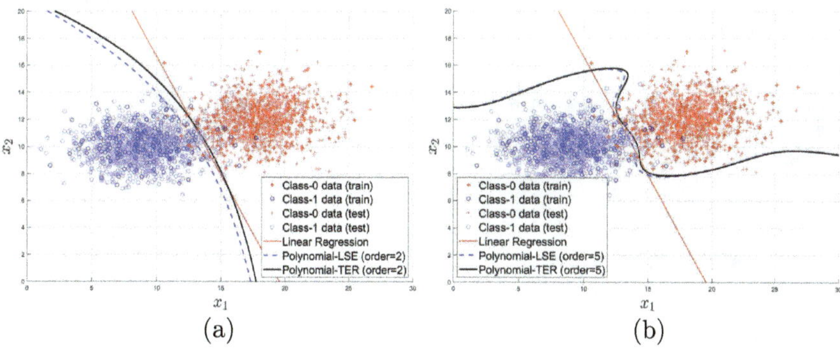

Fig. 10.7 Learning of two-dimensional Gaussian data points using a **a** second-order bivariate polynomial model, **b** fifth-order bivariate polynomial model

shows the decision boundaries of LR, fifth-order Polynomial-LSE, and fifth-order Polynomial-TER with respective test errors of 2.5, 2.75, and 2.5%. These results show Polynomial-LSE tends to show overfitting comparing with Polynomial-TER.

```
% Generate artificial single-Gaussian distributions for each class
% Train and test the FMP model using LSE and TER
clear all

order = 5;
b = 0;

%--- Generate Gaussian data points ---%
mu1 = [18 12];              % red cluster
sigma1 = [6 0.5; 0.5 2]; % red cluster
rng('default')
z1 = mvnrnd(mu1,sigma1,1300);
mu2= [9 10];               % blue cluster
sigma2 = [6 0.3; 0.3 1.5];  % blue cluster
rng('default')
z2 = mvnrnd(mu2,sigma2,1200);

%--- Form training and test sets ---%
Xtrn = [z1(1:300,:); z2(1:200,:)];        % red 300, blue 200
Xtst = [z1(301:1300,:); z2(201:1200,:)]; % red 1000, blue 1000
ytrn = [zeros(300,1); ones(200,1)];       % red 300, blue 200
ytst = [zeros(1000,1); ones(1000,1)];     % red 1000, blue 1000
[nsamp,nfeat] = size(Xtrn);

%--- Generate meshgrid for contour plots ---%
[X1, X2] = meshgrid(0:.1:30,0:.1:20);

%--- Linear Regression without intercept ---%
X = [ones(nsamp,1),Xtrn];
y = ytrn;
w1 = inv(X'*X)*X'*y;
%--- Test LR ---%
```

```
Xt = [ones(length(ytst),1),Xtst];
y_est1 = Xt*w1;
errs1 = sum(y_est1(ytst==0)>0.5) + sum(y_est1(ytst==1)<=0.5);
error_rate1 = 100*errs1/length(ytst); % percentage error count
%--- Grid test data for contour plot ---%
ONES = ones(size(X1));
y_grid1 = ONES.*w1(1) + X1.*w1(2) + X2.*w1(3);


%--- Train FMP using LSE ---%
[P, olist] = FMP(Xtrn,order);
alpha2 = inv(P'*P)*P'*y;
%--- Test FMP ---%
[Pt, olist] = FMP(Xtst,order,olist);
y_est2 = Pt*alpha2;
errs2 = sum(y_est2(ytst==0)>0.5) + sum(y_est2(ytst==1)<=0.5);
error_rate2 = 100*errs2/length(ytst); % percentage error count
%--- Grid test data for contour plot ---%
[nrow,ncol] = size(X1);
XX = [reshape(X1,nrow*ncol,1), reshape(X2,nrow*ncol,1)];
[Pgrid, olist] = FMP(XX,order,olist);
y_grid2_col = Pgrid*alpha2;
y_grid2 = reshape(y_grid2_col,nrow,ncol);


%--- Train FMP using TER ---%
Xp = Xtrn(y==1,:); Xm = Xtrn(y==0,:);
Yp = y(y==1,1); Ym = y(y==0,1);
[Pp, olist] = FMP(Xp,order,olist);
[Pm, olist] = FMP(Xm,order,olist);
[Mp,K] = size(Pp); [Mm,K] = size(Pm);
th = 0.5;
GG = inv((Pp'*Pp)/Mp + (Pm'*Pm)/Mm + b*eye(K));
FF = ((th-0.5)/Mm)*(sum(Pm,1))' + ((th+0.5)/Mp)*(sum(Pp,1))';
alpha8 = GG*FF;
%--- Test FMP ---%
[Pt, olist] = FMP(Xtst,order,olist);
y_est8 = Pt*alpha8;
errs8 = sum(y_est8(ytst==0)>0.5) + sum(y_est8(ytst==1)<=0.5);
error_rate8 = 100*errs8/length(ytst); % percentage error count
%--- Grid test data for contour plot ---%
[Pgrid, olist] = FMP(XX,order,olist);
y_grid8_col = Pgrid*alpha8;
y_grid8 = reshape(y_grid8_col,nrow,ncol);


figure(12), hold on
[c1,h1]=contour(X1,X2,y_grid1,[0.5 0.5],':');
[c2,h2]=contour(X1,X2,y_grid2,[0.5 0.5],'--');
[c8,h8]=contour(X1,X2,y_grid8,[0.5 0.5],'-');

hc01=plot(Xtst(1:1000,1),Xtst(1:1000,2),'+r','linewidth',1)
set(hc01,'Color',[1 0.5 0.5]),; % set the colour to light red
hc02=plot(Xtst(1001:2000,1),Xtst(1001:2000,2),'ob','linewidth',1)
set(hc02,'Color',[0.6 0.6 1]),; % set the colour to light blue
hc1=plot(Xtrn(1:300,1),Xtrn(1:300,2),'+r','linewidth',1.5)
hc2=plot(Xtrn(301:500,1),Xtrn(301:500,2),'ob','linewidth',1.0)
```

```
xlabel('$x_1$','Interpreter','latex','FontSize',28)
ylabel('$x_2$','Interpreter','latex','FontSize',28)
set(h1,'Color','r','LineWidth',1.5); % Linear Regression
set(h2,'Color','b','LineWidth',2); % LSE-FMP k=2
set(h8,'Color','k','LineWidth',2); % k-norm regression
set(gcf,'renderer','zbuffer');
string3 = sprintf('Polynomial-LSE (order=%d)',order);
string4 = sprintf('Polynomial-TER (order=%d)',order);
hLeg1 = legend([hc1, hc2, hc01, hc02, h1(1), h2(1), h8(1)], ...
'Class-0 data (train)', 'Class-1 data (train)', ...
'Class-0 data (test)', 'Class-1 data (test)', ...
'Linear Regression', string3, string4);
set(hLeg1,'FontSize',18);
grid
hold off
```

(iii) Two-Dimensional Mixture of Gaussians per Category

In this example, a data set with three categorial labels is generated. The data set contains 300 red labeled, 200 green labeled, and 300 blue labeled samples for training. Each category has another 1000 samples for testing. Each of the red and green labeled samples is generated by mixing two Gaussian distributed samples and the blue labeled samples are generated using a single-Gaussian distributed samples. A full multivariate (bivariate) polynomial (FMP) model and a reduced multivariate polynomial (RMP, see Example 3.8 for RM() function) model are adopted for TER learning the given training samples. There are 45 and 31 polynomial coefficients for the FMP and the RMP models respectively. Figure 10.8a–b show the decision regions for respectively TER-FMP model and TER-RMP, both of which at eighth order. The test accuracies are 89.3% and 88.2% respectively. This result shows higher model complexity of the FMP model has been preferred in this learning.

```
% Generate artificial single-Gaussian distributions for each category
% Train and test the RMP model using LSE and TER
clear all

order = 8;
b = 0; % no regularization

rng('default') % fix the random seed
%--- Generate mixture of Gaussian data points ---%
mu1a = [18 14];                                    % red cluster
sigma1a = [3 0.5; 0.5 2];                          % red cluster
z1a = repmat(mu1a,650,1) + randn(650,2)*sigma1a;% red cluster
mu1b = [20 10];                                    % red cluster
sigma1b = [3 0.1; 0.1 3];                          % red cluster
z1b = repmat(mu1b,650,1) + randn(650,2)*sigma1b;% red cluster
%-------------------------------------------%
mu2a= [9 12];                                      % green cluster
sigma2a = [3 0; 0 3];                              % green cluster
```

```matlab
z2a = repmat(mu2a,600,1) + randn(600,2)*sigma2a;% green cluster
mu2b= [13 12];                                  % green cluster
sigma2b = [2 0.3; 0.3 1.5];                     % green cluster
z2b = repmat(mu2b,600,1) + randn(600,2)*sigma2b;% green cluster
%----------------------------------------%
mu3= [15 6];                                    % blue cluster
sigma3 = [1.5 0.8; 0.2 2];                       % blue cluster
z3 = repmat(mu3,1300,1) + randn(1300,2)*sigma3; % blue cluster
%----------------------------------------%

%--- Form training and test sets ---%
Xtrn = [z1a(1:150,:); z1b(1:150,:); ...
z2a(1:100,:); z2b(1:100,:); ...
z3(1:300,:)]; % red 300, green 200, blue 300 points
Xtst = [z1a(151:650,:); z1b(151:650,:); ...
z2a(101:600,:); z2b(101:600,:); ...
z3(301:1300,:)]; % red 1000, green 1000, blue 1000 points
Ytrn = [repmat([1,0,0],300,1);
repmat([0,1,0],200,1);
repmat([0,0,1],300,1);];
Ytst = [repmat([1,0,0],1000,1);
repmat([0,1,0],1000,1);
repmat([0,0,1],1000,1);];
[nsamp,nfeat] = size(Xtrn);

%--- Generate meshgrid for contour plots ---%
range1=0:.1:30; range2=0:.1:20; [X1, X2] = meshgrid(range1,range2);

%--- Train RM model using TER ---%
P = RM(Xtrn,order); % Construct RM matrix
[alpha8] = TERtrain(P,Ytrn,b,1,1);

%--- Test RM model ---%
Pt = RM(Xtst,order); % Construct RM matrix
[Yest8, Yt_class_predict, TestAccuracy] = TERtest(Pt, Ytst, alpha8);
```

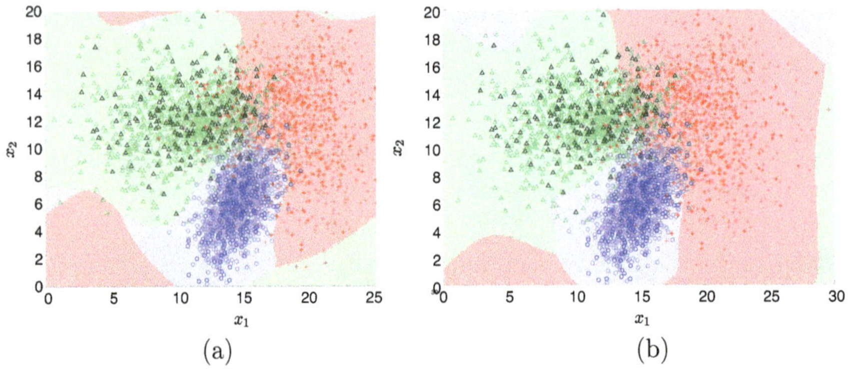

(a) (b)

Fig. 10.8 TER learning of two-dimensional Gaussian data points with three categories using an **a** eighth-order FMP model, **b** eighth-order RMP model

```
%--- Grid test data for contour plot ---%
[nrow,ncol] = size(X1);
XX = [reshape(X1,nrow*ncol,1), reshape(X2,nrow*ncol,1)];
Ygtst = zeros(length(XX),3); % dummy target for TERRMtest regions
PPt = RM(XX,order);
[Ygridout, Ygrid8est] = TERtest(PPt, Ygtst, alpha8);
% convert indicator matrix to label vector
y_region8 = zeros(length(Ygrid8est),1);
y_region8(Ygrid8est(:,1)==1)=1;
y_region8(Ygrid8est(:,2)==1)=2;
y_region8(Ygrid8est(:,3)==1)=3;
decisionmap8 = reshape(y_region8,nrow,ncol);

figure(1), hold on
%--- color the decision regions ---%
imagesc(range1,range2,decisionmap8);
set(gca,'ydir','normal');
% colormap: 1=light-red, 2=light-green, 3=light-blue
cmap = [1 0.8 0.8; 0.85 1 0.85; 0.9 0.9 1];
colormap(cmap);
%--- plot training and test data points ---%
hc01=plot(Xtst(1:1000,1),Xtst(1:1000,2),'+r','linewidth',1);
set(hc01,'Color',[1 0.5 0.5]); % set the colour to light red
hc02=plot(Xtst(1001:2000,1),Xtst(1001:2000,2),'^g','linewidth',1);
set(hc02,'Color',[0.3 1 0.3]); % set the colour to light green
hc03=plot(Xtst(2001:3000,1),Xtst(2001:3000,2),'ob','linewidth',1);
set(hc03,'Color',[0.6 0.6 1]); % set the colour to light blue
hc1=plot(Xtrn(1:300,1),Xtrn(1:300,2),'+r','linewidth',1.5);
hc2=plot(Xtrn(301:500,1),Xtrn(301:500,2),'^g','linewidth',1.0);
set(hc2,'Color',[0 0.3 0]); % set the colour to dark green
hc3=plot(Xtrn(501:800,1),Xtrn(501:800,2),'ob','linewidth',1.0);
xlabel('$x_1$','Interpreter','latex','FontSize',28);
ylabel('$x_2$','Interpreter','latex','FontSize',28);
hold off
axis([0 30 0 20])
TestAccuracy
```

(iv) Intertwined spirals: the two-spiral problem

In Fig. 10.9, a full bivariate polynomial model is used to fit the two-spiral data points
in two-dimensional space. A tenth-order polynomial and a thirteenth-order are used
to learn the given spiral data points as shown in Fig. 10.9a and b respectively. The
results show a slight underfitting of the tenth-order polynomial model and an exact
fitting of the thirteenth-order polynomial model for both LSE and TER methods. In
Fig. 10.10, a sixth-order polynomial model and a sixteenth-order polynomial model
are experimented. This example shows the departure of the fitting results between
LSE and TER for the sixteenth-order system.

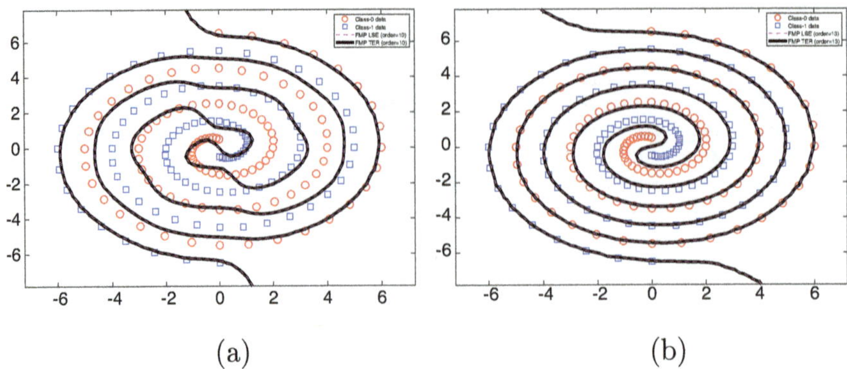

Fig. 10.9 Learning of the two-spiral problem: **a** using a tenth-order polynomial model, **b** using a thirteenth-order polynomial model

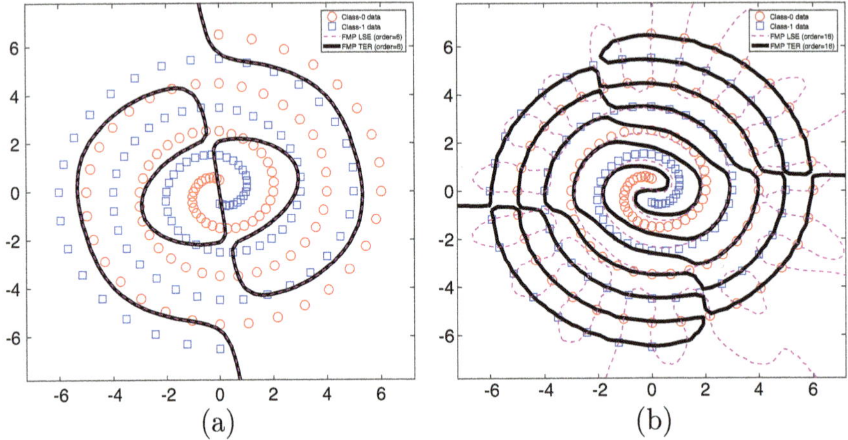

Fig. 10.10 Learning of the two-spiral problem: **a** using a sixth-order polynomial model, **b** using a sixteenth-order polynomial model

There are respectively 28, 66, 105, and 153 number of polynomial coefficients for the sixth-, tenth-, thirteenth-, and sixteenth-order polynomial models. Since there are 194 data samples for training, all the systems in Figs. 10.9 and 10.10 are over-determined. These results show the subtle difference between the LSE and the TER learning methods.

(v) Intertwined spirals: the three-spiral problem

In this example, three noisy intertwined spirals are generated with each spiral containing 500 samples for training and 500 samples for testing. In other words, there

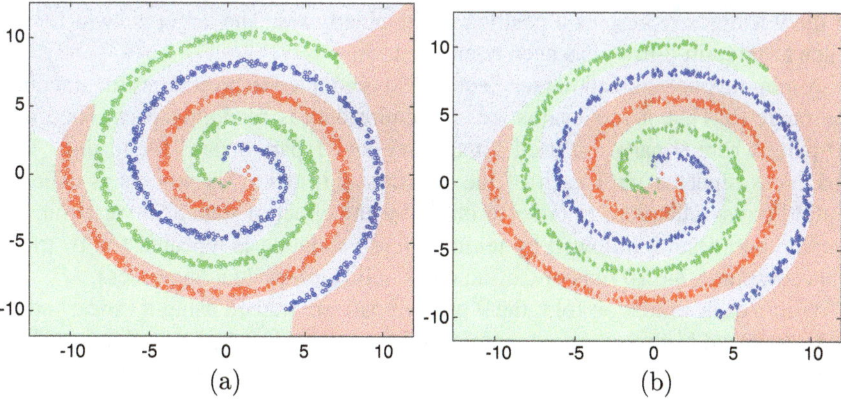

Fig. 10.11 Learning of the three-spiral problem using a tenth-order polynomial model: **a** training data points, **b** test data points and decision regions

are 3000 data samples altogether. A tenth-order full bivariate polynomial model with 66 parameters including the intercept term is used to learn the training of 1500 samples. Since there are three categories, an indicator matrix (5.15) is adopted as the training target for the over-determined system. The training is conducted without regularization using (5.18) where $\mathbf{P}$ is a tenth-order polynomial regressor matrix. Figure 10.11a and b show the training samples and the test samples respectively plotted on the learned decision regions.

10.2 Benchmark Examples

In this section, we showcase several real-world benchmark applications using data sets taken from the University of California Irvine (UCI) Machine Learning Repository (Kelly et al., 2024). Representative two-category, three-category and multiple-category applications are included.

10.2.1 Two-Category Problem: Tic-Tac-Toe Endgame

This game data set[1] encodes the end of the complete set of tic-tac-toe games into 958 instances with 9 features. The player "×" is assumed to start the first step. Each

---

of the 9 features represents a position of the nine boxes. The target is "win for ×" when a "three-in-a-row" has been completed.

A min-max normalization (see Sect. 2.3.3) is adopted to preprocess the data into the range [0, 1]. For test evaluation, the entire data is randomly partitioned into 10 parts where 9 parts are used for training and 1 part is used for testing. This process is repeated until each of the 10 parts is tested. This process is called a tenfold cross-validation. In order to minimize the effect of statistical variation, the random partitioning followed by tenfold cross-validation is repeated for 10 trials. This constitutes the 10 runs of tenfold cross-validation evaluation process.

Within each validation fold, the 9 parts of data are trained using a Linear model and a Reduced Multivariate polynomial (RM) model using the Least Squares Error (LSE), the Total Error Rate (TER), and the Area Under the ROC (AUC) based learning methods. The implemented training algorithms are thereby labeled as LSE-Linear, LSE-RM, TER-Linear, TER-RM, AUC-Linear and AUC-RM respectively.

The training data consists of 90% of the 958 samples which is around 862 samples. Among the 862 samples, another tenfold cross-validation is performed within the training set for model order selection. This process is known as tuning of the hyperparameter (model order). The effective size of training set during tuning is 90% of 862 samples, which is about 776 samples. These sample sizes (862 and 776) are much larger than the number of parameters of a sixth-order RM model (100 parameters including the parameter corresponding to the intercept). The system is thus over-determined for the experimented model order search from 1 to 6 for the RM model. The chosen model orders for RM are 2, 3, and 3 respectively for LSE, TER, and AUC learning methods. The average accuracy for each algorithm is shown in Fig. 10.12.

10.2.2 Three-Category Problem: Iris Flower Classification

This flower data set[2] contains 3 classes of 50 samples each. Each class refers to a type of iris plant namely the iris Setosa, iris Versicolour, and iris Virginica. Among the three classes, one class is linearly separable from the other two wherein the latter are not linearly separable from each other. The four attributes are summarized as follows.

1. Sepal length in *cm*
2. Sepal width in *cm*
3. Petal length in *cm*
4. Petal width in *cm*
5. Output class labels: Iris Setosa, Iris Versicolour, and Iris Virginica.

---

[2] This data set released by Fisher (1936) is among the best known database in the pattern recognition literature. The data set is licensed under a Creative Commons Attribution 4.0 International (CC BY 4.0) license https://creativecommons.org/licenses/by/4.0/legalcode and is available in the UCI Machine Learning Repository https://archive.ics.uci.edu/dataset/53/iris.

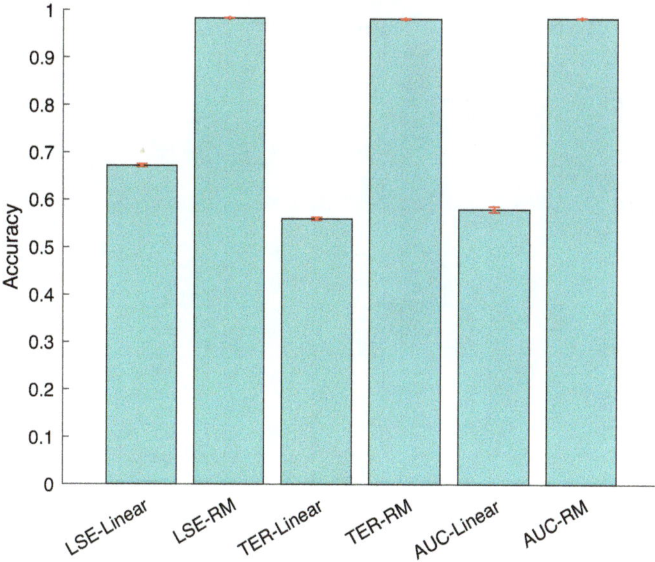

Fig. 10.12 Average accuracy for the Tic-tac-toe data set based on tenfold cross-validations for each algorithm. The standard deviation is shown as an error bar on top of each bar for each algorithm

The data is trained using the Linear model, the Reduced Multivariate polynomial (RM) model and the Full Multivariate Polynomial (FMP) model, using the Least Squares Error (LSE) and the Total Error Rate (TER) learning methods. For this data set of four dimensions, there are 70 and 29 polynomial terms for a fourth-order FMP and RM respectively. For the model selection training size of 121 (90–10% partitioning is for test evaluation and 90% of the 90% training set which is 81% of 150 samples is for training model selection), the system constitutes an over-determined system if the model order search is within the fourth order. The results show that a model beyond the linear one is necessary for this data set (Fig. 10.13).

10.2.3 Multiple-category Problem: Optical Recognition of Handwritten Digits

This digit data set[3] was collected based on a total of 43 people, wherein 30 contributed to the training set and the remaining 13 to the test set. For each image, the original 32×32 bitmap has been divided into non-overlapping blocks of 4×4 where the

---

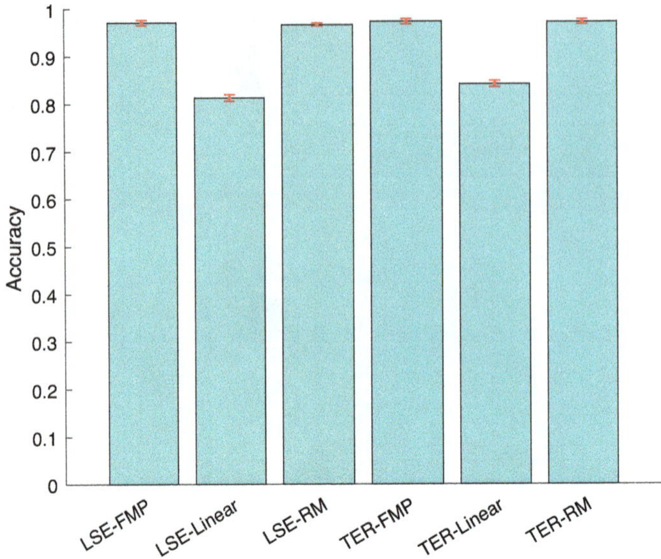

Fig. 10.13 Average accuracy for the Iris data set based on tenfold cross-validations for each algorithm. The standard deviation is shown as an error bar on top of each bar for each algorithm. The chosen model orders for LSE-RM, LSE-FMP, TER-RM, TER-FMP are respectively 2, 3, 3, 4

number of on pixels is counted within each block. This generated an input matrix of 8×8 where each element is an integer in the range [0, 16]. The dimension (64) is thus reduced (from 32×32) and the resulted image is invariant to minor distortions. The total number of samples collected for training and testing are respectively 3823 and 1797. Table 10.1 shows the sample size distribution in each output category. In our experiment, these two sets (training and test sets) of data are combined for running the 10 trials of tenfold cross-validation tests. Figures 10.14 and 10.15 show some samples of the image data taken from the training set and the testing set, respectively.

While noting that an expansion using the Full Multivariate Polynomial (FMP) model leads to an explosive feature size toward an ill-conditioned formulation, the data is trained using the Linear model and the Reduced Multivariate Polynomial (RMP) model using the Least Squares Error (LSE) and the Total Error Rate (TER) learning methods. The sample size for model order search is around 4552 ($= (3823 + 1797) \times 90 \times 90\%$). Since the number of parameters for RMP of tenth order is 1217 including the intercept term, the system is over-determined for RMP model within tenth order. The results in Fig. 10.16 show that a model more complex than the linear one is necessary toward a good classification of this data set.

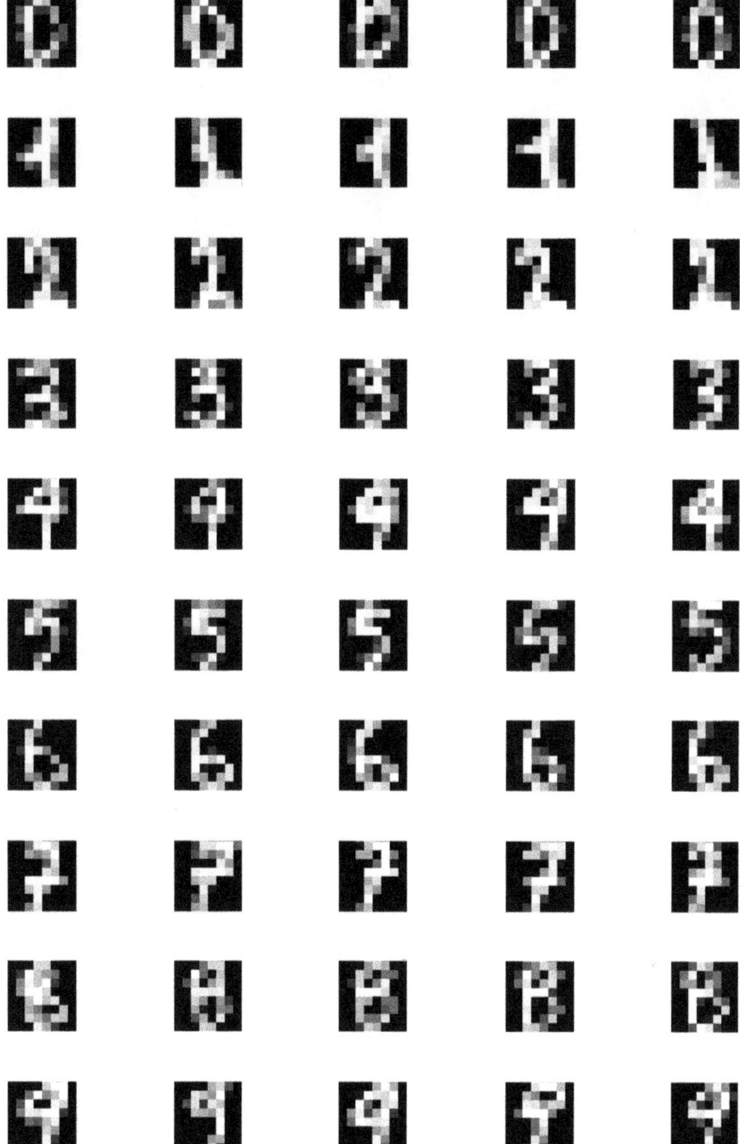

Fig. 10.14 Sample training images from the data set of optical recognition of handwritten digits (Kaynak, 1995)

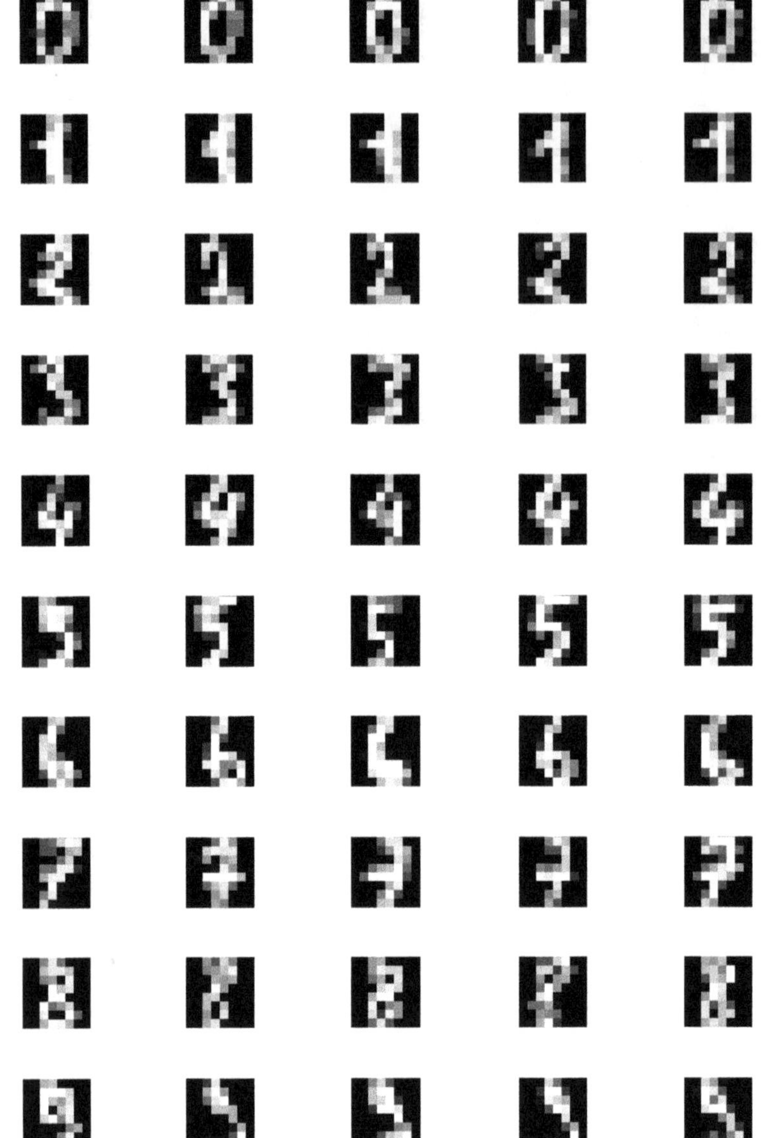

Fig. 10.15 Sample test images from the data set of optical recognition of handwritten digits (Kaynak, 1995)

Table 10.1 Sample size distribution in each category

Class number	Training set size	Test set size
0	376	178
1	389	182
2	380	177
3	389	183
4	387	181
5	376	182
6	377	181
7	387	179
8	380	174
9	382	180

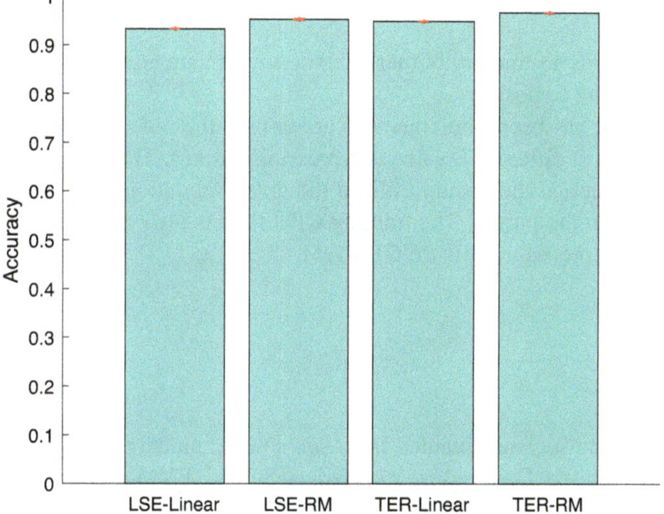

Fig. 10.16 Average accuracy for the optical digits data set based on tenfold cross-validations for each algorithm. The standard deviation is shown as error bar on top of each bar for each algorithm. The chosen model orders for LSE-RMP and TER-RMP are both the third-order

10.2.4 NIPS 2003 Feature Selection Data

Five feature selection data sets were created for the challenge organized along with the conference on Neural Information Processing Systems which was held in the year 2003 (NIPS 2003). The data sets were chosen to span a variety of domains namely, cancer prediction from mass-spectrometry data, handwritten digit recognition, text classification, and prediction of molecular activity. Among them, one data set is artificial. These data sets were formatted for the purpose of benchmarking variable selection algorithms in a controlled manner. The input variables are continuous or binary, sparse or dense. All problems are binary classification problems.

Table 10.2   NIPS 2003 feature selection challenge data sets

Data set	Domain	Type	# Feat.	# Trn	# Valid.	# Test (*org. not released)
Arcene	Mass spec.	Dense	10000	100	100	700
Dexter	Text categ.	Sparse	20000	300	300	2000
Dorothea	Drug discov.	Sp. bin.	100000	800	350	800
Gisette	Digit recog.	Dense	5000	6000	1000	6500
Madelon	Artifical data	Dense	500	2000	600	1800

* The test sets were not released during the challenge. They are now available in the UCI Machine Learning Repository

Table 10.2 presents a summary of these data sets which are now available in the UCI Machine Learning Repository.

The p-bridge has been implemented for learning these data to observe the prediction behavior at different k-value compression settings. The training utilized the given training set and the testing utilized the given validation set since the test set is not released to the public. The training CPU time is also recorded based on the runtime of an i9 processor with 96 GB RAM.

(i) Arcene

This Arcene[4] data set was obtained from two sources namely, the National Cancer Institute (NCI) and the Eastern Virginia Medical School (EVMS). The current version of NIPS 2003 was prepared by Isabelle Guyon. The data set consists of three subsets namely, the training set, the validation set, and the test set which have 100, 100, and 700 samples respectively. As the test set is withheld by the organizer, the training set and the validation set are combined in this experiment for a two-fold cross-validation evaluation. The task is to distinguish between cancer and normal patterns based on the mass-spectrometric data. The 10,000 attributes of input variables consist of continuous real values. According to the challenge organizer, the original features indicate the abundance of proteins in human sera having a given mass value. Among the 10,000 input features, 3000 are "distractor" features added by the organizer. These features have no predictive power. As a gauge regarding the difficulty of the data set, the "lambda" method in the sample code provided by the organizer yielded

---

[4] The data set was created by Isabelle Guyon, Steve Gunn, Asa Ben-Hur and Gideon Dror for the challenge and is licensed under a Creative Commons Attribution 4.0 International (CC BY 4.0) license https://creativecommons.org/licenses/by/4.0/legalcode and is available in the UCI Machine Learning Repository https://archive.ics.uci.edu/dataset/167/arcene.

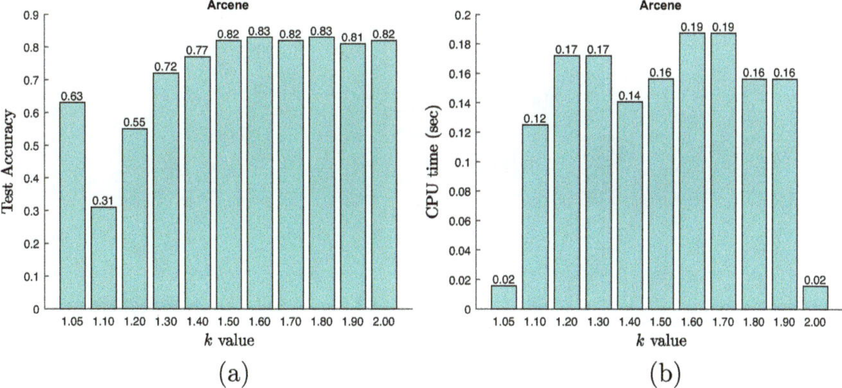

Fig. 10.17 **a** Test accuracy for the Arcene data set at different k values. **b** Training processing time for the compressive learning in seconds

an approximately 30% test error rate, while a linear SVM trained on all features achieved a significantly lower error rate of 3.5% under the original test protocol.

Figure 10.17a shows the results in terms of the test Accuracy at different k-values ([1.05, 1.10, 1.20, 1.30, 1.40, 1.50, 1.60, 1.70, 1.80, 1.90, 2.00]). These results show the accuracy peaks at $k = 1.60$ and $k = 1.80$. Figure 10.17b shows the variation of training processing time at different k values.

(ii) Dexter

This Dexter[5] data set was originally collected by the Carnegie Group, Inc. and Reuters, Ltd. in the course of developing the CONSTRUE text categorization system. The task is to distinguish between cancer and normal patterns from the mass-spectrometric data. The current version of NIPS 2003 was modified by Isabelle Guyon for the contest. The modified data has 20,000 attributes, within which 10,053 of them are random probes which carry no information for classification. The classification task is a binary one with equal number of positive and negative samples. The sample sizes for training, validation , and testing are respectively 300, 300, and 2,000. Since we do not have the test set, the following experiments are conducted based on multiple trials of two-fold tests using the combined training and validation sets. As a reference regarding the difficulty of the data, The "lambda" method in the sample code provided by the organizer had approximately a 20% test error rate and a linear SVM trained on all features a 5.8% error rate.

---

[5] The data set was created by Isabelle Guyon, Steve Gunn, Asa Ben-Hur, and Gideon Dror for the challenge and is licensed under a Creative Commons Attribution 4.0 International (CC BY 4.0) license https://creativecommons.org/licenses/by/4.0/legalcode and is available in the UCI Machine Learning Repository https://archive.ics.uci.edu/dataset/168/dexter.

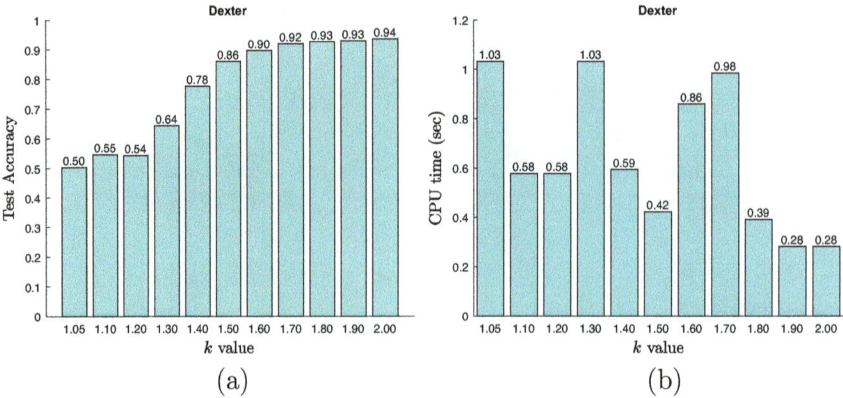

Fig. 10.18 **a** Test accuracy for the Dexter data set at different k values. **b** Training processing time for the compressive learning in seconds

Figure 10.18a shows the test accuracy at different k values. Based on these results, it is apparent that the performance peaks at $k = 2.00$ where a deterioration of prediction accuracy is observed for lower k values. The training processing time in Fig. 10.18b shows a higher processing cost when $k < 2.00$ in general.

(iii) Dorothea

The original data set with which Dorothea[6] was provided by DuPont Pharmaceuticals Research Laboratories for KDD Cup 2001. This data set was for drug discovery in identifying and isolating the receptor to which it should bind. The current version was modified by Isabelle Guyon for NIPS 2003 challenge. The data is very sparse with less than 1% of the entries containing non-zero values. They are respectively 800, 350 and 800 samples for training, validation, and testing. Since the test set is not available, the training and validation sets are combined for a two-fold test in our setting. The "lambda" method which comes with the sample code had a 21% test error rate. The linear SVM does not outperform this "lambda" method according to the organizer.

Figure 10.19a and b show respectively the accuracy and the CPU training time results using the p-bridge regression. These results do not show a significant change in prediction training processing performances.

---

[6] The data set was created by Isabelle Guyon, Steve Gunn, Asa Ben-Hur, and Gideon Dror for the challenge and is licensed under a Creative Commons Attribution 4.0 International (CC BY 4.0) license https://creativecommons.org/licenses/by/4.0/legalcode and is available in the UCI Machine Learning Repository https://archive.ics.uci.edu/dataset/169/dorothea.

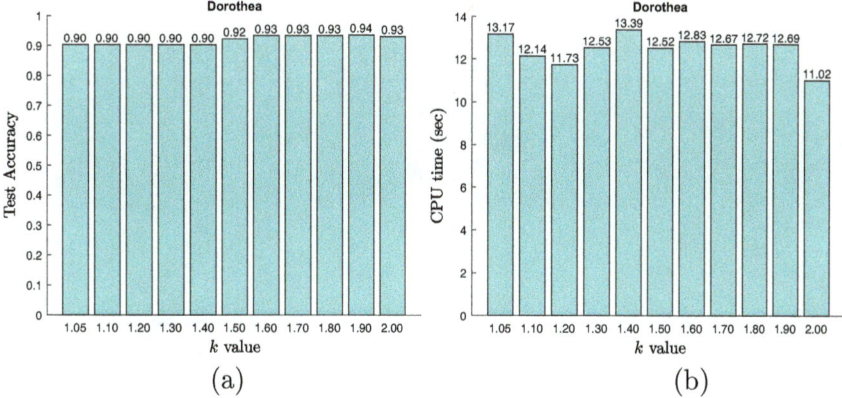

Fig. 10.19 **a** Test accuracy for the Dorothea data set at different k values. **b** Training processing time for the compressive learning in seconds

(iv) Gisette

The task of this Gisette[7] data set is to discriminate between two often confused handwritten Arabic numerals namely, the four and the nine. This is a binary classification problem with sparse continuous input variables. This data set was modified from the MNIST data by Isabelle Guyon for the NIPS 2003 challenge. There are 6,000, 1,000, and 6,500 samples for training, validation, and testing. The training set and the validation sets are combined in our two-fold experimental setting since the test set is not released. Among the 5,000 features, half of them are random probes. According to the organizer, the "lambda" method had approximately a 30% test error rate. A linear SVM trained on all features had a 3.5% error rate.

Figure 10.20a and b show respectively the prediction accuracy and the CPU training time results using the p-bridge regression. Similar to the above case, these results do not show a significant difference in prediction performance across different k values. However, the compression formulation shows a significantly high CPU cost compared with that of the ridge regression.

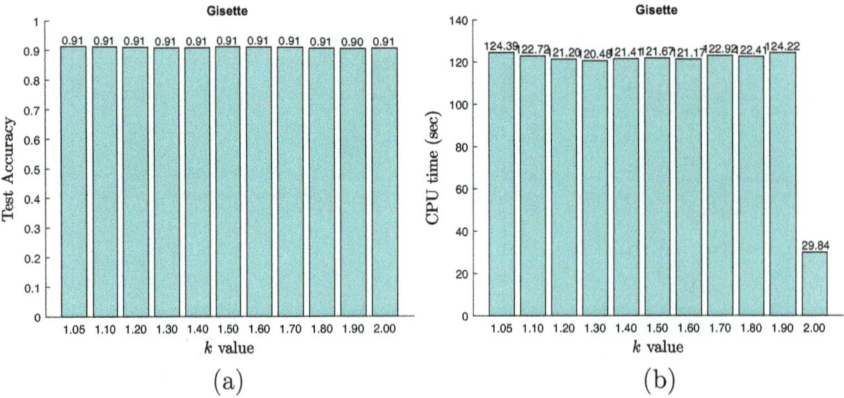

Fig. 10.20 **a** Test accuracy for the Gisette data set at different k values. **b** Training processing time for the compressive learning in seconds

(v) Madelon

This synthetic Madelon[8] data set was generated by the organizer of NIPS 2003 challenge. Among the 500 features, only a few of them are informative. There are 2,000, 600, and 1,800 samples being generated for training, validation, and testing. As in previous examples, the training set and the validation sets are combined for a two-fold cross-validation evaluation in our setting. The "lambda" method performs rather poorly on this highly nonlinear problem with a 41% error rate. According to the organizer, when the k-nearest neighbor method was used with $k = 3$ on 5 useful features, a 10% error rate was obtained.

Figure 10.21a and b show respectively the prediction accuracy and the CPU training time results using the p-bridge regression. These results suggest that the underlying decision boundary is way beyond the capability of a linear model. Similar to the above case, the training processing results also show a higher computational cost of p-bridge regression than that of the ridge regression.

10.3 Biometric Authentication

The field of *biometrics* refers to the automated measurement of physiological or behavioral characteristics of a person to determine or authenticate his or her identity. Among the various biometric traits, the face characteristics is easily acquired due

---

[8] The data set was created by Isabelle Guyon for the challenge and is licensed under a Creative Commons Attribution 4.0 International (CC BY 4.0) license https://creativecommons.org/licenses/by/4.0/legalcode and is available in the UCI Machine Learning Repository https://archive.ics.uci.edu/dataset/171/madelon.

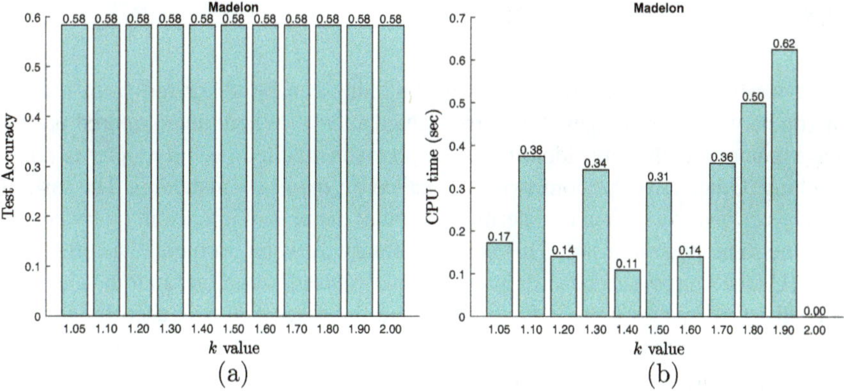

Fig. 10.21 **a** Test accuracy for the Madelon data set at different k values. **b** Training processing time for the compressive learning in seconds

to the pervasive deployment of CCD cameras. However, because of the limitation of current sensing technology and inherent issues, the acquired images of faces can vary significantly under changes due to illumination, head pose, facial expression (such as eye opening or closing), and occlusion (such as wearing glasses) factors. The subject of identity authentication based on face images is thus an interesting and important topic of study.

In face biometric, a comparison between two images of an identity is often performed on the extracted *feature templates* since the original images have sizes that can much burden the manipulation under current technology. Identity *verification* refers to a comparison or matching between a query template and the template of a known identity. On another hand, identity *identification* or *recognition* refers to a search in locating the identity of the query template. In a nutshell, biometric verification refers to a one-to-one matching of two templates whereas biometric identification/recognition refers to a one-to-many matching search.

Among face databases, the database of CMU face images (Mitchell, 2023)[9] is now considered an entry-level resource due to its relatively small-sample size. This database contains 640 images collected from 20 distinct subjects. Although there are 32 face images for each subject, some images have a ".bad" suffix and not utilized for our study. The total number of images utilized in our study is 624 images. The full size of each image is 120×128 pixels, with 256 gray levels per pixel. We shall use the quarter resolution images 30×32 pixels for our study. Due to the variation of illumination, facial expression, head pose and occlusion (with and without glasses), these images nevertheless constitute a representative example for our biometric study.

---

[9] This data set was created by Tom Mitchell (see Fig. 1.2 for some sample images). The data set is licensed under a Creative Commons Attribution 4.0 International (CC BY 4.0) license (https:// creativecommons.org/licenses/by/4.0/legalcode). The data set is available at https://archive.ics.uci. edu/dataset/124/cmu+face+images.

10.3.1   Face Verification Mode

In the verification mode, two groups of matching can be generated from a comparison between two images drawn from the database. When the compared pair of images came from the same identity, the comparison is called a *genuine-user* or *client* matching. Otherwise, the comparison is called an *impostor* matching. The task of *identity verification* is thus to determine whether the pair of images in query comes from the same person or not. This is thus a binary decision problem. The entire set of CMU face images can be split into two equal subsets based on random selection. One subset of images is used for training and the other subset is used for testing. Based on this organization, the set of genuine-user comparisons can be obtained by matching among images from the same identity, which is known as *intra-identity* matching. The set of impostor comparisons can be obtained by matching across the images of different identities, which is known as *inter-identity* matching. Due to a small number of bad images for certain subjects, there is an unequal distribution of images across subjects. Consequently, when randomly partitioning the total samples into training and testing sets, the resulting subsets may not contain an equal number of genuine-user and impostor matching pairs. Specifically, our random partitioning yielded 2,331 genuine-user matching pairs for training and 2,332 for testing. For impostor matches, there were 92,370 and 92,368 matching pairs in the training and testing sets, respectively. This results in an imbalanced category distribution, a common challenge in many real-world verification problems.

(i) A Baseline Method: Principal Component Analysis

The Principal Component Analysis (PCA) is a statistical procedure for converting a set of possibly correlated variables into a set of linearly uncorrelated variables. The principal modes of these converted variables are called principal components which are orthogonal to each other. Based on this property, the PCA is often adopted for extracting a set of uncorrelated features where less important features can be discarded without deteriorating the results heavily. The total number of features in each image sample from the adopted quarter resolution CMU face images is 960 ($30 \times 32 = 960$) pixels. After performing principal component analysis, the top 100 PCA features will be retained as feature templates for matching. This reduces the feature size suitable for implementation in a computational platform with low running memory. In a typical real-world application scenario, feature reduction may involve transforming millions of pixels into just a few hundred transformed features.

To facilitate computation using matrix-vector operations, each image sample is reshaped into a vector of 1×960 size. By stacking the 312 (half of 624) samples of the reshaped training data in matrix form, we can form a feature matrix $\mathbf{X}$ of 312×960 dimension. Let $\boldsymbol{\mu}^T$ (of 1×960 dimension) denotes the mean of $\mathbf{X}$ taken over the 312 sample rows. Then, the procedure for PCA starts with the formation of the mean centered data matrix given by $\boldsymbol{\Phi}^T = \mathbf{X}^T - [\boldsymbol{\mu}_1, \ldots, \boldsymbol{\mu}_{312}]$ where $\boldsymbol{\mu}_1 = \boldsymbol{\mu}_2 = \cdots = \boldsymbol{\mu}_{312} = \boldsymbol{\mu} \in \mathbb{R}^{960 \times 1}$. The mean centered data matrix $\boldsymbol{\Phi}$ is thus remained to be of 312×960 dimension. Next, compute the covariance matrix using

$$\mathbf{C}_1 = \boldsymbol{\Phi}^T \boldsymbol{\Phi} \in \mathbb{R}^{960 \times 960}, \tag{10.1}$$

and then compute the eigenvalues and eigenvectors based on the definition

$$\mathbf{C}_1 \mathbf{u}_i = \lambda_i \mathbf{u}_i, \quad i = 1, \ldots, 960. \tag{10.2}$$

However, it is more efficient to compute the eigenvalues and eigenvectors based on

$$\mathbf{C}_2 = \boldsymbol{\Phi} \boldsymbol{\Phi}^T \in \mathbb{R}^{312 \times 312}, \tag{10.3}$$

since the same set of non-zero eigenvalues can be obtained. This can be seen from the following relationship for $i = 1, \ldots, 312$:

$$
\begin{aligned}
\mathbf{C}_2 \mathbf{v}_i &= \mu_i \mathbf{v}_i \\
\Leftrightarrow \boldsymbol{\Phi} \boldsymbol{\Phi}^T \mathbf{v}_i &= \mu_i \mathbf{v}_i \\
\Rightarrow \boldsymbol{\Phi}^T \boldsymbol{\Phi} \boldsymbol{\Phi}^T \mathbf{v}_i &= \mu_i \boldsymbol{\Phi}^T \mathbf{v}_i \\
\Rightarrow \mathbf{C}_1 \boldsymbol{\Phi}^T \mathbf{v}_i &= \mu_i \boldsymbol{\Phi}^T \mathbf{v}_i \\
\Rightarrow \mathbf{C}_1 \mathbf{u}_i &= \lambda_i \mathbf{u}_i,
\end{aligned}
\tag{10.4}
$$

where $\mathbf{u}_i = \boldsymbol{\Phi}^T \mathbf{v}_i$ for $\lambda_i = \mu_i$, $i = 1, \ldots, 312$. Only the top 312 eigenvalues can be non-zero because the highest possible rank for both $\mathbf{C}_1$ and $\mathbf{C}_2$ is 312. To reduce the feature dimension, the top $p = 100$ largest eigenvalues with corresponding eigenvectors are used as the face template.

For illustration purpose and to match with the original image size for projection, we use the subset of trained eigenvectors $\mathbf{u}_i \in \mathbb{R}^{960 \times 1}, i = 1, \ldots, p$ taken from (10.2) for projecting the test images. Each test image (of size 30×32 is first reshaped as $\mathbf{x}_t \in \mathbb{R}^{960 \times 1}$) and then being projected onto the eigen-space using

$$\omega_i = \mathbf{u}_i^T (\mathbf{x}_t - \boldsymbol{\mu}), \quad i = 1, \ldots, p. \tag{10.5}$$

In other words, the face template for testing (projected test image) is taken as

$$\omega_{test} = \begin{bmatrix} \omega_1 \\ \omega_2 \\ \vdots \\ \omega_p \end{bmatrix}. \tag{10.6}$$

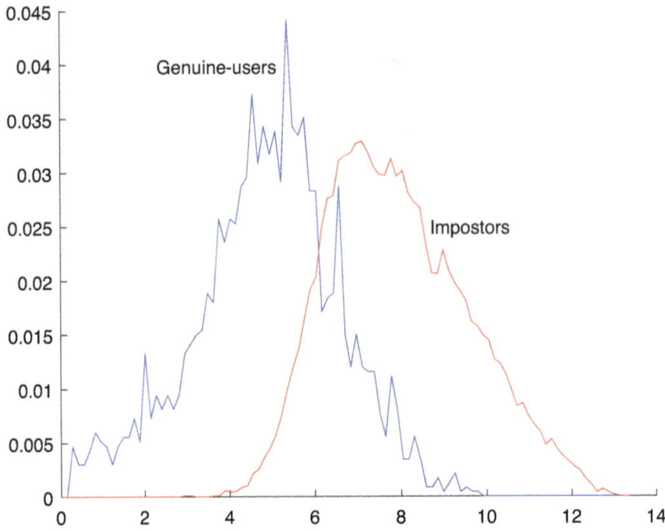

Fig. 10.22 PCA: the genuine-user and impostor distributions

For verification or authentication of identities, this test template ω_{test} is compared or matched with the template of a known identity (utilizing an image $\mathbf{x}_{known}$ with known identity) based on the Euclidean distance

$$\varepsilon = \|\omega_{test} - \omega_{known}\|, \tag{10.7}$$

where ε is the mismatch distance.

Based on the scores (10.7) of genuine-user matches (comparisons) and impostor matches (comparisons) within the test set, an Equal Error Rate (EER, see Sect. 4.3.3) of 17.93% is obtained for the PCA templates when $p = 100$. Figure 10.22 shows the distributions of the genuine-user and the impostor matchings.

(ii) Direct Pixel Matching

Without any dimension reduction, the entire quarter resolution (30×32) face images can be matched or compared directly using their pixel values. This is a non-learning-based method which can be achieved by vectorizing or reshaping each image to a column vector of 960×1 dimension where the Euclidean norm can be applied for matching. The distribution of genuine-user and impostor matchings within the test data set is shown in Fig. 10.23 where an EER of 18.18% is obtained. When each image is sub-sampled by skipping a few pixels row-wise and column-wise, the matching accuracy can vary. Figure 10.24 shows the EERs for image sizes namely,

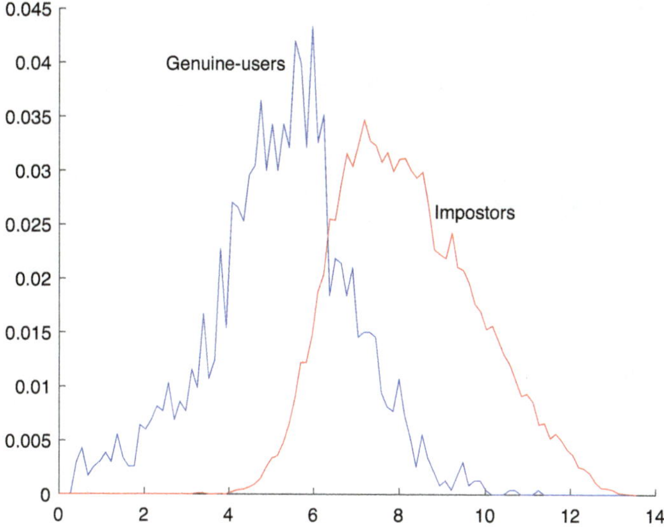

Fig. 10.23 Pixel matching: the genuine-user and impostor distributions

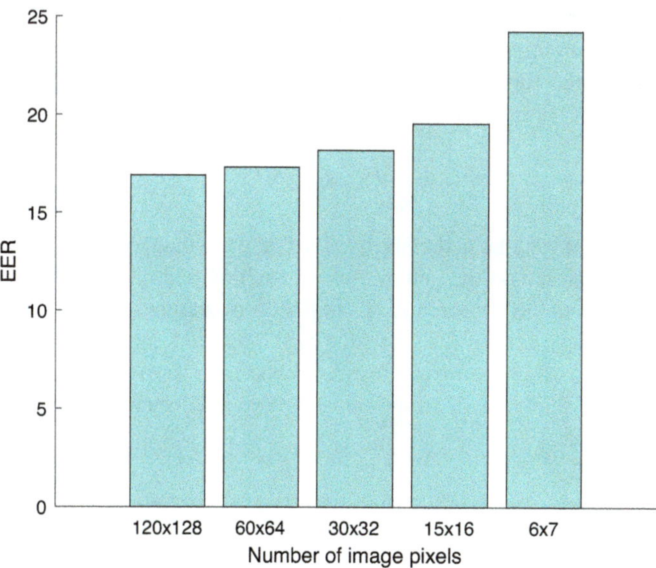

Fig. 10.24 Sub-sampled pixel matchings

120×128 (full), 60×64 (half), 30×32 (quarter), 15×16 and 6×7 pixels. Due to the discarding of pixel information, the matching accuracy deteriorates when more useful information is discarded than the non-useful one.

(iii) Linear Pixel Regression

Besides the PCA-based method for reducing the feature dimension, a simple regression can be applied to the pixel values to learn the more influential pixels. The size of the quarter-sampled image is of $30 \times 32 = 960$ pixels. Suppose the reshaped sub-sampled images for training is denoted as $\mathbf{x}_i \in \mathbb{R}^{960 \times 1}, i = 1, \ldots, 312$ (half of the 624 images). Then, the absolute difference between each pair (i.e., $\tilde{\mathbf{x}} = \mathrm{abs}(\mathbf{x}_i - \mathbf{x}_j) \in \mathbb{R}^{960 \times 1}$) of images within the training set (which has $2,332$ genuine-user pairs and $92,370$ impostor pairs) can be stacked in matrix form as $\mathbf{X} \in \mathbb{R}^{(2332+92370) \times (960+1)}$. To speed up the computation time, a subset of the impostor matching scores has been utilized giving $\mathbf{X} \in \mathbb{R}^{(2332+9360) \times (960+1)}$.

Using this training data set $\mathbf{X}$, a weight vector $\mathbf{w}$ can be adopted to differentiate the importance among the 960 features. Fitting a linear prediction model to the target $\mathbf{y} \in \{0, 1\}$ can then be written as $\mathbf{y} = \mathbf{X}\hat{\mathbf{w}} + \boldsymbol{\varepsilon}$ where "0" indicates a genuine-user and "1" indicates an impostor. The error of fit is given by $\boldsymbol{\varepsilon}$ which can be minimized based on its squared form given by $\boldsymbol{\varepsilon}^T \boldsymbol{\varepsilon}$ (see SSE minimization in Sect. 5.1). The solution for an SSE optimized weight vector for an over-determined system[10] is

$$\hat{\mathbf{w}} = \left(\mathbf{X}^T \mathbf{X}\right)^{-1} \mathbf{X}^T \mathbf{y}. \tag{10.8}$$

This trained weight vector can then be used with the test data matrix $\mathbf{X}_t$ for prediction using $\hat{\mathbf{y}}_t = \mathbf{X}_t \hat{\mathbf{w}}$. Based on the genuine-user matching pairs and impostor matching pairs generated from the test set, an EER of 3.09% is obtained for the CMU face data set. In this result, the size of the impostor matching pairs has been sub-sampled to 9360 and 9371 pairs for training and testing respectively to speed up the computation time. Figure 10.25 shows the distributions of the genuine-user and impostor generated from the prediction.

(iv) Compressive Regression

When the full dimensions of the images are used without subsampling, the p-bridge regression can be applied to stretch the parameters for possible sparse learning. For illustration purposes, we use the quarter resolution of the CMU face images for compressive regression in this example.

---

[10] We have $(2332 + 9360)$ training samples of 961 dimension. This is an over-determined system since we have more training samples than the feature dimension.

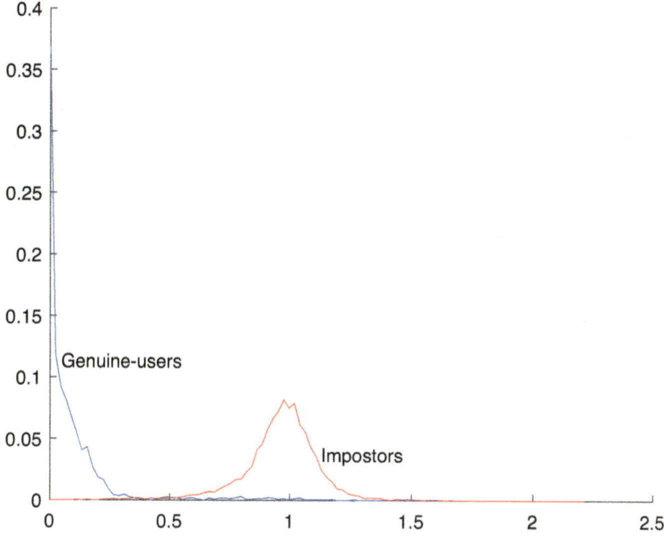

Fig. 10.25   Linear regression on the pixels: the genuine-user and impostor distributions

Figures 10.26 and 10.27 show respectively the estimated genuine-user and impostor distributions for $k \in \{1.05, 1.30, 1.60, 1.90\}$ and the estimated parameters sorted in ascending order. The test EER for this compressive estimation is shown in Fig. 10.28. The top parameters of the compressive regression can be subsequently exploited for sparse estimation.

10.3.2   Face Identification Mode

In the *identification* or *recognition* mode, the query face image is compared with images from the database of known identities. This turns out to be a *one-to-many* comparisons. Similar to the verification mode, the comparison is frequently compared based on the extracted templates or features in order to have a reduced computational cost as well as having the unwanted noise removed.

(i) PCA in Identification Mode

The principal component analysis (PCA) is again adopted here for recognition of identities from the training database using a template with reduced size. Similar to that in the verification mode, all the images of the CMU database are projected onto the eigenvectors (using equation (10.5)–(10.6)) to produce the face templates for both

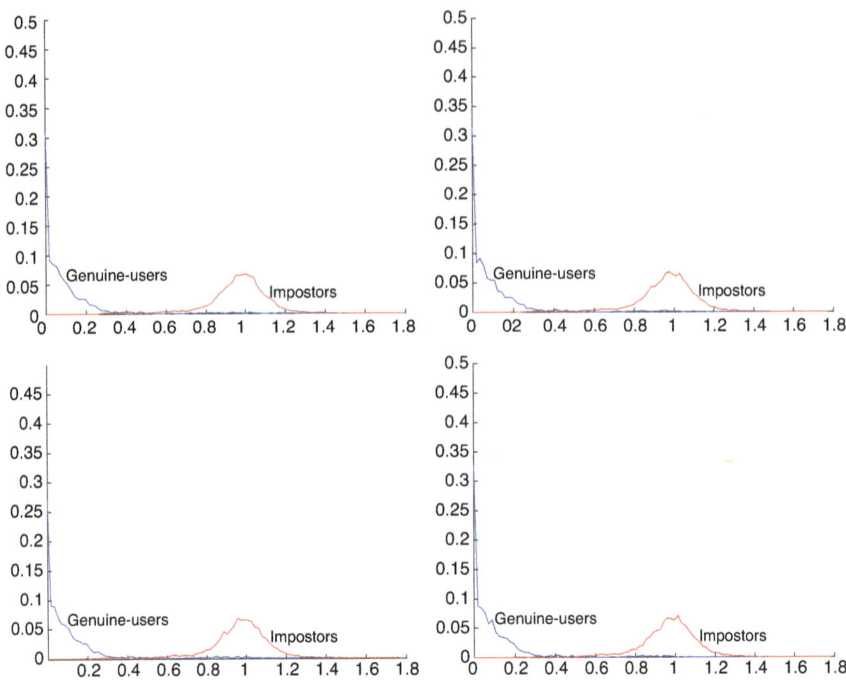

Fig. 10.26 The estimated genuine-user and impostor distributions for $k = 1.05$ (upper left), $k = 1.30$ (upper right), $k = 1.60$ (lower left), and $k = 1.90$ (lower right)

the training and test sets. Since there are 28–32 images for each identity in the training set, an average training template over these samples is adopted for representation of each identity in the training set. Subsequently, the identification mode is conducted by matching each sample of the test set with each of those averaged training templates from the 20 identities. Figure 10.29 shows the distribution plots for the genuine-user and impostor scores. When only the top $p = 100$ eigenvectors (with the remaining discarded) are utilized for template projection, an EER of 9.94% is achieved.

For identification mode, the matching rank with respect to each identity is an important measure to gauge the accuracy performance. In other words, we wish to know the success rate for identifying the correct person from the training set for each test image. For this experimental setup, the rank-1 identification rate is found to be 85%.

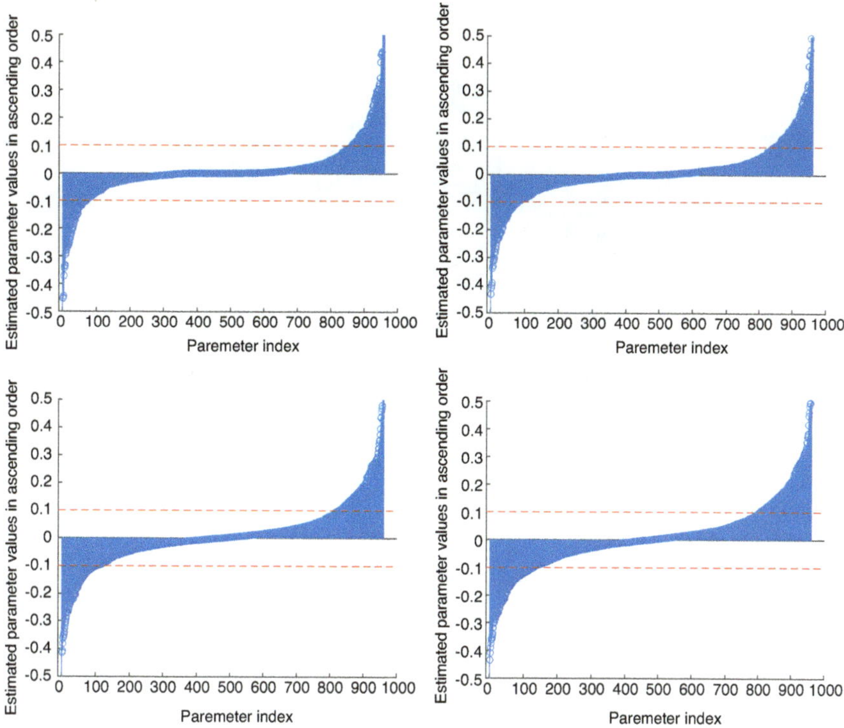

Fig. 10.27 The estimated parameter values sorted in ascending order for $k = 1.05$ (upper left), $k = 1.30$ (upper right), $k = 1.60$ (lower left), and $k = 1.90$ (lower right)

(ii) Direct Pixel Matching in Identification Mode

The averaged template taken from the training image samples for each identity is utilized as the identity search template. The adopted template uses all the image pixels which are reshaped as a vector template for comparison. Figure 10.30 shows the distributions of the genuine users and the impostors with an EER of 10.47%. The rank-1 identification rate is observed to be 85%.

(iii) Linear Regression in Identification Mode

Apart from using the mean value of the training samples to represent each identity as the matching template, the training samples can be used for training a set of weights that correspond to each template feature (in this case each pixel is one template feature). The outcome of training returns a weight vector for each identity which indicates the importance of individual pixel feature.

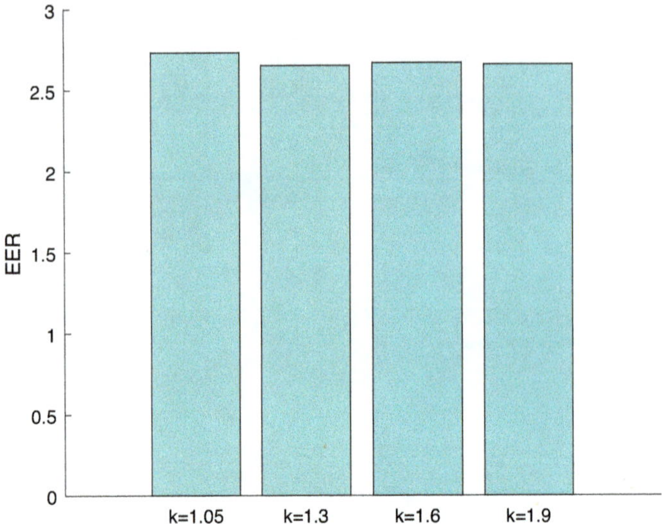

Fig. 10.28 Linear regression on the pixels: the genuine-user and impostor distributions

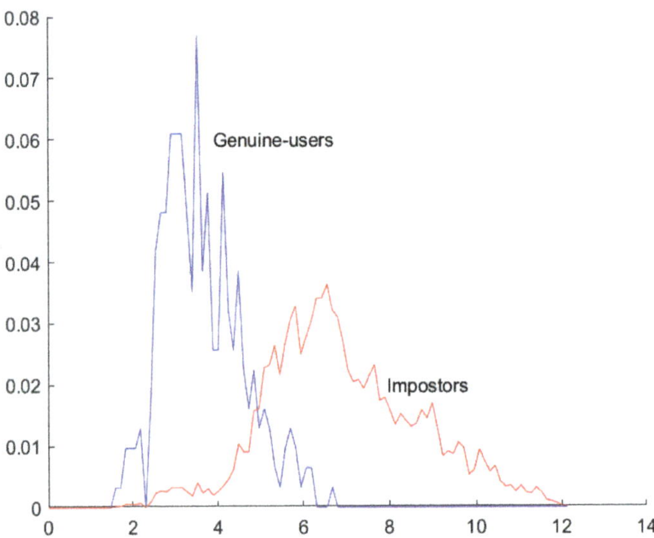

Fig. 10.29 PCA in identification mode: the genuine-user and impostor distributions

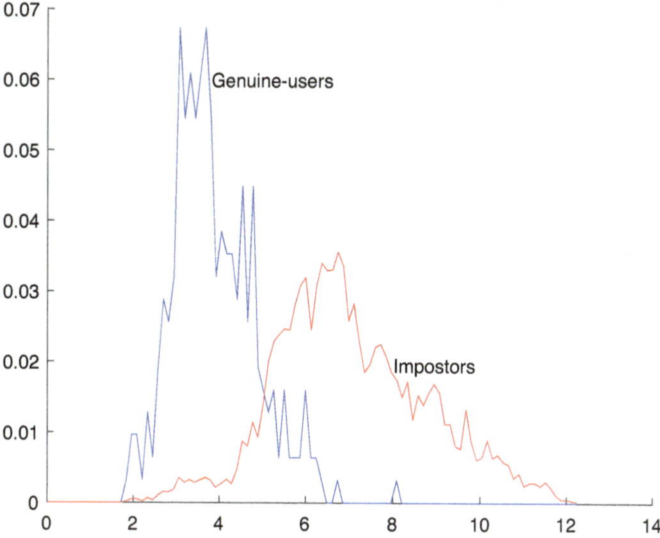

Fig. 10.30 Pixel matching in identification mode: the genuine-user and impostor distributions

Specifically, the pixels of the training samples can be packed as a regressor matrix $\mathbf{X}$ with its corresponding target matrix $\mathbf{Y}$ as follows:

$$\mathbf{X} = \begin{bmatrix} 1 & x_{1,1} & x_{1,2} & \cdots & x_{1,d} \\ 1 & x_{2,1} & x_{2,2} & \cdots & x_{2,d} \\ \vdots & \vdots & \vdots & \cdots & \vdots \\ \vdots & \vdots & \vdots & \cdots & \vdots \\ \vdots & \vdots & \vdots & \cdots & \vdots \\ \vdots & \vdots & \vdots & \cdots & \vdots \\ 1 & x_{m,1} & x_{m,2} & \cdots & x_{m,d} \end{bmatrix}_{d+1}, \quad \mathbf{Y} = \begin{bmatrix} 1 & 0 \\ \vdots & \vdots \\ 1 & 0 \\ & \ddots \\ 0 & 1 \\ \vdots & \vdots \\ 0 & 1 \end{bmatrix}_{20}, \quad (10.9)$$

where each row of $\mathbf{X}$ represents a face image, with pixel values reshaped into a horizontal vector. The matrix $\mathbf{Y}$ is an indicator matrix (see Eq. (5.15) in Sect. 5.1). Each row of $\mathbf{Y}$ identifies the corresponding sample row by placing a "1" in the relevant column, while all other columns containing "0". In other words, the column position in each row of the matrix $\mathbf{Y}$ corresponds to a unique identity. Hence, there are 20 columns in $\mathbf{Y}$ to represent the 20 persons in the CMU face data set. Here, $m = 312$ and d is the total number of pixels in the image. The ones at the left-most column of $\mathbf{X}$ are bias terms added in order to provide a shifting capability for fitting. Since we have $m = 312$ training samples and $d = 960$ pixel features, the system is under-determined. Hence, the dual space is adopted for training the weights of the multi-category (20 classes for the 20 people) problem (see (5.30) in Sect. 5.1):

$$\hat{\mathbf{W}} = \mathbf{X}^T \left(\mathbf{X}\mathbf{X}^T \right)^{-1} \mathbf{Y}. \tag{10.10}$$

For evaluation, the test regressor matrix $\mathbf{X}_t$ is packed in the same way as that in (10.9) but using images from the test set. The output estimation for test samples is then computed using

$$\hat{\mathbf{Y}}_t = \mathbf{X}_t \hat{\mathbf{W}}. \tag{10.11}$$

The determination of the identity is finally obtained using the *one-versus-all* technique. In other words, for each row of the estimated $\hat{\mathbf{Y}}_t$, the identity category is determined by the column position of the largest row elements. Under this mode of operation, the accuracy of recognition for the CMU face data set is 98.72%. Apart from using the original pixels in $\mathbf{X}$, the PCA templates can also be used for constructing the regressor $\mathbf{X}$ under this mode.

(iv) Compressive Regression in Identification Mode

When all the elements of $\mathbf{X}$ are normalized within $[0, 1]$, then the above regression protocol can be adopted for simplied p-bridge estimation using

$$\hat{\mathbf{W}} = (\mathbf{X}^T)^{\circ \frac{1}{k-1}} \left(\mathbf{X}(\mathbf{X}^T)^{\circ \frac{1}{k-1}} \right)^{-1} \mathbf{Y}. \tag{10.12}$$

Depending on the k-value setting, the parameters are stretched at different extent. After having the weight matrix estimated, it can be applied to test samples for prediction using (10.11). Figure 10.31 shows the test recognition accuracies obtained for various k-values ($k \in \{1.2, 1.3, 1.4, 1.6, 2.0\}$). The results show deterioration of accuracy at low k-values. When a sparse set of weight parameters is needed, the top ranking weight parameters in terms of their absolute magnitude can be selected for a re-estimation using the ℓ_2-norm ($k = 2$).

10.3.3 Decision Fusion for Multiple Biometrics

When a multiple number of biometric traits are available, it has been argued that adoption of multiple traits not only provides a higher accuracy of authentication but also a higher level of security. However, the ultimate effectiveness of using multiple biometric traits depends significantly on the implementation and how the weakest trait is addressed. In this section, we shall treat fusion of multiple biometrics at score

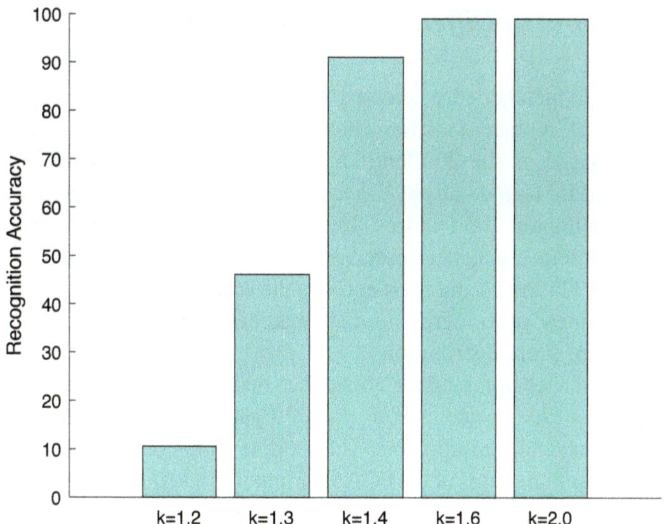

Fig. 10.31 Recognition accuracy plotted over k values

level as a binary classification problem and show that the authentication accuracy can be improved with appropriate classifier design.

Before delving into the fusion design, we shall briefly introduce the concept of biometric information fusion across different levels. Essentially, information fusion for multiple biometrics can be conducted before or after the matching process. Fusion before matching can be performed at sensor level and feature level. Fusion after matching can be performed at score level (or confidence level), rank level, and abstract level. For fusion at sensor level, the original data collected from the sensors such as visual camera and fingerprint sensor are fused directly. However, due to the high dimensionality of the original data, which contains a large amount of noise or unwanted signals, achieving effective fusion can be challenging. For fusion at feature level, although much of the unwanted signals can be mitigated, the extracted features still remain at high dimension and this continues to post difficulty toward good fusion performance. Fusion at score level right after matching is the most popular in multibiometrics since it comes with low dimension and yet retains the necessary resolution for discrimination. Fusion at rank and abstract level losses much detailed information for discrimination.

The fusion of multiple biometrics can involve combining different biometric traits, such as face, fingerprint, palmprint, and palm-vein. Alternatively, it can entail merging scores from different biometric matchers within the same biometric trait. In the following example, we illustrate the fusion of scores from the PCA face matcher (Sect. 10.3.1(i)) and the linear pixel regression face matcher (Sect. 10.3.1(iii)) under the verification mode.

Fusing the Scores of Different Face Matchers

Two different face matchers are selected for scores fusion. The first face matcher selected is the PCA face matcher under the verification mode described in Sect. 10.3.1(i). Based on the 20 identities with each identity having 28–32 image samples in the CMU face database, the test matching accuracy in terms of EER is 17.93% based on the top 100 features. The second face matcher selected is the linear pixel regression face matcher described in Sect. 10.3.1(iii) which has an EER = 3.09%. Figure 10.32a and b depict respectively the score distributions of the PCA face matcher and the linear pixel regression face matcher from Sect. 10.3.1, highlighting the contrast in their score distributions

The main challenges of fusing the above face matchers include imbalance in their accuracies (EER = 17.93% versus EER = 3.09%) and imbalance in the sample sizes of the genuine-users and the impostors. The score level fusion has been implemented by bundling the training scores of the PCA face matcher and the linear pixel regression matcher to form the training input features. By denoting the scores of the PCA face matcher and the linear pixel regression matcher by $\mathbf{s}_{PCA}$ and $\mathbf{s}_{Regression}$ respectively, the input features for scores fusion can be written as $[\mathbf{s}_{PCA}, \mathbf{s}_{Regression}]$. Here, we note that both $\mathbf{s}_{PCA}$ and $\mathbf{s}_{Regression}$ are of length $2331 + 9359$ as there are 2331 genuine-user scores and 9359 impostor scores. The 9359 impostor scores has been obtained by subsampling to reduce the class size imbalance.

By putting $\mathbf{X} = [\mathbf{s}_{PCA}, \mathbf{s}_{Regression}]$, the reduced multivariate polynomial model (RM) (Sect. 3.3.2) has been employed to expand the features (including an intercept term) for LSE-based[11] (Sect. 5.1.3) and TER-based (Sect. 5.2.1) fusion. Figure 10.33a

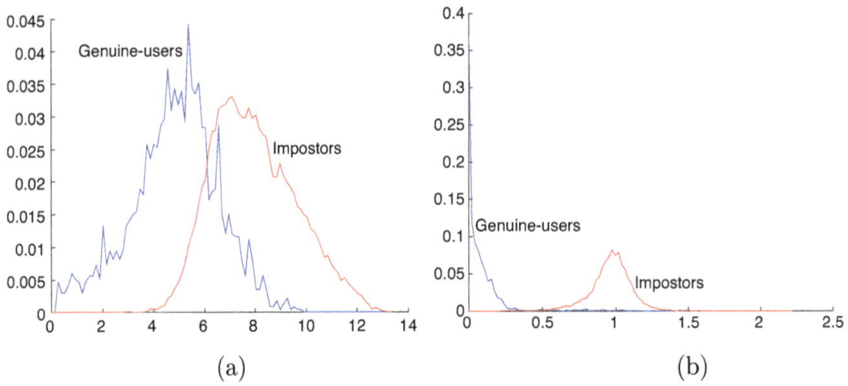

Fig. 10.32 **a** The score distribution of the PCA face matcher, **b** The score distribution of the Pixel regression face matcher

---

[11] In general, the methods in Sects. 5.1.1–5.1.4 can be considered as the Least Squares Error based method as the error distance has been minimized. The primal ridge regression, as a specific case of LSE, is adopted here due to the over-determined nature of the formulation.

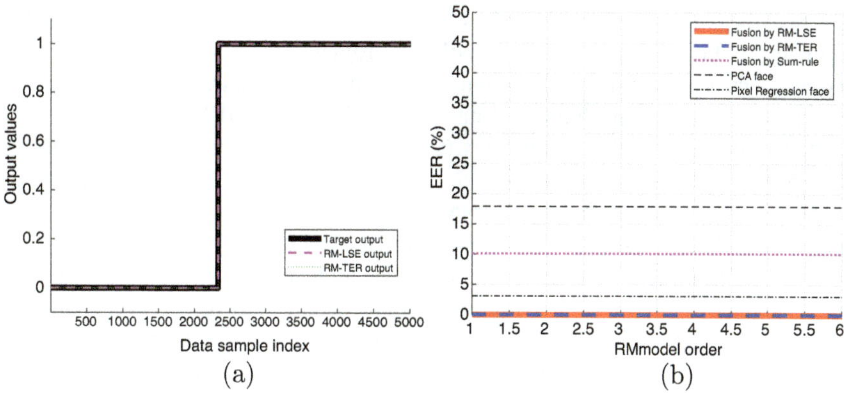

Fig. 10.33 **a** The outputs of the second-order fusion, **b** The Equal Error Rates (EERs in %) of fusion with those before fusion

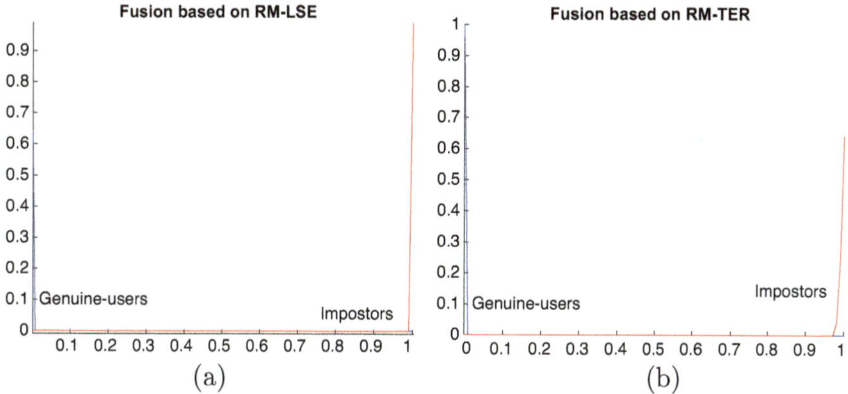

Fig. 10.34 **a** The score distribution of LSE-based fusion, **b** The score distribution of TER-based fusion

shows the learned outputs of RM-LSE (RM model with LSE learning) and RM-TER (RM model with TER learning) and Fig. 10.33b shows the EER values at different RM orders. Figure 10.34a and b show the score distributions of LSE-based fusion and TER-based fusion respectively. These results show the enhancement of EER values for both RM-LSE fusion and RM-TER fusion.

10.4 Bibliographic Notes

The **Archimedean spiral** (also known as arithmetic spiral) (Wikipedia, 2017) starts in the origin and makes a curve with three rounds. Essentially, the distances of intersection points along a line through the origin are the same. In other words, the distances between the spiral branches are the same. Archimedes described such a spiral in his book "On Spirals". In polar coordinates (r, θ) where r is the radial distance and θ is the angle in radian, the Archimedean spiral can be written as $r = a + b\theta$ where a and b are real numbers (Wikipedia, 2017). The spiral can be written in the rectangular coordinates (Cartesian coordinates) form as $b\theta \cos(\theta + a) + b\theta \sin(\theta + a)$ where a controls its orientation and b controls its "tightness" or the distance between the spiral branches. The adopted intertwined spirals were created based on the Archimedean spiral equation in rectangular coordinates.

The **UCI machine learning repository** is an archive to share databases[12] for the empirical analysis of machine learning algorithms. The archive was first created as an ftp archive in 1987 by David Aha and subsequently managed by fellow librarians at UC Irvine (Kelly et al., 2024). The data sets are classified in terms of their default tasks. Since then, the repository has accumulated over 666 data sets which cover various fields.

The **NIPS feature selection challenge** (Guyon, 2003; Guyon et al., 2004) was introduced in 2003 with five high-dimensional data sets. The goal was to find feature selection algorithms that significantly outperform methods using all features. Each data set was partitioned into training, validation, and test sets where the users were provided only the labels of training and validation sets together with their respective input features at the initial stage of the challenge. Although the challenge ended in December 2003, the data sets are still available for evaluation and benchmarking of algorithms with only the test set labels withheld by the organizer. These data sets are also available in the UCI machine learning repository (Kelly et al., 2024).

The database of **CMU face images** (Mitchell, 2023) consists of 640 grayscale images with varying pose, expression, and eye conditions. The poses include straight, left, right, and up orientations. The expressions include neutral, happy, sad, and angry. The eye conditions include wearing sunglasses or not. Each identity has 32 images, each with a full resolution of 120×128 pixels. The images also come in half (60×64) and quarter (30×32) resolutions. As some of the images have a ".bad" suffix, only the remaining 624 images are utilized in our experiments. This database is available in the UCI machine learning repository (Kelly et al., 2024).

According to Wikipedia, the **Principal Component Analysis (PCA)** was invented in 1901 by Pearson (1901), as an analog of the principal axis theorem in mechanics. It was later independently developed and named "principal components" by Harold Hotelling in the 1930s (Hotelling, 1933, 1936). In the field of signal processing, it is also called the discrete Karhunen–Loéve Transform (KLT). Apart from this, several other terms such as "Hotelling transform", "Proper Orthogonal Decomposition (POD)", "Singular Value Decomposition (SVD)", "Eigenvalue Decomposition (EVD)", "Empirical Eigenfunction Decomposition", "Empirical Component Analysis", and "Spectral Decomposition" are also related to PCA.

---

[12] Under a Creative Commons Attribution 4.0 International (CC BY 4.0) license (https://creativecommons.org/licenses/by/4.0/legalcode). The web link to the UCI machine learning repository is https://archive.ics.uci.edu/.

Solution Manual

Chapter 1

Introduction

Exercises

  1.1 What is the relationship among AI (Artificial Intelligence), ML (Machine Learning), and DL (Deep Learning)?

1.1 Ans: The relationship among AI (Artificial Intelligence), ML (Machine Learning), and DL (Deep Learning) can be represented in Fig. A.1. Essentially, AI is an umbrella term for all forms of automation including learning and reasoning. ML is a subset of AI focusing on learning. DL is a specific tool of learning based on the artificial neural networks.

  1.2 A computer program is said to learn from experience E with respect to some task T and some performance measure P, if its performance on T, as measured by P, improves with experience E. Suppose we feed a learning algorithm a lot of historical weather data, and have it learn to predict weather. In this setting what is T?

  (a) The historical weather data
  (b) The probability of it correctly predicting a future data's weather
  (c) The weather prediction task
  (d) None of these

1.2 Ans: (c)

© The Editor(s) (if applicable) and The Author(s), under exclusive license to Springer   327
Nature Singapore Pte Ltd. 2025
K.-A. Toh et al., *Analytic Learning Methods for Pattern Recognition*,
https://doi.org/10.1007/978-981-96-2151-4

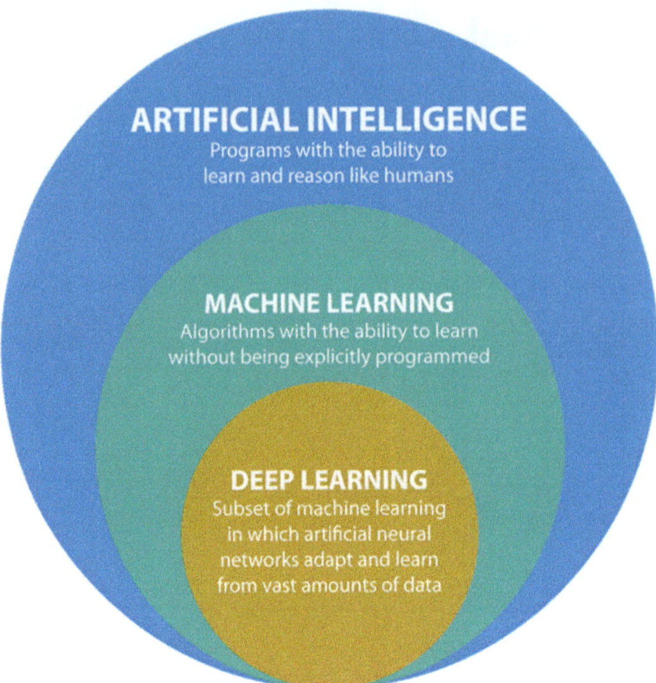

Fig. A.1 Relationship among artificial intelligence, machine learning and deep learning

   1.3 What is data preprocessing and what are the typical processing tasks?

1.3 Ans: Data preprocessing is the process of cleaning and preparing data for learning and analysis. This can involve tasks such as removing invalid data, data imputation, and data normalization to prevent anomalies from dominating the learning and analysis.

   1.4 In standardization, the data features will be re-scaled with

     (a) Mean 1 and Variance 1
     (b) Mean 0 and Variance 1
     (c) Minimum 0 and maximum 1
     (d) Minimum -1 and maximum $+1$

1.4 Ans: (b)

   1.5 Which of the following is/are example(s) of feature extraction?

     (a) Measure the length, the width, the texture, and the number of fins of fishes for classification using computer vision
     (b) Imputation of missing data
     (c) Principal component analysis
     (d) Construct histogram of data values for each sample

1.5 Ans: (a,c,d) Reason:

(a) The length, the width, the texture, and the number of fins of fishes can be considered data extracted from fishes.

(c) The largest principal component subsets can be used as extracted features.

(d) The histogram provides a statistical view of data from the binning perspective.

1.6 Suppose you are working on stock market prediction. Typically tens of millions of shares of a company's stock are traded each day. You would like to predict the number of shares that will be traded tomorrow for the company.

   (i) Would you treat this as a classification or a regression problem?
      (a) Regression
      (b) Classification
      (c) Clustering
      (d) None of these
   (ii) If the data you have collected involved millions of attributes, what would you do?

1.6 Ans: (i)(a), (ii)(Extract relevant features or perform a dimension reduction.)

1.7 Suppose you are working on weather prediction and use a learning algorithm to predict tomorrow's temperature (in degrees Centigrade/ Fahrenheit).

   (i) Would you treat this as a classification or a regression problem?
      (a) Regression
      (b) Classification
      (c) Clustering
      (d) None of these.
   (ii) What kind of data should you gather?

1.7 Ans: (i) (a)

   (ii) Weather forecasts are made by collecting quantitative data (e.g., changes in barometric pressure, current weather conditions, and sky condition or cloud cover) about the current state of the atmosphere at a given place and using meteorology to project how the atmosphere will change.

1.8 What is/are the differences between supervised and unsupervised learning?

1.8 Ans: The algorithms of supervised learning use labeled data to get trained. The algorithms of unsupervised learning use unlabeled data for training purposes. In unsupervised learning, the models identify hidden data trends and do not take any feedback.

1.9 In a report, a research team claimed that their method is the best based solely on the low training error evaluation. The claim in the report is trustworthy. True or false?

1.9 Ans: The evaluation should use the unseen test error instead.

1.10 A research team analyzed information from several earlier studies involving more than 188 million of adults. They observed the link between sitting and risk of mortality over 20–30 years. In the studies, about 28% of deaths could be attributed to sitting, 21% to television viewing, the researchers said. The researchers concluded the analysis with the following statement: "Sit less than 5 years, add 2 years to your life". This report is trustworthy as it involved a huge database. True or false?

1.10 Ans: This can be considered a causality creep since there is no proof for the causation.

Chapter 2

Data Preprocessing

Exercises

2.1 There is no difference between information and data. True or False?

2.1 Ans: False: Information is data that is processed in such a way that it becomes meaningful to the person who receives it, so information is not only a collection of raw facts.

2.2 For categorical data such as the type of fruits, the one-hot encoding is preferred over an arbitrary number assignment such as one for orange, two for apple, three for mango, etc. True or False?

2.2 Ans: Reason: Yes, this is because the higher values in arbitrary number assignment could have a higher influence for large numbers than that for small numbers. One-hot encoding does not have this problem.

2.3 Which of the following represents the characteristics of ordinal data?

(a) Ordinal data has a median
(b) Value of interval is unknown
(c) Measures non-numeric traits such as satisfaction, happiness, etc.
(d) Establish a relative rank
(e) None of these

2.3 Ans: (a, b, c, d) and reasons:

(a) Ordinal data has a median: Median is the value in the middle but not the middle value of a scale and can be calculated with data which has an innate order.

(b) Value of interval is unknown: The ranking order does not have any scale or values in-between.

(c) Measures non-numeric traits such as satisfaction, happiness, etc.: Measures quantitative levels of intensity qualitatively.

(d) Establish a relative rank: Provides an order of intensity, strength, etc.

2.4 Determine whether the following items belong to interval or ratio data.

(a) The number of hours one spent studying each day
(b) The score of an intelligence test
(c) The temperatures in cities throughout the country that are listed in most newspapers
(d) The birth weights of babies who were born at Samsung Hospital last week
(e) Ages of students in a machine learning class
(f) Number of pages in your statistic book
(g) The annual salaries for all teachers in a town
(h) The cost of a pair of shoes

2.4 Ans: (a) The number of hours one spent studying each day (ratio scale)
(b) The score of an intelligence test (ratio scale)
(c) The temperatures in Fahrenheit (interval scale)
(d) The birth weights of babies who were born at Samsung Hospital last week (ratio scale)
(e) Ages of students in a machine learning class (ratio scale)
(f) Number of pages in your statistic book (ratio scale)
(g) The annual salaries for all teachers in a town (ratio scale)
(h) The cost of a pair of shoes (ratio scale)

2.5 Which of the following is incorrect regarding the properties of distance metrics?

(a) The distance from $\mathbf{x}$ to $\mathbf{y}$ is the same as the distance from $\mathbf{y}$ to $\mathbf{x}$
(b) The metric must satisfy the triangular inequality
(c) The distance between two identical vectors, $\mathbf{x}$ and $\mathbf{y}$, is a non-zero value
(d) The distance must never be negative
(e) None of these

2.5 Ans: (c) The distance between two identical vectors, $\mathbf{x}$ and $\mathbf{y}$, is a non-zero value. Reason: The distance between two identical vectors, $\mathbf{x}$ and $\mathbf{y}$, is zero.

2.6 Which of the following distance measures calculates the distance between two binary vectors?

(a) Euclidean distance
(b) Manhattan distance
(c) Minkowski distance
(d) Hamming distance
(e) None of these

2.6 Ans: (d) Hamming distance. Reason: The Hamming distance between two strings of equal length is the number of positions at which the corresponding symbols are different. In other words, it measures the minimum number of errors that could have transformed one string into the other.

2.7 What is the preferred way to handle missing or corrupted data in a small data set?

(a) Drop the entire data sample that contains the corrupted features
(b) Replace missing values with mean/median/mode
(c) Replace missing features with zeros
(d) Do not change the corrupted features
(e) None of these

2.7 Ans: (b) Reason: Dropping the data sample would reduce the small data size further. Replace missing features with zeros or other arbitrary values would introduce bias to the data sample.

2.8 When the bounds of data are not known, the z-score standardization is preferred over min-max normalization. True or False?

2.8 Ans: True. Reason: This is because min-max normalization needs the boundary values for computing the standardized values of data.

2.9 You are given a set of data for supervised learning. A sample block of data looks like this:

1.0234,	0.3202,	25,	0.0081,	60033.81,	3
1.1356,	0.3308,	21,	0.0067,	69283.18,	-1
0.9988,	0.2526,	23,	0.0093,	66034.33,	2
1.1858,	0.3001,	22,	0.0077,	64037.35,	1
1.1533,	0.3853,	21,	0.0066,	62033.58,	-13
1.0755,	0.3102,	21,	0.0098,	60183.65,	1
1.0045,	0.2901,	22,	0.0065,	61093.98,	-2
1.1031,	0.3912,	21,	0.0088,	69033.23,	-1

Each row corresponds to a sample data measurement with 5 input features and 1 response. What kind of undesired effect can you anticipate if this set of raw data is used for learning?

(a) Those features with nominal values may create numerical instability
(b) Those features with very large values may overshadow those with very small values
(c) Those features with integer values may add bias to the system
(d) The sample with output response value -13 should be imputed.
(e) None of these

2.9 Ans: (b) Those features with very large values may overshadow those with very small values. We can either use min-max or z-score normalization to resolve the problem. (The output value of -13 does not seem abnormal for a regression problem)

2.10 Suppose you are given the following data regarding the monthly income of participants of a yoga learning group.

Participants	a	b	c	d	e	f	g	h	i
Income \$	5000	3000	4000	4500	6000	3500	5000	800000	5500

Among mean, mode, and median, which is the most appropriate statistic to describe the data? Provide the value of this statistic based on the given table.

2.10 Ans: The median has better representation for this set of data.

Median: 3000, 3500, 4000, 4500, **5000**, 5000, 5500, 6000, 800000

Chapter 3

Linear Parametric Models

Exercises

3.1 Given $\mathbf{Xw} = \mathbf{y}$ where $\mathbf{X} = \begin{bmatrix} 1 & 1 \\ 5 & 6 \end{bmatrix}, \mathbf{y} = \begin{bmatrix} 0 \\ 1 \end{bmatrix}$.

(a) What kind of system is this? (even-, over- or under-determined?)
(b) Is $\mathbf{X}$ invertible? Why?
(c) Solve for $\mathbf{w}$ if it is solvable.

3.1 Ans: (a) This is an even-determined system because it has the same the number of equations and the number of unknowns.

(b) $\mathbf{X}$ is invertible since $\det(\mathbf{X}) = 1 \times 6 - 1 \times 5 = 1 \neq 0$.
$$\mathbf{X}^{-1} = \begin{bmatrix} 6 & -1 \\ -5 & 1 \end{bmatrix}.$$

(c) $\hat{\mathbf{w}} = \mathbf{X}^{-1}\mathbf{y} = \begin{bmatrix} 6 & -1 \\ -5 & 1 \end{bmatrix}\begin{bmatrix} 0 \\ 1 \end{bmatrix} = \begin{bmatrix} -1 \\ 1 \end{bmatrix}.$

3.2 The linear system given by $\begin{bmatrix} 1 & 2 \\ 3 & 4 \\ 5 & 6 \end{bmatrix}\begin{bmatrix} w_1 \\ w_2 \end{bmatrix} = \begin{bmatrix} 1 \\ 2 \\ 3 \end{bmatrix}$ is over-determined and thus has no exact solution. True or False?

3.2 Ans: False. Reason: since the `rank(X)` = `rank([X|y])` = 2, it has an exact solution.

3.3 Consider the following system:

$$w_1 + w_2 = 0$$
$$w_1 - w_2 + w_3 - w_4 = 1$$
$$w_3 + w_4 = 2$$

What kind of system is this? Can the system be solved? Select one or more from below:

(a) The system is even-determined and can be solved with an exact solution.
(b) The system is over-determined and can be solved with an approximated solution.
(c) The system is over-determined and it does not have an exact solution.
(d) The system is under-determined and can be solved with an exact least norm solution.
(e) The system is under-determined and it has infinite number of solutions.

3.3 Ans: (d, e) An under-determined system with full row rank has infinite number of solutions. The least norm solution is unique and is given by

$$\hat{\mathbf{w}} = \mathbf{X}^T (\mathbf{X}\mathbf{X}^T)^{-1}\mathbf{y} = \begin{bmatrix} 1 & 1 & 0 \\ 1 & -1 & 0 \\ 0 & 1 & 1 \\ 0 & -1 & 1 \end{bmatrix} \begin{bmatrix} 1 & 1 & 0 & 0 \\ 1 & -1 & 1 & -1 \\ 0 & 0 & 1 & 1 \end{bmatrix} \begin{bmatrix} 0 \\ 1 \\ 2 \end{bmatrix} = \begin{bmatrix} 0.25 \\ -0.25 \\ 1.25 \\ 0.75 \end{bmatrix}.$$

3.4 Identify those polynomial models shown below which include at least a second-order term:

(a) $1 + x_1 + x_2 + x_1 x_2 - x_1^2 - x_2^2$
(b) $1 + x_1 + x_2 + x_3 + x_4 + x_5$
(c) $1 + x_1 + x_2 + x_1 x_2$
(d) $1 + x_1 + x_2 + x_1(x_1 - x_2)$
(e) None of these

3.4 Ans: (a, c, d) Reason: The degree of a polynomial is the highest degree of its monomials. You can find the degree by adding up the exponents of the variables that appear in it. For example, the polynomial $x^2 y^3 + xy + 4x^3 y^4$ has 3 terms. The first term has a degree of $2 + 3 = 5$; The second term has a degree of $1 + 1 = 2$; The third term has a degree of $4 + 3 = 7$; Therefore the polynomial has a degree of 7 which is the highest degree of any term.

3.5 Given the input feature x and the corresponding response y as follows.

x	−10	−7	−3	−1	4	8
y	10	11	8	6	3	4

(a) Perform a third-order polynomial regression and plot the result of line fitting.

(b) Given a test point $x = -5$ predict y using the learned polynomial model.

3.5 Ans: (a) Polynomial of third-order for 1D feature: $g(\mathbf{x}) = w_0 + w_1 x + w_2 x^2 + w_3 x^3$. Pack all the data points to form the polynomial matrix

$$\mathbf{P} = \begin{bmatrix} 1 & -10 & (-10)^2 & (-10)^3 \\ 1 & -7 & (-7)^2 & (-7)^3 \\ 1 & -3 & (-3)^2 & (-3)^3 \\ 1 & -1 & (-1)^2 & (-1)^3 \\ 1 & 4 & (4)^2 & (4)^3 \\ 1 & 8 & (8)^2 & (8)^3 \end{bmatrix} \text{ with } \mathbf{y} = \begin{bmatrix} 10 \\ 11 \\ 8 \\ 6 \\ 3 \\ 4 \end{bmatrix}.$$

This is an over-determined system with solution given by

$$\hat{\mathbf{w}} = (\mathbf{P}^T \mathbf{P})^{-1} \mathbf{P}^T \mathbf{y} = \begin{bmatrix} 5.3280 \\ -0.8977 \\ 0.0330 \\ 0.0075 \end{bmatrix}$$

(b) Prediction: $\hat{y} = \mathbf{p}_t^T \hat{\mathbf{w}} = [1, -5, (-5)^2, (-5)^3]^T \hat{\mathbf{w}} = 9.7081$ (Fig. A.2).

3.6 This question is also about polynomial models. Which of the following models has discriminative feature for the data given by

$$\begin{bmatrix} x_1 \\ x_2 \end{bmatrix} : \begin{bmatrix} -1 \\ -1 \end{bmatrix}, \begin{bmatrix} +1 \\ +1 \end{bmatrix}, \begin{bmatrix} -1 \\ +1 \end{bmatrix}, \begin{bmatrix} +1 \\ -1 \end{bmatrix}$$

with corresponding target values $y \in \{+1, +1, -1, -1\}$?

(a) $g(\mathbf{x}) = 1 + w_1 x_1 + w_2 x_2$

(b) $g(\mathbf{x}) = 1 + w_1 x_1 + w_2 x_2 + w_3 x_1 x_2$

(c) $g(\mathbf{x}) = 1 + w_1 x_1 + w_2 x_2 + w_3 x_1^2$

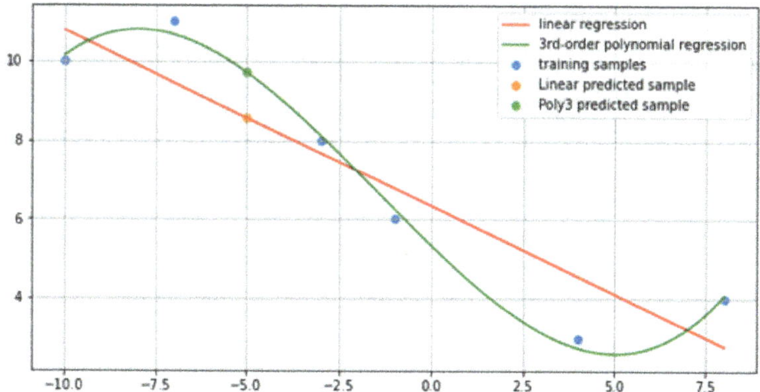

Fig. A.2 The fitting curve

(d) $g(\mathbf{x}) = 1 + w_1 x_1 + w_2 x_2 + w_3 x_2^2$

(d) $g(\mathbf{x}) = 1 + w_1 x_1 + w_2 x_2 + w_3 x_1^2 + w_4 x_2^2$

(e) None of these

3.6 Ans: (b)

Reason: By substituting the $\mathbf{x}$ values into the monomial terms of the given functions, we have

- $x_1 x_2 : \{+1, +1, -1, -1\}$ (This term is discriminative since it has good alignment with the target values)
- $x_1^2 : \{+1, +1, +1, +1\}$ (This term is not discriminative)
- $x_2^2 : \{+1, +1, +1, +1\}$ (This term is not discriminative)

Hence the only discriminative function is (b) which contains discriminative feature $x_1 x_2$.

3.7 This question is also about multivariate polynomials.

(a) Write down the expression for a third-order polynomial model having a three-dimensional input.

(b) Write down the $\mathbf{P}$ matrix for this polynomial given

$$\mathbf{X} = \begin{bmatrix} x_{11} & x_{12} & x_{13} \\ x_{21} & x_{22} & x_{23} \end{bmatrix} = \begin{bmatrix} 1 & 0 & 1 \\ 1 & -1 & 1 \end{bmatrix}.$$

(c) Given $\mathbf{y} = \begin{bmatrix} 0 \\ 1 \end{bmatrix}$, can a unique solution be obtained for this under-determined system of $\mathbf{P}$ matrix?

3.7 Ans: (a) A polynomial model of third order can be written as $g(\mathbf{x}) = w_0 + w_1 x_1 + w_2 x_2 + w_3 x_3 + w_4 x_1 x_2 + w_5 x_2 x_3 + w_6 x_1 x_3 + w_7 x_1^2 + w_8 x_2^2 + w_9 x_3^2 + w_{10} x_2 x_1^2 + w_{11} x_3 x_1^2 + w_{12} x_1 x_2^2 + w_{13} x_3 x_2^2 + w_{14} x_1 x_3^2 + w_{15} x_2 x_3^2 + w_{16} x_1 x_2 x_3 + w_{17} x_1^3 + w_{18} x_2^3 + w_{19} x_3^3$.

(b) The polynomial expansion with data is written in matrix form as

$$\mathbf{P} = \begin{bmatrix} 1\,1 & 0\,1 & 0 & 0\,1\,1\,0\,1 & 0\,1\,0\,0\,1 & 0 & 0\,1 & 0\,1 \\ 1\,1 & -1\,1 & -1 & -1\,1\,1\,1\,1 & -1\,1\,1\,1\,1 & -1 & -1\,1 & -1\,1 \end{bmatrix}.$$

(c) Yes, a unique solution can be obtained (using Python) for the under-determined system as follows:
$\hat{\mathbf{w}} = \mathbf{P}^T (\mathbf{P}\mathbf{P}^T)^{-1} \mathbf{y} = [0,\ 0,\ -0.1,\ 0,\ -0.1,\ -0.1,\ 0,\ 0,\ 0.1,\ 0,\ -0.1,\ 0,\ 0.1,\ 0.1,\ 0,\ -0.1,\ -0.1,\ 0,\ -0.1,\ 0]$.

3.8 This question is about the reduced multivariate polynomials.

(a) Write down the expression for a third-order RM model having a three-dimensional input.

(b) Write down the $\mathbf{P}$ matrix for this polynomial given

$$\mathbf{X} = \begin{bmatrix} x_{11} & x_{12} & x_{13} \\ x_{21} & x_{22} & x_{23} \end{bmatrix} = \begin{bmatrix} 1 & 0 & 1 \\ 1 & -1 & 1 \end{bmatrix}.$$

(c) Given $\mathbf{y} = \begin{bmatrix} 0 \\ 1 \end{bmatrix}$, can a unique solution be obtained for this under-determined system of $\mathbf{P}$ matrix?

3.8 Ans: (a) A RM model of third order can be written as $g(\mathbf{x}) = w_0 + w_1 x_1 +$
$w_2 x_2 + w_3 x_3 \quad + w_4 x_1^2 + w_5 x_2^2 + w_6 x_3^2 + w_7 x_1^3 + w_8 x_2^3 + w_9 x_3^3 \quad +$
$w_{10} x_1 (x_1 + x_2 + x_3) + w_{11} x_2 (x_1 + x_2 + x_3) + w_{12} x_3 (x_1 + x_2 + x_3)$
$+ w_{13} x_1 (x_1 + x_2 + x_3)^2 + w_{14} x_2 (x_1 + x_2 + x_3)^2 + w_{15} x_3 x_3 (x_1 + x_2 +$
$x_3)^2$.

(b) The reduced polynomial expansion with data is written in matrix form
as
$$\mathbf{P} = \begin{bmatrix} 1 & 1 & 0 & 1 & 1 & 0 & 1 & 1 & 0 & 1 & 2 & 0 & 2 & 4 & 0 & 4 \\ 1 & 1 & -1 & 1 & 1 & 1 & 1 & 1 & -1 & 1 & 1 & -1 & 1 & 1 & -1 & 1 \end{bmatrix}.$$

(c) Yes, a unique solution can be obtained for the under-determined system as follows:
$\hat{\mathbf{w}} = \mathbf{P}^T (\mathbf{P}\mathbf{P}^T)^{-1} \mathbf{y} = [0.0716, 0.0716, -0.1202, 0.0716, 0.0716, 0.1202,$
$0.0716, 0.0716, -0.1202, 0.0716, 0.0230, -0.1202, 0.0230,$
$-0.0742, -0.1202, -0.0742]$.

3.9 The two-layer perceptron network with linear output layer can be written in two ways, namely

$$g(\mathbf{W}_o, \mathbf{W}_h, \mathbf{x}) = \mathbf{W}_o L(\mathbf{W}_h \mathbf{x}), \quad \mathbf{W}_h \in \mathbb{R}^{h \times d}, \ \mathbf{W}_o \in \mathbb{R}^{o \times h},$$

and

$$g(\mathbf{W}_o, \mathbf{W}_h, \mathbf{x}) = L(\mathbf{x}^T \mathbf{W}_h) \mathbf{W}_o, \quad \mathbf{W}_h \in \mathbb{R}^{d \times h}, \ \mathbf{W}_o \in \mathbb{R}^{h \times o},$$

where $L(\cdot) = \max(0, \cdot)$ for $\mathbf{x} \in \mathbb{R}^{d \times 1}$. Let's use the second form for our computation and assume $\mathbf{W}_h$ has been fixed at

$$\mathbf{W}_h = \begin{bmatrix} 2 & 2 & 3 \\ 3 & 0 & 2 \end{bmatrix}.$$

Suppose you are given several training samples $\mathbf{x}$ with corresponding target y:
$$\mathbf{x} : \left\{ \begin{bmatrix} 1 \\ 0 \end{bmatrix}, \begin{bmatrix} 0 \\ 1 \end{bmatrix}, \begin{bmatrix} 0 \\ 0 \end{bmatrix}, \begin{bmatrix} 1 \\ 1 \end{bmatrix} \right\}, \quad y : \{0, 0, 1, 1\}.$$

Can the output weights $\mathbf{W}_o$ be solved deterministically?

3.9 Ans: Yes, the solution is over-determined which is given by

$$\hat{\mathbf{W}}_o = (\mathbf{P}^T \mathbf{P})^{-1} \mathbf{P}^T \mathbf{y} = [0.25, 0, 0.5]^T$$

where $\mathbf{P} = L(\mathbf{x}^T \mathbf{W}_h)$ and $\mathbf{y} = [0, 0, 1, 1]^T$.

1. Codes:
```
X = [1 0; 0 1; 0 0; 1 1]
y = [0; 0; 1; 1]
Wh = [ 2 2 3; 3 0 2]
P = ReLU(X*Wh)
```

```
w = inv(P'*P)*P'*y

function p = ReLU(x)
    p = max(0,x);
end
```

3.10 Code a function to build a regressor matrix for the Reduced Multivariate
      Polynomials Reciprocal Sigmoid features within MATLAB/Octave.

3.10 Ans: Codes:

```
function P = RMRsigmoid(X,order)
%  Build regressor matrix P (mxK):
%     order = desired order of approximation,
%     X = input matrix (mxl), K = number of parameters to be est.
%     m = number of data samples, l = input dimension.
[m,l] = size(X); MM1=[]; MM3=[]; Msum=sum(X,2);
X = 1 + exp(-X); % Reciprocal Sigmoid features
for i=1:order
    for k=1:l
        M1(:,k)=X(:,k).^i;
        if (i>1)
            M3(:,k)=X(:,k).*Msum.^(i-1);
        end
    end
    MM1=[MM1,M1];
    if (i>1)
        MM3=[MM3,M3];
    end
end
P = [ones(m,1),MM1,MM3];
return;
```

Chapter 4

Learning Score Functions

Exercises

4.1 What is the difference between a regression loss and a classification loss?

4.1 Ans: The regression loss involves prediction of real valued, continuous tar-
           get values. The classification loss involves prediction of discrete target

category or class and label. While regression loss predicts values, classification loss predicts target classes or labels.

4.2 What is the difference between a loss function and an error function?

4.2 Ans: An error function measures the deviation of an observable value from a prediction. A loss function works on an error to quantify its negative consequences of error. For example, the regression loss measures the squared error distance between the predicted value and the target value. The classification loss measures the number of samples being classified incorrectly based on the error observed or deviation of an observable value from a prediction.

4.3 What is the difference between a loss function and a cost function?

4.3 Ans: A loss function quantifies the negative consequences of error for a single training sample, and the cost function aggregates these losses from the training samples to form a measurement objective. The cost function is also known as the objective function, which is a function specified for either minimization or maximization. The objective function is a mathematical optimization term that embodies the set of functions being optimized and the constraints involved in achieving the optimization.

4.4 What are the common loss functions for regression and that for classification?

4.4 Ans: Common losses for regression include the Mean-Squared Error (MSE) and Mean Absolute Error (MAE). Common losses for classification include the Margin Loss, Cross-entropy loss, and 0–1 loss.

4.5 The plot below shows the decision boundary of Classifier A. Write down the confusion matrix for Classifier A below.

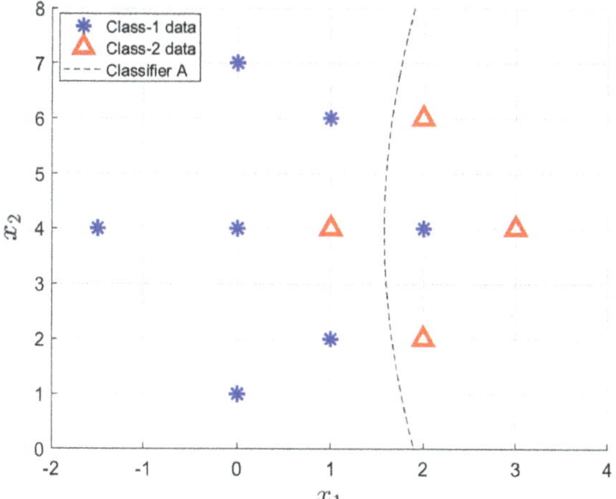

4.5 Ans: Confusion matrix

	Class-1	Class-2
Predicted Class-1	6	1
Predicted Class-2	1	3

4.6 Consider the experimental observations of the Training error rate Tr and the Validation error rate Va for a machine learning algorithm for each setting of the parameter θ. Choose a parameter (P) based on Tr and Va.

θ	Tr	Va
0.1	0.08	0.23
0.2	0.33	0.38
0.3	0.21	0.18
0.4	0.13	0.26
0.5	0.16	0.18

Which value of parameter θ will you choose based on the above observation?

4.6 Ans: $\theta = 0.5$

Reason: There are two cases with lowest Va value: at $\theta = 0.3$ and at $\theta = 0.5$. The former case shows more underfit in training than that of the latter case.

4.7 Consider the binary classification problem, which you are dealing with, has highly imbalanced classes. The majority class has 99 hundred samples and the minority class has 1 hundred samples. Which of the following metric(s) would you choose for assessing the classification performance? (Select all relevant metric(s))

(a) Accuracy
(b) Cost sensitive accuracy
(c) Precision and recall
(d) Type-I and type-II errors
(d) None of these

4.7 Ans: The answers are (b, c, d).

Reason: A cost sensitive accuracy is needed here for imbalance classes. The choices (b, c, d) show distinct results for each of the two classes (such as minority/majority, type-I/type-II or positive/negative classes).

4.8 According to the plots of Figs. 4.8b and 4.10 (reproduced below), what is the relationship between the Gini Coefficient and the Area Under the ROC (AUC)?

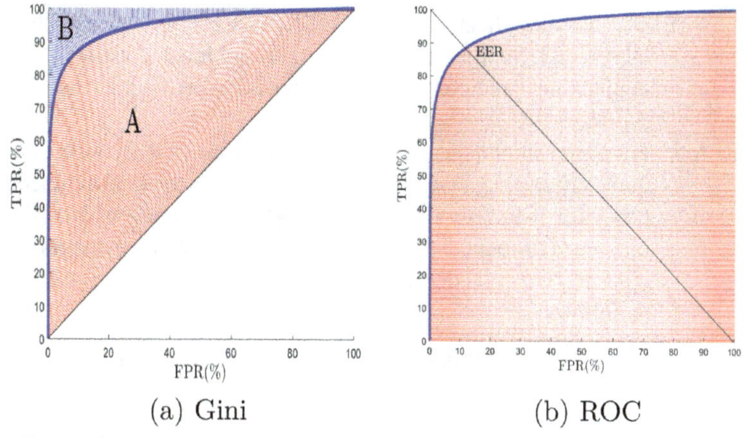

(a) Gini   (b) ROC

4.8 Ans: From plot (a), the Gini-coefficient = A/(A+B). Since the area (A+B) is 1/2, Gini = A/(1/2) = 2A.

From the ROC area in plot (b), AUC = A +1/2 $\Rightarrow$ A = AUC −0.5.

Substitute A into the Gini above: Gini = 2(AUC −0.5).

4.9 Given the following confusion matrix for a binary classifier:

	Class(+1)	Class(−1)
Predicted Class(+1)	990	20
Predicted Class(−1)	10	380

Let n^+ and n^- denote the numbers of Class(+1) and Class(−1) samples respectively. Calculate the

(a) True Positive rate;
(b) True Negative rate;
(c) Accuracy at $s = n^-/n^+$ (this is the most widely adopted accuracy);
(d) Accuracy at $s = 1$;
(e) Precision at $s = n^-/n^+$.

4.9 Ans: (a) True Positive rate: $TPR = TP/n^+ = 990/(990 + 10) = 0.99$.

(b) True Negative rate: $TNR = TNP/n^- = 380/(380 + 20) = 0.95$.

(c) Accuracy at $s = n^-/n^+$: $(TP + TN)/(n^+ + n^-) = (990 + 380)/(1000 + 400) = 0.9786$.

(d) Accuracy at $s = 1$: $(TPR + TNR)/2 = (0.99 + 0.95)/2 = 0.97$.

(e) Precision at $s = n^-/n^+$: $TP/(TP + FP) = 990/(990 + 20) = 0.9802$.

4.10 The Misclassification Error (MCE) is the count of samples classified incorrectly. Suppose the binary targets are labeled as $\{-1, +1\}$ and $g(x)$ denotes the predictor output. Express the MCE in terms of the margin $y_i g(x_i)$.

4.10 Ans: For each sample, the margin $y_i g(x_i)$ is positive if the sample is classified correctly and negative vice versa. The MCE is the total number of samples classified incorrectly. Using $\mathrm{sgn}(y_i g(x_i))$ where $\mathrm{sgn}(\cdot) \in \{-1, +1\}$, we only need to count those samples which give negative margins. To obtain a counting with positive numbers, we need to invert the sign of the counting as follows:

$$\mathrm{MCE} = -\sum_{i=1}^{n} \min\{0, \mathrm{sgn}(y_i g(x_i))\}.$$

Chapter 5

Analytic Learning

Exercises

5.1 The polynomial function given by $g(x, w) = w_0 + w_1 x_1 + w_2 x_2 + w_3 x_1 x_2 + w_4 x_1^2 + w_5 x_2^2$ is linear with respect to the input feature vector $x = [x_1, x_2]^T$, and hence the learning of its parameters can be analytic. True or False?

5.1 Ans: False. Reason: $g(x, w)$ is NOT linear with respect to the input feature vector $x = [x_1, x_2]^T$, but it is linear with respect to w.

5.2 The contours of decision surface in the following plot have been produced by learning a second-order polynomial model based on four data points drawn from two categories. Which of the following statement(s) is/are true?

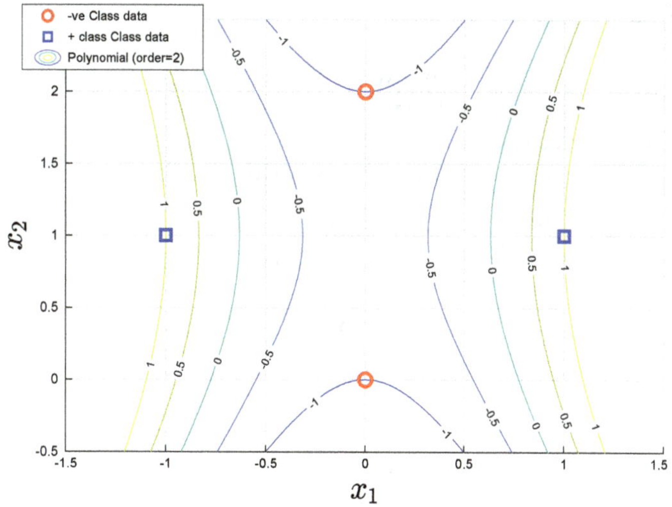

(a) The system is an over-determined one
(b) The system is an under-determined one
(c) The system is an even-determined one
(d) The learned solution is exact with no approximation
(e) None of these

5.2 Ans: (b) and (d).

Reason: The second-order polynomial expansion on two features (x_1 and x_2 of the plot) give rise to model $g(x, w) = w_0 + w_1 x_1 + w_2 x_2 + w_3 x_1 x_2 + w_4 x_1^2 + w_5 x_2^2$ which has 6 parameters. Using this polynomial to learn 4 data points in the plot constitutes an under-determined system. The learning is thus exact with no approximation.

5.3 Consider the optimization problem given by

$$\min_{w} J(w) = \min_{w} \left\{ \sum_{i=1}^{m} (f(w, x) - y_i)^2 + \lambda w^T w \right\}$$

where $f(w, x_i) = x_i^T w$ with each x_i being non-zero independent entry, $w \in \mathbb{R}^{(d+1) \times 1}$ is the parameter, each $y_i \in \mathbb{R}$ is the learning target. When the covariance $X^T X$ is invertible, a least squares solution exists in the form of $\hat{w} = (X^T X)^{-1} X^T y$. Which of the following statement(s) is/are correct?

(a) The first derivative of $f(w, x_i)$ with respect to w at $\hat{w}$ must be zero.
(b) The first derivative of $f(w, x_i)$ with respect to w at $\hat{w}$ need not be zero.
(c) The second derivative of $f(w, x_i)$ with respect to w at $\hat{w}$ must be zero.
(d) The second derivative of $f(w, x_i)$ with respect to w at $\hat{w}$ must be positive definite.
(e) Non of these.

5.3 Ans: (b) Reason:

- For setting the derivative to 0, it concerns the cost function J and not the learning model $f(\boldsymbol{w}, \boldsymbol{x}_i)$!
- Nevertheless, the first derivative of the learning model $f(\boldsymbol{w}, \boldsymbol{x}_i) = \boldsymbol{x}_i^T \boldsymbol{w}$ wrt $\boldsymbol{w}$ is $\boldsymbol{x}_i$ and it is not necessarily 0.

5.4 Consider the following data samples with x_1 and x_2 being the data features, and y being the learning target output class label.

x_1	0	1	1	0
x_2	0	1	0	1
y	-1	-1	+1	+1

Without regularization, construct a polynomial model of second-order to learn these data and then predict the test output labels of $(x_1, x_2) \in \{(0.1, 0.1), (0.9, 0.9), (0.9, 0.1), (0.1, 0.9)\}$.

5.4 Ans: The second-order polynomial model can be written as $g(x_1, x_2) = w_0 + w_1 x_1 + w_2 x_2 + w_3 x_1 x_2 + w_4 x_1^2 + w_5 x_2^2$ and the corresponding polynomial regressor matrix is

$$\mathbf{P} = \begin{bmatrix} 1 & 0 & 0 & 0 & 0 & 0 \\ 1 & 1 & 1 & 1 & 1 & 1 \\ 1 & 1 & 0 & 0 & 1 & 0 \\ 1 & 0 & 1 & 0 & 0 & 1 \end{bmatrix}.$$

This is an under-determined system since there are more parameters than sample size. The solution for the parameters is given by

$$\hat{\boldsymbol{w}} = \mathbf{P}^T (\mathbf{P}\mathbf{P}^T)^{-1} \mathbf{y}$$

$$= \begin{bmatrix} 1 & 1 & 1 & 1 \\ 0 & 1 & 1 & 0 \\ 0 & 1 & 0 & 1 \\ 0 & 1 & 0 & 0 \\ 0 & 1 & 1 & 0 \\ 0 & 1 & 0 & 1 \end{bmatrix} \begin{bmatrix} 1 & 0 & 0 & 0 & 0 & 0 \\ 1 & 1 & 1 & 1 & 1 & 1 \\ 1 & 1 & 0 & 0 & 1 & 0 \\ 1 & 0 & 1 & 0 & 0 & 1 \end{bmatrix}^{-1} \begin{bmatrix} -1 \\ -1 \\ +1 \\ +1 \end{bmatrix}$$

$$= \begin{bmatrix} -1 \\ 1 \\ 1 \\ -4 \\ 1 \\ 1 \end{bmatrix}$$

The test label predictions are

$$\hat{g}_{class}(x_1, x_2) = \text{sgn}(\mathbf{P}_t \hat{\mathbf{w}}) = \text{sgn}\left(\begin{bmatrix} -0.82 \\ -0.82 \\ 0.46 \\ 0.46 \end{bmatrix}\right) = \begin{bmatrix} -1 \\ -1 \\ +1 \\ +1 \end{bmatrix}.$$

5.5 Consider the following data samples with x_1 and x_2 being the data features, and y being the learning target output value.

x_1	0	0	1	2	0	1	3	2	4	5	3	6
x_2	0	1	0	1	2	2	1	3	5	3	6	8
y	−2.1	−1.3	−1.5	1.0	0.5	0.8	2.1	1.8	3.3	2.8	3.8	5.0

First construct a polynomial model of third-order without regularization to learn these data and then predict the test outputs of $(x_1, x_2) \in \{(0.5, 0.3), (0.8, 0.8)\}$. Next learn the same third-order polynomial model with regularization at $\lambda = 0.0001$ for predicting the same test points.

5.5 Ans: The third-order polynomial model can be written as $g(x_1, x_2) = w_0 + w_1 x_1 + w_2 x_2 + w_3 x_1 x_2 + w_4 x_1^2 + w_5 x_2^2 + w_6 x_1^2 x_2 + w_7 x_2^2 x_1 + w_8 x_1^3 + w_9 x_2^3$. The corresponding $\mathbf{P}$ matrix incorporating the training data can be written as

$$\mathbf{P} = \begin{bmatrix} 1 & 0 & 0 & 0 & 0 & 0 & 0 & 0 & 0 & 0 \\ 1 & 0 & 1 & 0 & 0 & 1 & 0 & 0 & 0 & 1 \\ 1 & 1 & 0 & 0 & 1 & 0 & 0 & 0 & 1 & 0 \\ 1 & 2 & 1 & 2 & 4 & 1 & 4 & 2 & 8 & 1 \\ 1 & 0 & 2 & 0 & 0 & 4 & 0 & 0 & 0 & 8 \\ 1 & 1 & 2 & 2 & 1 & 4 & 2 & 4 & 1 & 8 \\ 1 & 3 & 1 & 3 & 9 & 1 & 9 & 3 & 27 & 1 \\ 1 & 2 & 3 & 6 & 4 & 9 & 12 & 18 & 8 & 27 \\ 1 & 4 & 5 & 20 & 16 & 25 & 80 & 100 & 64 & 125 \\ 1 & 5 & 3 & 15 & 25 & 9 & 75 & 45 & 125 & 27 \\ 1 & 3 & 6 & 18 & 9 & 36 & 54 & 108 & 27 & 216 \\ 1 & 6 & 8 & 48 & 36 & 64 & 288 & 384 & 216 & 512 \end{bmatrix}.$$

Based on this setting, the solutions without and with regularization are respectively $\hat{\mathbf{w}}_a = (\mathbf{P}^T \mathbf{P})^{-1} \mathbf{P}^T \mathbf{y}$ and $\hat{\mathbf{w}}_b = (\mathbf{P}^T \mathbf{P} + \lambda \mathbf{I})^{-1} \mathbf{P}^T \mathbf{y}$ where $\mathbf{y} = [-2.1, -1.3, -1.5, 1.0, 0.5, 0.8, 2.1, 1.8, 3.3, 2.8, 3.8, 5.0]^T$.

The learned parameters for non-regularized and regularized cases are

$\hat{\mathbf{w}}_a = [-2.4096, 0.8002, 1.6881, -0.4563, 0.4047, -0.2161, 0.0959, -0.0433, -0.0913, 0.0371]^T$

and

$\hat{\mathbf{w}}_b = [-2.4093, 0.8001, 1.6879, -0.4563, 0.4047, -0.2160, 0.0958, -0.0433, -0.0912, 0.0371]^T$

respectively.

The prediction the two test points are

$\hat{\mathbf{y}}_a = \mathbf{P}_t \hat{\mathbf{w}}_a = [-1.4949, -0.5910]^T$ and $\hat{\mathbf{y}}_b = \mathbf{P}_t \hat{\mathbf{w}}_b = [-1.4948, -0.5910]^T$ for the non-regularized case and the regularized case respectively.

Depending on the sensitivity of inverse computation, the non-regularized case might encounter "matrix close to singular or badly scaled, results may be inaccurate." The two learning results also show that the regularized case gives an estimation with a smaller norm value.

5.6 You are given a collection of 6 training data points of two features ($\mathbf{x} = [x_1, x_2]^T$) and their class labels ($y \in \{-1, +1\}$) as follows:

$$y = -1 : \ \mathbf{x} \in \left\{ \begin{bmatrix} 0 \\ 0 \end{bmatrix}, \begin{bmatrix} 0 \\ 0.2 \end{bmatrix}, \begin{bmatrix} 0.2 \\ 0.2 \end{bmatrix}, \begin{bmatrix} 0.2 \\ 0 \end{bmatrix} \right\},$$

$$y = +1 : \ \mathbf{x} \in \left\{ \begin{bmatrix} 0.2 \\ 0.1 \end{bmatrix}, \begin{bmatrix} 0.3 \\ 0.1 \end{bmatrix} \right\}.$$

These data have been used for training a partial polynomial model of third-order given by equation (1) utilizing the criterion function given by equation (2) below.

$$g(\mathbf{w}, \mathbf{x}) = w_0 + \sum_{i=1}^{2}\sum_{j=1}^{2} w_{ij}x_i x_j + \sum_{i=1}^{2}\sum_{j=1}^{2}\sum_{k=1}^{2} w_{ijk}x_i x_j x_k, \ (1)$$

$$\min_{w} J(\mathbf{w}) = \min_{w} \sum_{i=1}^{m} (g(\mathbf{w}, \mathbf{x}) - y_i)^2 . (2)$$

How many of the above samples are incorrectly classified by the partial polynomial model incorporating a signum threshold?

5.6 Ans: The dual solution without regularization is $\hat{\mathbf{w}} = \mathbf{P}^T (\mathbf{P}\mathbf{P}^T)^{-1}\mathbf{y}$ and the classified output is $\hat{\mathbf{y}} = \text{sgn}(\mathbf{P}\hat{\mathbf{w}}) = [-1, -1, -1, -1, +1, +1]^T$.

Utilizing Python for the computation:

```python
import numpy as np
from numpy.linalg import inv
from sklearn.preprocessing import PolynomialFeatures

X = np.array([[0,0],[0,0.2],[0.2,0.2],[0.2,0],[0.2,0.1],[0.3,0.1]])
y = np.array([-1, -1, -1, -1, 1, 1])
poly = PolynomialFeatures(3)
Pfull = poly.fit_transform(X)
Ppart = Pfull[:,3:10]
P = np.column_stack((np.ones((len(Ppart), 1)),Ppart))
if(P.shape[1] > P.shape[0]): # dual
    w = P.T @ inv(P @ P.T) @ y
else: # primal
    w = (inv(P.T @ P) @ P.T) @ y
```

```
y_trained = P @ w
y_class = np.sign(y_trained)
print(y_class)
```

5.7 Generate the data points using the MATLAB codes below:

```
x = [0.1:.1:10];
rand('state',8);
y = [0.6+0.05*rand(1,length(x)/2),0.3+0.05*rand
(1,length(x)/2)]';
```

Based on the knowledge that a kernel measures some kind of distance between two data points, construct the kernel matrix corresponding to $\mathbf{K}(i,j) = e^{\frac{1}{2\sigma^2}\|x_i - x_j\|_2^2}$ for two σ values given by 0.1 and 1. Learn the above data by kernel ridge regression with the two given σ values and use the two learning results to predict test samples given by

```
x_ts = [0.1:.01:10.05];
```

What do you observe regarding the difference between the two predictions?

5.7 Ans: Firstly the learning kernel matrix can be constructed by

```
% Form Training Kernel
sigma = 1;
for i = 1:length(x)
    for j = 1:length(x)
        K(i,j) = exp(-(x(i)-x(j))^2/(2*sigma^2));
    end
end
```

The value of `sigma` can be changed to 1 to form another kernel matrix.

Next, learn the data using the kernel ridge regression as follows.

```
% Training
lambda = 0.001;
alpha = inv(K + lambda*eye(size(K)))*y;
```

For the test prediction, form the test kernel matrix and then use it for prediction as follows.

```
% Form Test Kernel
x_ts = [0.1:.01:10.05]; % Test data
for i = 1:length(x_ts)
    for j = 1:length(x)
        Kts(i,j) = exp(-(x_ts(i)-x(j))^2/(2*sigma^2));
    end
end
```

```
% Test output
y_ts = Kts*alpha;
```

Plotting the results:

```
figure(1), hold on
h1=plot(x,y,'xr','MarkerSize',8,'LineWidth',2);
h2=plot(x_ts,y_ts,'-k','LineWidth',1);
xlabel('$x$','Interpreter','latex','FontSize',18)
```

```
ylabel('$y$','Interpreter','latex','FontSize',18)

% title('$\lambda=0.001,\ \sigma=1$','Interpreter','latex','FontSize',18)

title('$\lambda=0.001,\ \sigma=0.1$','Interpreter','latex','FontSize',18)

lgd=legend([h1, h2], 'Data', 'Test prediction');

axis([0 10.05 0.28 0.68])

grid

hold off
```

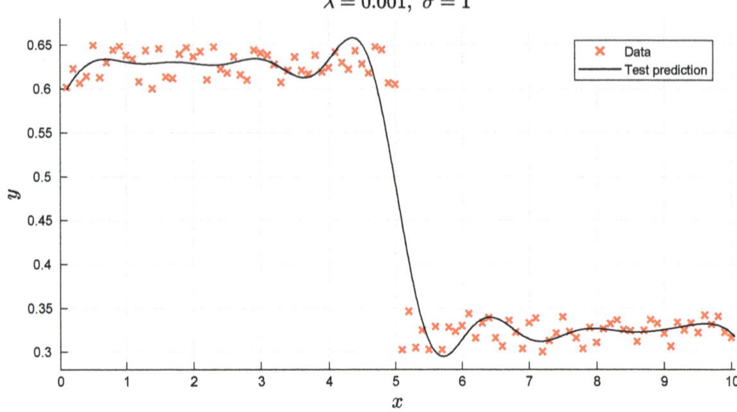

5.8 Consider the following training data points for two categories,
$\begin{bmatrix} x_1 \\ x_2 \end{bmatrix}$: $\begin{bmatrix} 3 \\ 1 \end{bmatrix}$, $\begin{bmatrix} 2 \\ 3 \end{bmatrix}$, $\begin{bmatrix} 0 \\ 1 \end{bmatrix}$, $\begin{bmatrix} 2 \\ 1 \end{bmatrix}$, $\begin{bmatrix} 1 \\ 3 \end{bmatrix}$, which correspond to labels y : $\{0, 0, 0, 1, 1\}$ respectively.

(a) The first goal is to predict the class label of $\mathbf{x}_{t1} = \begin{bmatrix} 3 \\ 3 \end{bmatrix}$ and $\mathbf{x}_{t2} = \begin{bmatrix} 1 \\ 1 \end{bmatrix}$ based on the primal TER method utilizing a linear model with a single output (i.e., utilize y : $\{0, 0, 0, 1, 1\}$ as the single column of learning target).

(b) The second goal is to predict the class label of $\mathbf{x}_{t1} = \begin{bmatrix} 3 \\ 3 \end{bmatrix}$ and $\mathbf{x}_{t2} = \begin{bmatrix} 1 \\ 1 \end{bmatrix}$ based on the dual TER method utilizing a third-order polynomial model.

5.8 Ans: (a) Primal TER learning utilizing a linear model of single output:
By packing the data in matrix form:

$$
\mathbf{X} = \begin{bmatrix} 1, & 3, & 1 \\ 1, & 2, & 3 \\ 1, & 0, & 1 \\ 1, & 2, & 1 \\ 1, & 1, & 3 \end{bmatrix}, \quad \mathbf{y} = \begin{bmatrix} 0 \\ 0 \\ 0 \\ 1 \\ 1 \end{bmatrix}. \tag{1.1}
$$

The TER learning solution for over-determined systems is given by

$$\hat{\mathbf{w}} = (\frac{1}{m^+}\mathbf{X}_+^T\mathbf{X}_+ + \frac{1}{m^-}\mathbf{X}_-^T\mathbf{X}_-)^{-1}(\frac{1}{m^+}\mathbf{X}_+^T\mathbf{y}_+ + \frac{1}{m^-}\mathbf{X}_-^T\mathbf{y}_-),$$

where data has been split into the $+$ve-class and $-$ve-class matrices as follows:
$$\mathbf{X}_+ = \begin{bmatrix} 1, & 2, & 1 \\ 1, & 1, & 3 \end{bmatrix}, \mathbf{y}_+ = \begin{bmatrix} 1 \\ 1 \end{bmatrix} \text{ and } \mathbf{X}_- = \begin{bmatrix} 1, & 3, & 1 \\ 1, & 2, & 3 \\ 1, & 0, & 1 \end{bmatrix}, \mathbf{y}_- = \begin{bmatrix} 0 \\ 0 \\ 0 \end{bmatrix}.$$
Finally, the prediction can be conducted sample-wise or in multiple samples:

(i) Single sample: $\hat{\mathbf{y}}^T = \mathbf{x}_t^T\hat{\mathbf{w}}$ for unseen data sample $\mathbf{x}_t$ i.e.,
$$\hat{\mathbf{y}}_1^T = \mathbf{x}_{t1}^T\hat{\mathbf{w}} = 0.5484 > 0.5 \Rightarrow \text{class1},$$
$$\hat{\mathbf{y}}_2^T = \mathbf{x}_{t2}^T\hat{\mathbf{w}} = 0.4516 < 0.5 \Rightarrow \text{class0}$$

(ii) Multiple samples: $\hat{\mathbf{Y}} = \mathbf{X}_t\hat{\mathbf{w}}$ for unseen data matrix $\mathbf{X}_t$ packed in similar manner.

(b) Dual TER learning utilizing a third-order polynomial model:
By expanding the data matrix (1.1) into the third-order polynomial $\mathbf{P}$ matrix for training (consisting of 5 rows/samples of $\boldsymbol{p}^T$) and the corresponding weight parameter matrix $\mathbf{W}$, we have

$$\mathbf{P} = \begin{bmatrix} \overbrace{1 \ x_1 \ x_2 \ x_1x_2 \ x_1^2 \ x_2^2 \ x_1^2x_2 \ x_1x_2^2 \ x_1^3 \ x_2^3} \\ 1 \ 3 \ 1 \ (3)(1) \ 3^2 \ 1^2 \ 3^2(1) \ (3)1^2 \ 3^3 \ 1^3 \\ 1 \ 2 \ 3 \ (2)(3) \ 2^2 \ 3^2 \ 2^2(3) \ (2)3^2 \ 2^3 \ 3^3 \\ 1 \ 0 \ 1 \ (0)(1) \ 0^2 \ 1^2 \ 0^2(1) \ (0)1^2 \ 0^3 \ 1^3 \\ 1 \ 2 \ 1 \ (2)(1) \ 2^2 \ 1^2 \ 2^2(1) \ (2)1^2 \ 2^3 \ 1^3 \\ 1 \ 1 \ 3 \ (1)(3) \ 1^2 \ 3^2 \ 1^2(3) \ (1)3^2 \ 1^3 \ 3^3 \end{bmatrix}, \mathbf{w} = \begin{bmatrix} w_0 \\ w_1 \\ w_2 \\ w_3 \\ w_4 \\ w_5 \\ w_6 \\ w_7 \\ w_8 \\ w_9 \end{bmatrix}.$$

$$(1.2)$$

Next, construct a diagonal matrix $\mathbf{M}$ using either $\frac{1}{m^+}$ or $\frac{1}{m^-}$ at row positions corresponding to their respective category. In other words, each row of $\mathbf{P}$, $\mathbf{M}$ and $\mathbf{y}$ should have their binary category labels matched.

- TER learning: $\hat{\mathbf{w}} = \mathbf{P}^T(\mathbf{MPP}^T)^{-1}\mathbf{My}$ for under-determined systems.
- Prediction:

(i) Single sample: $\hat{\mathbf{y}}^T = \boldsymbol{p}(\mathbf{x}_t)^T\hat{\mathbf{w}}$ for unseen data sample $\mathbf{x}_t$ i.e.,
$$\hat{\mathbf{y}}_1^T = \boldsymbol{p}(\mathbf{x}_{t1})^T\hat{\mathbf{w}} = -2.5376 < 0.5 \Rightarrow \text{class0},$$
$$\hat{\mathbf{y}}_2^T = \boldsymbol{p}(\mathbf{x}_{t2})^T\hat{\mathbf{w}} = 0.6530 > 0.5 \Rightarrow \text{class1}$$

(ii) Multiple samples: $\hat{\mathbf{Y}} = \mathbf{P}_t\hat{\mathbf{W}}$ for polynomial matrix $\mathbf{P}_t$ of unseen data packed in similar manner.

5.9 Solve the above question based on one-hot encoding of the target instead of utilizing a single target output.

5.9 Ans: (a) Primal TER learning utilizing multiple target outputs:

By packing the data in matrix form and considering a one-hot encoding for the learning target given by

$$\mathbf{X} = \begin{bmatrix} 1, & 3, & 1 \\ 1, & 2, & 3 \\ 1, & 0, & 1 \\ 1, & 2, & 1 \\ 1, & 1, & 3 \end{bmatrix}, \quad \mathbf{Y} = \begin{bmatrix} 0 & 1 \\ 0 & 1 \\ 0 & 1 \\ 1 & 0 \\ 1 & 0 \end{bmatrix}. \tag{1.3}$$

The TER learning for over-determined systems is given by

$$\hat{\mathbf{w}}_k = (\frac{1}{m^+}\mathbf{X}_+^T\mathbf{X}_+ + \frac{1}{m^-}\mathbf{X}_-^T\mathbf{X}_-)^{-1}(\frac{1}{m^+}\mathbf{X}_+^T\mathbf{y}_+ + \frac{1}{m^-}\mathbf{X}_-^T\mathbf{y}_-),$$
$$k = 1, 2$$

where data is split into the $+$ve-class and $-$ve-class matrices:

$$\mathbf{X}_+ = \begin{bmatrix} 1, & 2, & 1 \\ 1, & 1, & 3 \end{bmatrix}, \mathbf{y}_+ = \begin{bmatrix} 1 \\ 1 \end{bmatrix} \text{ and } \mathbf{X}_- = \begin{bmatrix} 1, & 3, & 1 \\ 1, & 2, & 3 \\ 1, & 0, & 1 \end{bmatrix}, \mathbf{y}_- = \begin{bmatrix} 0 \\ 0 \\ 0 \end{bmatrix} \text{ for } k = 1.$$

$$\mathbf{X}_- = \begin{bmatrix} 1, & 2, & 1 \\ 1, & 1, & 3 \end{bmatrix}, \mathbf{y}_- = \begin{bmatrix} 0 \\ 0 \end{bmatrix} \text{ and } \mathbf{X}_+ = \begin{bmatrix} 1, & 3, & 1 \\ 1, & 2, & 3 \\ 1, & 0, & 1 \end{bmatrix}, \mathbf{y}_+ = \begin{bmatrix} 1 \\ 1 \\ 1 \end{bmatrix} \text{ for } k = 2.$$

Finally, the prediction can be conducted sample-wise or in multiple samples using $\hat{\mathbf{W}} = [\hat{\mathbf{w}}_1, \hat{\mathbf{w}}_2]$:

(i) Single sample: $\hat{\mathbf{y}}^T = \mathbf{x}_t^T\hat{\mathbf{W}}$ for unseen data sample $\mathbf{x}_t \Rightarrow$ The column number of the row vector $\hat{\mathbf{y}}^T$ with the largest value is the class number. ($\hat{\mathbf{y}}_1^T = \mathbf{x}_{t1}^T\hat{\mathbf{W}} = [\mathbf{0.5484}, 0.4516]$, $\hat{\mathbf{y}}_2^T = \mathbf{x}_{t2}^T\hat{\mathbf{w}} = [0.4516, \mathbf{0.5484}]$)

(ii) Multiple samples: $\hat{\mathbf{Y}} = \mathbf{X}_t\hat{\mathbf{W}}$ for unseen data matrix $\mathbf{X}_t$ packed in similar manner.

(b) Dual TER learning utilizing a third-order polynomial model:

By expanding the data matrix (1.1) into the third-order polynomial $\mathbf{P}$ matrix for training (consisting of 6 rows/samples of $\mathbf{p}^T$) and the corresponding weight parameter matrix $\mathbf{W}$, we have

$$
\mathbf{P} = \begin{bmatrix}
1 & 1 & 2 & (1)(2) & 1^2 & 2^2 & 1^2(2) & (1)2^2 & 1^3 & 2^3 \\
1 & 2 & 0 & (2)(0) & 2^2 & 0^2 & 2^2(0) & (2)0^2 & 2^3 & 0^3 \\
1 & 1 & 1 & (1)(1) & 1^2 & 1^2 & 1^2(1) & (1)1^2 & 1^3 & 1^3 \\
1 & 3 & 1 & (3)(1) & 3^2 & 1^2 & 3^2(1) & (3)1^2 & 3^3 & 1^3 \\
1 & 2 & 3 & (2)(3) & 2^2 & 3^2 & 2^2(3) & (2)3^2 & 2^3 & 3^3 \\
1 & 3 & 3 & (3)(3) & 3^2 & 3^2 & 3^2(3) & (3)3^2 & 3^3 & 3^3
\end{bmatrix}, \quad
\mathbf{W} = \begin{bmatrix}
w_{0,1} & w_{0,2} \\
w_{1,1} & w_{1,2} \\
w_{2,1} & w_{2,2} \\
w_{3,1} & w_{3,2} \\
w_{4,1} & w_{4,2} \\
w_{5,1} & w_{5,2} \\
w_{6,1} & w_{6,2} \\
w_{7,1} & w_{7,2} \\
w_{8,1} & w_{8,2} \\
w_{9,1} & w_{9,2}
\end{bmatrix}.
$$

where the columns of $\mathbf{P}$ correspond to $1 \; x_1 \; x_2 \; x_1 x_2 \; x_1^2 \; x_2^2 \; x_1^2 x_2 \; x_1 x_2^2 \; x_1^3 \; x_2^3$.

$$(1.4)$$

Next, construct a diagonal matrix $\mathbf{M}_k$ using either $\frac{1}{m^+}$ or $\frac{1}{m^-}$ at row positions corresponding to their respective category for each column k. $\mathbf{y}_k$, $k = 1, 2$ correspond to the k-th column of the indicator matrix $\mathbf{Y}$ in (1.3). In other words, each row of $\mathbf{P}_k$, $\mathbf{M}_k$ and $\mathbf{y}_k$ for $k = 1, 2$ should have their binary category labels matched. Note: $\mathbf{P}_k$ is the same matrix as $\mathbf{P}$ but with each $+$ve and $-$ve sample aligned with that in $\mathbf{y}_k$.

- TER learning: $\hat{\mathbf{w}}_k = \mathbf{P}_k^T (\mathbf{M}_k \mathbf{P}_k \mathbf{P}_k^T)^{-1} \mathbf{M}_k \mathbf{y}_k$, $k = 1, 2$ for under-determined systems. Pack parameters as $\hat{\mathbf{W}} = [\hat{\mathbf{w}}_1, \hat{\mathbf{w}}_2]$.
- Prediction:

(i) Single sample: $\hat{\mathbf{y}}^T = p(\mathbf{x}_t)^T \hat{\mathbf{W}}$ for unseen data sample $\mathbf{x}_t \Rightarrow$ The column number of the row vector $\hat{\mathbf{y}}^T$ with the largest value is the class number.
$(\hat{\mathbf{y}}_1^T = p(\mathbf{x}_{t1})^T \hat{\mathbf{W}} = [-2.5376, \mathbf{3.7500}], \hat{\mathbf{y}}_2^T = p(\mathbf{x}_{t2})^T \hat{\mathbf{W}} = [\mathbf{0.6530}, 0.3232])$

(ii) Multiple samples: $\hat{\mathbf{Y}} = \mathbf{P}_t \hat{\mathbf{W}}$ for polynomial matrix $\mathbf{P}_t$ of unseen data packed in similar manner.

5.10 Given the data points of two categories which are labeled by circles and squares as shown in the following figure. The decision boundaries of Classifier A, Classifier B, and Classifier C are shown as dotted line, dashed line, and solid line respectively.

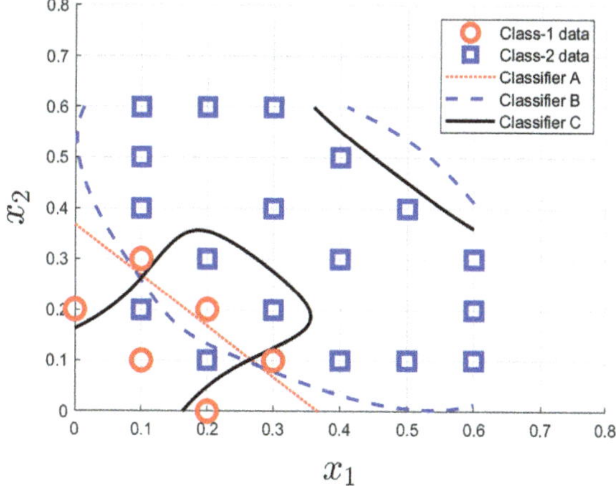

(i) Write out the confusion matrix for each classifier.
(ii) Calculate the Total Error Rate for each Classifier.
(iii) Which classifier can predict better for unseen data? Why?

5.10 Ans: (i) Classifier A

	C1	C2
C1	3	2
C2	3	16

Classifier B

	C1	C2
C1	3	2
C2	3	16

Classifier C

	C1	C2
C1	2	4
C2	4	14

(ii) Total Error Rate for each Classifier:
A: $3/6 + 2/18 = 11/18$
B: $3/6 + 2/18 = 11/18$
C: $4/6 + 4/18 = 16/18$

(iii) Which classifier can predict better for unseen data?
Classifier A and Classifier B show better generalization than Classifier C. They have simpler decision boundary than that of Classifier C.

Chapter 6

Penalized Learning

Exercises

6.1 The solution for a primal ridge regression problem with feature vectors in $\mathbf{X}$ and targets in $\mathbf{y}$ can be written as $\hat{\mathbf{w}} = (\mathbf{X}^T\mathbf{X} + \lambda\mathbf{I})^{-1}\mathbf{X}^T\mathbf{y}$ for $\lambda > 0$. As λ increases, $\hat{\mathbf{w}}^T\hat{\mathbf{w}}$ increases. True or False?

6.1 Ans: False. Reason: As λ increases, the penalty to $\hat{\mathbf{w}}^T\hat{\mathbf{w}}$ increases and hence $\hat{\mathbf{w}}^T\hat{\mathbf{w}}$ decreases.

6.2 Which technique in classifier learning involves minimizing an error cost function while incorporating an L_p-norm penalty on the model parameters?

(a) Ridge regression
(b) Lasso regression
(c) Elastic net
(d) Bridge regression
(e) None of these.

6.2 Ans: (d)

(a) Ridge regression: incorporating an L_2-norm penalty on the model parameters.
(b) Lasso regression: incorporating an L_1-norm penalty on the model parameters.
(c) Elastic net: varies between L_1-norm and L_2-norm penalty on the model parameters.
(d) Bridge regression: incorporating an L_p-norm penalty on the model parameters.
(e) None of these.

6.3 Consider the penalized learning formulated by

$$\min_{w} J(x, w) = \min_{w} \left\{ \sum_{i=1}^{m} (f(x_i, w) - y_i)^2 + \lambda \|w\|_p^p, \right\}, \quad \lambda > 0$$

where $f(x, w)$ is a learning model with augmented input $x \in \mathbb{R}^{d+1}$, learning parameter $w \in \mathbb{R}^{d+1}$ and y_i, $i = 1, ..., m$ being the learning target. Which of the following statement(s) is/are true?

(a) When $p \leq 1$, it performs both variable selection and regularization.
(b) When $p > 2$, it does not perform coefficient shrinkage.
(c) When $p < 2$, it performs coefficient shrinkage.
(d) When $p = 2$, the solution is sparse.
(e) None of these.

6.3 Ans: (a) and (c)

Reason: the method performs coefficient shrinkage for all values of p since the coefficient norm term is also minimized along minimization of the error term when $\lambda > 0$.

6.4 Consider the criterion

$$\hat{w} = \min_{w} \left\{ \sum_{i=1}^{m} \left(y_i - w_0 - \sum_{j=1}^{d} x_{ij} w_j \right)^2 + \lambda \sum_{j=1}^{d} |w_j|^q \right\}, \quad q \geq 0,$$

where x and y are the input and target values, $\mathbf{w}$ is the parameter vector and λ is a penalty factor. The contours of constant value of $\sum_j |\mathbf{w}_j|^q$ are shown in figure below for the case of two inputs. Write down the common name/term (for those items marked with "?") for each value of q and state whether there is coefficient shrinkage and sparseness for each estimation (by circling the answer) (Note: the answer for LASSO is already provided):

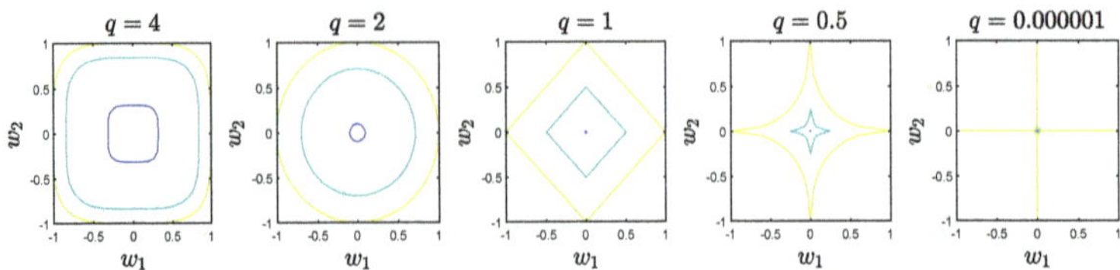

Value of q	Common Names	Coefficient shrinkage?	Sparse?
$q = 0$	?	(Yes/No)?	(Yes/No)?
$q = 0.5$	–NA–	(Yes/No)?	(Yes/No)?
$q = 1$	lasso	Yes	Yes
$q = 2$	?	(Yes/No)?	(Yes/No)?
$q = 4$	–NA–	(Yes/No)?	(Yes/No)?

Value of q	Common Terms	Coefficient shrinkage?	Sparse?
$q = 0$	variable subset selection	(Yes)	(Yes)
$q = 0.5$	–	(Yes)	(Yes)
$q = 1$	lasso	(Yes)	(Yes)
$q = 2$	ridge regression	(Yes)	(No)
$q = 4$	–	(Yes)	(No)

6.4 Ans:

6.5 Consider the XOR problem with four training data samples given by $(x_1, x_2) \in \{(0, 1), (2, 1), (1, 0), (1, 2)\}$ with their corresponding target outputs $y \in \{0, 0, 1, 1\}$. Use the following quadratic polynomial model to learn the XOR data based on the proximal bridge regression at $k = 1.1$ and $k = 2$ using $\lambda = 10$.

$$p(x_1, x_2) = w_0 + w_1 x_1 + w_2 x_2 + w_3 x_1^2 + w_4 x_2^2 + w_5 x_1 x_2$$

Rank the coefficients in terms of their magnitudes. Comment on any difference in the rankings.

6.5 Ans: This is an under-determined case. The MSEs at $k = 1.1$ and $k = 2$ are respectively 0.4478 and 0.3295. This shows compromise of fitting accuracy to favor sparseness for $k = 1.1$.

The estimated parameter vectors are respectively

$$\boldsymbol{w}_{k=1.1} = \begin{bmatrix} 0.0001 \\ 0.0005 \\ -0.0005 \\ -0.0416 \\ 0.0516 \\ 0.0018 \end{bmatrix}, \quad \boldsymbol{w}_{k=2} = \begin{bmatrix} 0.0278 \\ 0.0213 \\ 0.0366 \\ -0.0223 \\ 0.0656 \\ 0.0300 \end{bmatrix}.$$

The ranking orders for the learned coefficients are

$$\boldsymbol{w}_{k=1.1} : w_4, w_3, w_5, w_1/w_2, w_0,$$

$$\boldsymbol{w}_{k=2.0} : w_4, w_2, w_5, w_0, w_3, w_1.$$

The above magnitude ranking shows the different importance of the coefficients between the learned parameters of $\boldsymbol{w}_{k=1.1}$ and $\boldsymbol{w}_{k=2}$. The coefficient w_4 for x_2^2 in both cases shows the highest influence in determining the learning decision.

Codes in MATLAB:

```matlab
% XOR training data
X = [ 0  1; 2  1; 1  0; 1  2]; % XOR inputs (training data)
ytr = [0; 0; 1; 1];            % XOR outputs (training data)
%--- 2nd-order Polynomial features ---%
Ptr = [ones(4,1),X(:,1), X(:,2), X(:,1).^2, X(:,2).^2, X(:,1).*X(:,2)];
Lamval = [10];
kval = [1.1, 2];
for nn = 1:length(Lamval)
    for ll=1:length(kval)
        alpha(:,nn,ll) = p_bridge(Ptr,ytr,kval(ll),Lamval(nn));
        ytr_bridge = Ptr*alpha(:,nn,ll);
        MSE(nn,ll) = mean((ytr - ytr_bridge).^2);
    end
end


function [alpha] = p_bridge(P,y,k,lambda,primal)
% primal = "1", "0" for primal, dual respectively
%    exclusion of "primal" -> default choice based on over- or under-determined system
if (nargin==4) primal = 2; end % default choice if no input for primal
if (primal==1) Prow = 2; Pcol = 1;
elseif (primal==0) Prow = 1; Pcol = 2;
else [Prow,Pcol] = size(P); end
if (Prow<Pcol) % under-determined case
    if (k==2)
        Pwarp = P';
    else
        Pwarp = abs(P').^(1/(k-1));
    end
```

```matlab
    ID = eye(size(P*P'));

    III = (P'/(P*P' + lambda*ID))*P;

    VVV = (Pwarp/(P*Pwarp + lambda*ID))*y;

    alpha = sign(VVV).*abs( (III*(VVV.^(k-1))).^(1/(k-1)) );
else % over-determined case
    pseudoinv = (P'*P + lambda*eye(size(P'*P)))\P';

    alpha = pseudoinv*y; % initial estimation

    for kk=1:5
        pseudoinv = (diag((lambda*k/2)*(abs(alpha(:,1))).^(k-2))+P'*P)\P';

        alpha = pseudoinv*y;

    end

end

end % end function
```

6.6 Consider the simulated data with the true model given by $y = \boldsymbol{w}^T \boldsymbol{x} + \sigma \epsilon$ with $\epsilon \sim N(0, 1)$. Generate a data set consisting of 100/100/400 (training/validation/test) samples with $\sigma = 15$ and

$$\boldsymbol{w} = [0, 3, 6, 8, 0, 0, 0, 0, 0, 0]^T.$$

The correlation between x_i and x_j is zero.
(Hint: the data can be generated using the following MATLAB code:
`Xall = mvnrnd(zeros(1,10),diag(ones(1,10)),600);`
which uses a fixed random seed such as `rng(8)`.)
Compute the Mean-Squared Error (MSE) of learning using ordinary least-sqaures, ridge regression and proximal bridge regression based on the generated data. Compare their learned parameters too.
(Hint: since the true model is known, the MSE can be computed using $\text{MSE} = (\hat{\boldsymbol{w}} - \boldsymbol{w})^T \frac{\mathbf{X}_t^T \mathbf{X}_t}{n} (\hat{\boldsymbol{w}} - \boldsymbol{w})$ where $\mathbf{X}_t$ and n are respectively the matrix containing the test samples and the test sample size.)

```matlab
sig = 15; beta = [0,3,6,8,0,0,0,0,0,0]';

for kk = 1:1 %for a single trial

  %--- Generate data ---%

  rng(kk*8); Xall = mvnrnd(zeros(1,10),diag(ones(1,10)),600);

  %--------------------%

  [nrow,ncol,ntrial] = size(Xall);

  X = Xall; % no bias term

  rng(kk*18); epsilon = normrnd(0,1,600,1);

  y = X*beta + sig*epsilon; % target, exclude bias term

  rng(kk*68); Idx = randperm(600);

  Xrand = X(Idx,:); yrand = y(Idx);

  Xtr = Xrand(1:100,:); ytr = yrand(1:100);

  Xva = Xrand(101:200,:); yva = yrand(101:200);

  Xts = Xrand(201:600,:); yts = yrand(201:600);
```

```matlab
    %--- OLS regression ---%
    alpha_ols = ridge(ytr,Xtr,0);
    yts_ols = Xts*alpha_ols;
    MSE_ols(kk) = (alpha_ols - beta)'*(Xts'*Xts)*(alpha_ols - beta)/length(yts);
    ALPHA_ols(:,kk) = alpha_ols;


    %--- ridge regression ---%
    clear Lam; clear kval; clear minVal; clear minIdx;
    Lam = [0:.01:1,2:1:10,20:10:100,200:100:1000,2000:1000:10000];
    for jj=1:length(Lam)
        alpha_ridge = ridge(ytr,Xtr,Lam(jj));
        yva_ridge = Xva*alpha_ridge;
        MSE_ridgeval(jj) = mean((yva - yva_ridge).^2);
    end
    [minVal,minIdx] = min(MSE_ridgeval);
    alpha_ridge = ridge(ytr,Xtr,Lam(minIdx));
    yts_ridge = Xts*alpha_ridge;
    MSE_ridge(kk) = (alpha_ridge - beta)'*(Xts'*Xts)*(alpha_ridge - beta)/length(yts);
    ALPHA_ridge(:,kk) = alpha_ridge;
    Lam_ridge = Lam(minIdx);


    %--- pbrdige regression k=min ---%
    clear Lam; clear kval; clear minVal; clear minIdx; clear MSE_bridgeval;
    Lam = [0:.01:1,2:1:10,20:10:100,200:100:1000,2000:1000:10000];
    kval = [1:.01:2];
    clear alpha_bridge; clear yva_bridge; clear MSE_bridgeval;
    for jj=1:length(Lam)
        [kk,2,jj]
        for ll=1:length(kval)
            alpha_bridge = bridge(Xtr,ytr,kval(ll),Lam(jj));
            yva_bridge = Xva*alpha_bridge;
            MSE_bridgeval(jj,ll) = mean((yva - yva_bridge).^2);
        end
    end
    [minVal1,minIdx1] = min(MSE_bridgeval,[],1);
    [minVal2,minIdx2] = min(minVal1);
    alpha_bridgekmin = bridge(Xtr,ytr,kval(minIdx2),Lam(minIdx1(minIdx2)));
    yts_bridgekmin = Xts*alpha_bridgekmin;
    MSE_bridgekmin(kk) = (alpha_bridgekmin - beta)'*(Xts'*Xts)*(alpha_bridgekmin - beta)/length(yts);
    ALPHA_bridgekmin(:,kk) = alpha_bridgekmin;
    k_bridge = kval(minIdx2);
    Lam_bridge = Lam(minIdx1(minIdx2));
end


median_MSE_ols = median(MSE_ols);
```

```
std_MSE_ols = std(MSE_ols);

OLS_result = [median_MSE_ols, std_MSE_ols/sqrt(length(MSE_ols))]

median_MSE_ridge = median(MSE_ridge);

std_MSE_ridge = std(MSE_ridge);

Ridge_result = [median_MSE_ridge, std_MSE_ridge/sqrt(length(MSE_ridge))]

median_MSE_bridgekmin = median(MSE_bridgekmin);

std_MSE_bridgekmin = std(MSE_bridgekmin);

Bridgekmin_result = [median_MSE_bridgekmin, std_MSE_bridgekmin/sqrt(length(MSE_bridgekmin))]

% display learned parameters

[mean(ALPHA_ols,2), mean(ALPHA_ridge,2), mean(ALPHA_bridgekmin,2)]
```

The selected k and λ values for proximal bridge are respectively 1.01 and 300. The selected λ for ridge regression is 40. The learned parameters are

$$
[\hat{\boldsymbol{w}}_{\text{ols}},\ \hat{\boldsymbol{w}}_{\text{ridge}},\ \hat{\boldsymbol{w}}_{\text{pbridge}}] =
\begin{bmatrix}
-0.6719 & 0.0770 & 0.0002 \\
5.0816 & 3.2727 & 3.1247 \\
7.2959 & 5.2663 & 5.9396 \\
9.6343 & 6.7390 & 7.8913 \\
0.8932 & 0.5700 & 0.0073 \\
0.4527 & -0.4748 & -0.0009 \\
-3.6663 & -2.4003 & -1.5043 \\
-1.6277 & -0.6048 & -0.0005 \\
-0.6747 & -0.7199 & -0.0127 \\
-0.8095 & -0.7749 & -0.0370
\end{bmatrix}.
$$

The MSE values for $\hat{\boldsymbol{w}}_{\text{ols}}$, $\hat{\boldsymbol{w}}_{\text{ridge}}$, $\hat{\boldsymbol{w}}_{\text{pbridge}}$ are respectively 25.9375, 9.9451, 2.2504. These results show the compressive capability of proximal bridge.

6.7 For the over-determined and under-determined solutions of the proximal bridge regression, are the estimations biased?

6.7 Ans: We shall work on the single output case only since the multiple outputs case is a direct stacking of the single output case. Assume the data is generated according to $\mathbf{y} = \mathbf{X}\boldsymbol{\alpha} + \boldsymbol{\epsilon}$ with $\mathbf{X} \in \mathbb{R}^{M \times D}$, $\boldsymbol{\alpha} \in \mathbb{R}^{D}$ where $\boldsymbol{\epsilon}$ is a zero mean noise with covariance matric $\mathbf{C}$. For the over-determined case, we have $M \geq D$. Suppose the estimation is initialized by $\boldsymbol{\alpha}_0$, then the expectation of $\hat{\boldsymbol{\alpha}}$ is

$$
\begin{aligned}
E[\hat{\boldsymbol{\alpha}}] &= E\left[\left(\frac{\lambda k}{2}\,\text{diag}\{|\boldsymbol{\alpha}_0|^{\circ(k-2)}\} + \mathbf{X}^T\mathbf{X}\right)^{-1}\mathbf{X}^T(\mathbf{X}\boldsymbol{\alpha} + \boldsymbol{\epsilon})\right] \\
&= \left(\frac{\lambda k}{2}\,\text{diag}\{|\boldsymbol{\alpha}_0|^{\circ(k-2)}\} + \mathbf{X}^T\mathbf{X}\right)^{-1}\mathbf{X}^T\mathbf{X}\boldsymbol{\alpha} \\
&\neq \boldsymbol{\alpha},\ \ \forall k > 1, \lambda > 0.
\end{aligned}
\tag{1.5}
$$

This shows that the estimation is biased for the ranges of our working k values and λ values. For the special case when $\lambda = 0$, we have an unbiased ols estimation since $E[\hat{\alpha}] = E\left[\left(\mathbf{X}^T\mathbf{X}\right)^{-1}\mathbf{X}^T\left(\mathbf{X}\alpha + \epsilon\right)\right] = E\left[\left(\mathbf{X}^T\mathbf{X}\right)^{-1}\mathbf{X}^T\mathbf{X}\alpha\right] = \alpha$.

For the under-determined system where $M < D$, the expectation of estimate is

$$E[\hat{\alpha}] = E\left[\mathrm{sgn}\left(\boldsymbol{\theta}\right) \circ \left|\mathbf{X}^T\left(\mathbf{X}\mathbf{X}^T\right)^{-1}\mathbf{X}\boldsymbol{\theta}^{\circ(k-1)}\right|^{\circ\frac{1}{k-1}}\right],$$

$$\boldsymbol{\theta} = |\mathbf{X}^T|^{\circ\frac{1}{k-1}}\left[\mathbf{X}|\mathbf{X}^T|^{\circ\frac{1}{k-1}}\right]^{-1}\left(\mathbf{X}\alpha + \epsilon\right),$$

$$\neq \alpha. \tag{1.6}$$

The inequality holds because the rank of

$$\left|\mathbf{X}^T\left(\mathbf{X}\mathbf{X}^T\right)^{-1}\mathbf{X}\left\{|\mathbf{X}^T|^{\circ\frac{1}{k-1}}\left[\mathbf{X}|\mathbf{X}^T|^{\circ\frac{1}{k-1}}\right]^{-1}\mathbf{X}\alpha\right\}^{\circ(k-1)}\right|^{\circ\frac{1}{k-1}}$$

is at most M (due to $\left(\mathbf{X}\mathbf{X}^T\right)^{-1}$) which is smaller than D. The above analyses for the over- and the under-determined cases show that both the estimates are biased, and this is consistent with the compressed estimation where some ideal parameters have been suppressed.

6.8 The penalty term in ridge regression can be considered as an augmentation data set on the ordinary least squares. By augmenting the data matrix $\mathbf{X}$ with d additional rows and the target $\mathbf{y}$ with d zeros, show that such relationship between data augmentation and regularization can be established.

6.8 Ans: Consider $\mathbf{X} \in \mathbb{R}^{d \times d}$ with corresponding target $\mathbf{y} \in \mathbb{R}^{d \times 1}$. By appending the identity matrix $\mathbf{I}$ to $\mathbf{X}$ with appropriate scaling and appending zeros $\mathbf{0}$ to $\mathbf{y}$, we have

$$\tilde{\mathbf{X}} = \begin{bmatrix} \mathbf{X} \\ \sqrt{\lambda}\mathbf{I} \end{bmatrix}, \quad \tilde{\mathbf{y}} = \begin{bmatrix} \mathbf{y} \\ \mathbf{0} \end{bmatrix}.$$

The least squares solution to this augmented problem is

$$\mathbf{w} = (\tilde{\mathbf{X}}^T\tilde{\mathbf{X}})^{-1}\tilde{\mathbf{X}}^T\tilde{\mathbf{y}}$$

where the matrices can be expanded in terms of their composing blocks:

$$\tilde{\mathbf{X}}^T\tilde{\mathbf{X}} = \begin{bmatrix} \mathbf{X}^T & \sqrt{\lambda}\mathbf{I} \end{bmatrix}\begin{bmatrix} \mathbf{X} \\ \sqrt{\lambda}\mathbf{I} \end{bmatrix} = \mathbf{X}^T\mathbf{X} + \lambda\mathbf{I},$$

$$\tilde{\mathbf{X}}^T\tilde{\mathbf{y}} = \begin{bmatrix} \mathbf{X}^T & \sqrt{\lambda}\mathbf{I} \end{bmatrix}\begin{bmatrix} \mathbf{y} \\ \mathbf{0} \end{bmatrix} = \mathbf{X}^T\mathbf{y}.$$

This solution to the augmented problem can thus be written as

$$\mathbf{w} = (\mathbf{X}^T\mathbf{X} + \lambda\mathbf{I})^{-1}\mathbf{X}^T\mathbf{y}.$$

6.9 By utilizing the singular value decomposition of the centered regressor matrix $\mathbf{X}$, derive an expression to show that $\|\mathbf{w}\|$ increases as its tuning parameter $\lambda \to 0$.

6.9 Ans: The singular value decomposition of the centered regressor matrix $\mathbf{X} \in \mathbb{R}^{m \times d}$ can be writtend as $\mathbf{X} = \mathbf{UDV}^T$ where $\mathbf{U}$ is a $m \times d$ matrix with orthonormal columns that span the column space of $\mathbf{X}$, and $\mathbf{V}$ is a $d \times d$ orthogonal matrix, and finally $\mathbf{D}$ is a $d \times d$ diagonal matrix with elements d_j ordered such that $d_1 \geq d_2 \geq \cdots d_d \geq 0$. Based on this decomposition, we have

$$\mathbf{X}^T\mathbf{X} = \mathbf{VDU}^T\mathbf{UDV}^T = \mathbf{VD}^2\mathbf{V}^T.$$

Based on the above, we have

$$\begin{aligned}
\mathbf{w} &= (\mathbf{X}^T\mathbf{X} + \lambda\mathbf{I})^{-1}\mathbf{X}^T\mathbf{y} \\
&= (\mathbf{VD}^2\mathbf{V}^T + \lambda\mathbf{VV}^T)^{-1}\mathbf{VDU}^T\mathbf{y} \\
&= (\mathbf{V}(\mathbf{D}^2 + \lambda\mathbf{I})\mathbf{V}^T)^{-1}\mathbf{VDU}^T\mathbf{y} \\
&= (\mathbf{V}(\mathbf{D}^2 + \lambda\mathbf{I}))^{-1}\mathbf{DU}^T\mathbf{y}.
\end{aligned}$$

Utilizing this result, we then have

$$\begin{aligned}
\|\mathbf{w}\|_2^2 &= \mathbf{y}^T\mathbf{UD}(\mathbf{D}^2 + \lambda\mathbf{I}))^{-1}\mathbf{V}^T\mathbf{V}(\mathbf{D}^2 + \lambda\mathbf{I}))^{-1}\mathbf{DU}^T\mathbf{y} \\
&= \mathbf{y}^T\mathbf{UD}(\mathbf{D}^2 + \lambda\mathbf{I})^{-2}\mathbf{DU}^T\mathbf{y} \\
&= (\mathbf{U}^T\mathbf{y})^T[\mathbf{D}(\mathbf{D}^2 + \lambda\mathbf{I})^{-2}\mathbf{D}](\mathbf{U}^T\mathbf{y}).
\end{aligned}$$

Since the middle matrix is diagonal with elements $\frac{d_j^2}{(d_j^2+\lambda)^2}$, we thus have

$$\|\mathbf{w}\|_2^2 = \sum_{j=1}^{d} \frac{d_j^2(\mathbf{U}^T\mathbf{y})_j^2}{(d_j^2 + \lambda)^2},$$

with $(\mathbf{U}^T\mathbf{y})_j$ indicating the jth component of the vector $(\mathbf{U}^T\mathbf{y})$. Therefore, as $\lambda \to 0$, the fraction $\frac{d_j^2}{(d_j^2+\lambda)^2}$ increases.

6.10 Match among the types (ℓ_1-norm, ℓ_2-norm, $\ell_{1.5}$-norm) of penalized learning with their corresponding parameter profile below:

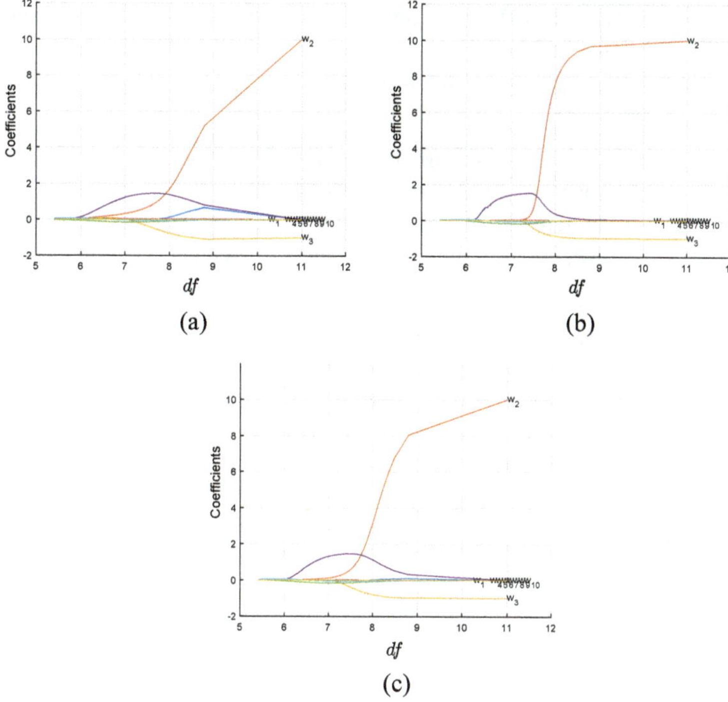

6.10 Ans: (a) ℓ_2-norm: the most gentle compression of parameters

(b) ℓ_1-norm: the steepest compression of parameters with some go to zero

(c) $\ell_{1.5}$-norm: moderate compression

Chapter 7

Network Learning

Exercises

7.1 Solving the parameters of the system of linear equations under the kernel and the range spaces result in the minimum ℓ_2-norm of the parameters. True or False?

7.1 Ans: True. See Lemma 7.1.

7.2 Utilization of the ℓ_1 loss in network residual learning leads to sparseness of the network weights. True or False?

7.2 Ans: False. This is because the learning residual will be forced to be sparse but not the weights.

    7.3 Select from below any ways that are effective to reduce overfitting in neural networks.

      (a) Including an L_2 regularization.
      (b) Increasing the number of hidden nodes.
      (c) Increasing the learning rate.
      (d) Augmenting the training data with synthetic samples.

7.3 Ans: (a) and (d).

    7.4 Consider two networks using the same activation function in all their layers. Then, when trained appropriately, a neural network with 3 layers will not have a higher training error than that of a single-layer neural network.

7.4 Ans: True. This is because a single-layer network does not have the universal approximation capability possessed by a 3-layer network.

    7.5 Suppose you have trained a network for a huge number of iterations with high training and test errors. Which of the following, done in isolation, has a better chance of reducing both the training and test errors?

      (a) Add more hidden layer(s).
      (b) Add more nodes to hidden layers.
      (c) Add more training data.
      (d) Reduce connections of hidden layer(s).
      (e) None of these.

7.5 Ans: (a) Add more hidden layer(s): this might not improve the situation when data is not increased.
      (b) Add more nodes to hidden layers: tends to overfit when overly trained.
      (c) **Add more training data: both training and test errors decrease if added data is representative.**
      (d) Reduce connections of hidden layer(s): this might increase the training error while decreasing the test error marginally.

    7.6 Consider a three-layer neural network utilizing the sigmoid function given by $1/(1 + \exp(-\beta x))$ as the activation in all layers. The parameter β of the sigmoid controls the slope of the activation function. A training error of 89.35 and a testing error of 91.8 are observed. Which of the following is true?

      (a) Increasing β will most likely increase the test error.
      (b) Increasing β will most likely reduce the test error.
      (c) Decreasing β will most likely reduce the test error.
      (d) Decreasing β will most likely increase the test error.
      (e) Not enough information is provided to determine how β should be changed.
      (f) None of these.

7.6 Ans: (e)

7.7 A fully connected network of 3 layers can be written as

$$\mathbf{G} = f([\mathbf{1},\, f([\mathbf{1},\, f(\mathbf{XW}_1)]\mathbf{W}_2)]\mathbf{W}_3)$$

where

$$\mathbf{X} = \begin{bmatrix} 1 & 1 & 3.0 \\ 1 & 2 & 2.5 \end{bmatrix}, \quad \mathbf{W}_1 = \begin{bmatrix} -1 & 0 & 1 \\ 0 & -1 & 0 \\ 1 & 0 & 1 \end{bmatrix},$$

$$\mathbf{W}_2 = \mathbf{W}_3 = \begin{bmatrix} -1 & 0 & 1 \\ 0 & -1 & 0 \\ 1 & 0 & 1 \\ 1 & -1 & 1 \end{bmatrix}.$$

Suppose the Rectified Linear Unit (ReLU) has been used as the activation function (f) for all the nodes. Compute the network output matrix $\mathbf{G}$ based on the given network weights and data.

7.7 Ans: Utilizing MATLAB to compute the network output, we have

```
x = [1 3; 2 2.5];
X = [ones(2,1),x];
W1 = [-1 0 1; 0 -1 0; 1 0 1];
W2 = [-1 0 1; 0 -1 0; 1 0 1; 1 -1 1];
W3 = W2;
F = ReLU([ones(2,1),ReLU([ones(2,1),ReLU(X*W1)]
*W2)]*W3)
*[2mm] function g = ReLU(a)
g = max(a,0);
end
```

The network output is $\begin{bmatrix} 4.0 & 0 & 6.0 \\ 3.5 & 0 & 5.5 \end{bmatrix}$.

7.8 The following graph shows the structure of a simple neural network with a single hidden layer. The input layer consists of three dimensions $x = [x_1, x_2, x_3]^T$. The hidden layer includes three units $y = [y_1, y_2, y_3]^T$. The output layer includes one unit z. The bias terms are ignored for simplicity.

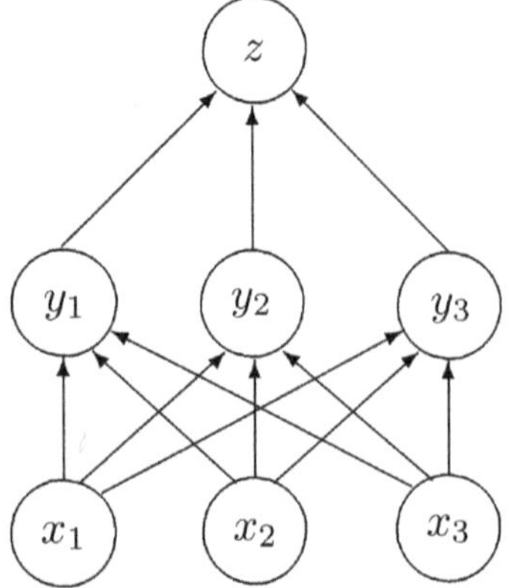

A linear rectified unit $\mathrm{ReLU}(a) = \max(0, a)$ is adopted as activation function for the hidden layer and the output layer. Given the loss function $L(z, t) = \frac{1}{2}(z - t)^2$ with t being the target value for learning. Denote by W_1 and W_2 weight matrices connecting input and hidden layer, and hidden layer and output respectively. They are initialized as follows:

$$W_1 = \begin{bmatrix} 1 & 0 & 1 \\ 0 & 1 & 0 \\ 1 & 0 & 1 \end{bmatrix}, \quad W_2 = \begin{bmatrix} 1 & -1 & 1 \end{bmatrix}.$$

Write out symbolically (no need to put any specific values of W_1 and W_2) the feedforward functional mapping $x \to z$ using ReLU, W_1, W_2. In other words, express y in terms of ReLU, W_1, W_2 and x.

7.8 Ans: $z = \mathrm{ReLU}(W_2\mathrm{ReLU}(W_1 x))$ or $z = \mathrm{ReLU}(\mathrm{ReLU}(x^T W_1^T)W_2^T)$.

7.9 Consider the network N below. The input layer of the network consists of three dimensions $x = [x_1, x_2, x_3]^T$ and the bias. The hidden layer includes three units $h = [h_1, h_2, h_3]^T$ and the bias. The output layer includes one unit y.

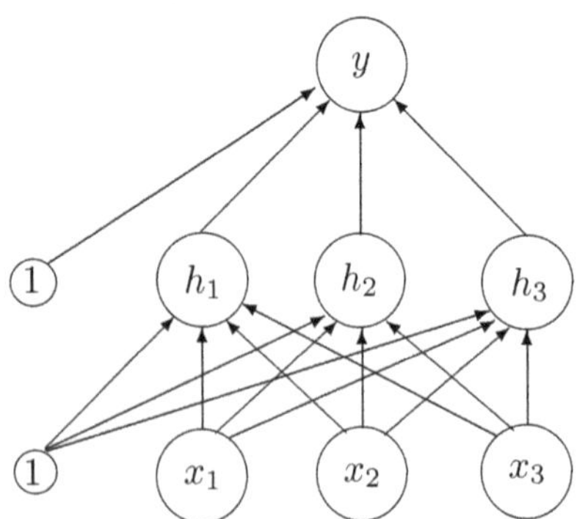

After training this network, the values of the weights (including the bias weights) are

$$\mathbf{W}_1 = \begin{bmatrix} 1 & -1 & 0.5 & 1 \\ 0.5 & -1 & 0.2 & -1 \\ 1 & -0.5 & 1 & 1 \end{bmatrix}, \quad \mathbf{w}_2 = \begin{bmatrix} 1 & 0.5 & 1 & -1 \end{bmatrix}^T.$$

(a) Suppose the same activation function σ has been adopted for all the hidden and output nodes. Write out symbolically (no need to plug in the specific values of $\mathbf{W}_1$ and $\mathbf{w}_2$ just yet) the feedforward functional mapping $N : x \mapsto y$ using σ, $\mathbf{W}_1$, $\mathbf{w}_2$. In other words, express y in terms of σ, $\mathbf{W}_1$, $\mathbf{w}_2$ and x.

(b) Write down the gradient of the loss function with respect to the weights. In other words, compute the following terms symbolically:

(i) The gradient with respect to $\mathbf{w}_2$.
(ii) The gradient with respect to $\mathbf{W}_1$.

(c) Suppose you input a three-dimensional sample $x = [1, 3, 1]^T$ to the above network N. You tried four different types of activation functions $(\phi_1, \phi_2, \phi_3, \phi_4)$ (on all the hidden and output nodes) which respectively output the values $(y_1, y_2, y_3, y_4) = (0, \ 0.6818, \ -3.1500, \ -0.9963)$. What is a valid guess for the activation functions $(\phi_1, \phi_2, \phi_3, \phi_4)$?

7.9 Ans: (a)

$$h = \sigma(\mathbf{W}_1 x)$$
$$y = \sigma(\mathbf{w}_2^T h)$$

(b)

(i) The gradient relative to $\mathbf{w}_2$.

$$\frac{\partial L}{\partial \mathbf{w}_2} = \left(\frac{\partial y}{\partial \mathbf{w}_2} \right)^T \frac{\partial L}{\partial y}$$

(ii) The gradient relative to $\mathbf{W}_1$.

$$\frac{\partial L}{\partial \mathbf{W}_1} = \left(\frac{\partial h}{\partial \mathbf{W}_1} \right)^T \left(\frac{\partial y}{\partial h} \right)^T \frac{\partial L}{\partial y}$$

(c) MATLAB codes:

```
φ₁: function output = ReLU(x)
        output = max(0,x);
    end

φ₂: function output = sig(x)
        output = 1./(1+exp(-x));
    end

φ₃: function output = lin(x)
        output = x;
    end

φ₄: tanh()
```

7.10 Consider a binary classification task of classifying images as apple versus non-apple. You design a deep network with a single output neuron z for the learning. To make the binary decision, this output neuron goes through an activation given by $\hat{y} = \sigma(\text{ReLU}(z))$ where $\sigma(\cdot)$ denotes the sigmoid function and $\text{ReLU}(\cdot)$ denotes the linear rectified unit. Identify all correct setting(s) below to get full credit.

(a) We can set the target as $y \in \{-1, +1\}$ for apple and non-apple respectively and then classify the output as apple when $\hat{y} \geq 0$ and non-apple when $\hat{y} < 0$.

(b) We can set the target as $y \in \{0, 1\}$ for apple and non-apple respectively and then classify the output as apple when $\hat{y} \geq 0.5$ and non-apple when $\hat{y} < 0.5$.

(c) We can set the target as $y \in \{-1, +1\}$ for non-apple and apple respectively and then classify the output as non-apple when $\hat{y} \geq 0$ and apple when $\hat{y} < 0$.

(d) We can set the target as $y \in \{0, 1\}$ for non-apple and apple respectively and then classify the output as non-apple when $\hat{y} \geq 0.5$ and apple when $\hat{y} < 0.5$.

(e) None of these

7.10 Ans: (e) None of these
(Since the output of ReLU is always positive, the output of sigmoid is always $\sigma(\text{ReLU}(z)) \geq 0.5$. Hence it is impossible to make decision based on the listed settings)

Chapter 8

Ensemble Learning

Exercises

8.1 Explain how the correlation between base learners in a bagging ensemble affects the overall variance of the ensemble prediction.

8.1 Ans: The variance of the ensemble prediction is affected by the correlations between the predictions of individual base learners. To understand this, consider the ensemble prediction $F(x)$ which is the average of predictions from M base learners $h_m(x)$, i.e.,

$$F(x) = \frac{1}{M} \sum_{m=1}^{M} h_m(x).$$

Assuming each learner has the same variance σ^2 and the pairwise correlation between different learners' predictions is ρ, the overall variance of the ensemble's prediction can be derived as follows:

$$\mathrm{Var}(F(x)) = \mathrm{Var}\left(\frac{1}{M} \sum_{m=1}^{M} h_m(x) \right) = \frac{1}{M^2} \mathrm{Var}\left(\sum_{m=1}^{M} h_m(x) \right).$$

Given that the variance of the sum of correlated variables is

$$\mathrm{Var}\left(\sum_{m=1}^{M} h_m(x) \right) = \sum_{m=1}^{M} \mathrm{Var}(h_m(x)) + \sum_{m \neq n} \mathrm{Cov}(h_m(x), h_n(x)),$$

where $\mathrm{Var}(h_m(x)) = \sigma^2$ for all m and $\mathrm{Cov}(h_m(x), h_n(x)) = \rho\sigma^2$, the equation becomes

$$\mathrm{Var}\left(\sum_{m=1}^{M} h_m(x) \right) = M\sigma^2 + M(M-1)\rho\sigma^2.$$

Substituting back, we get

$$\mathrm{Var}(F(x)) = \frac{\sigma^2}{M} + \frac{M-1}{M}\rho\sigma^2.$$

Rearranging, this yields

$$\text{Var}(F(x)) = \rho\sigma^2 + \frac{1-\rho}{M}\sigma^2.$$

This formulation shows that the variance of the ensemble decreases as the number of learners M increases, but the reduction is limited by the correlation ρ. If learners are completely uncorrelated ($\rho = 0$), variance reduction is maximal. Conversely, if learners are fully correlated ($\rho = 1$), multiple learners do not reduce variance.

8.2 Explain the concept of "Irreducible Error" and how it impacts the performance of ensemble learning methods like bagging and boosting.

8.2 Ans: Irreducible error refers to the error inherent in the problem itself, due to noise or other data-related issues that cannot be eliminated by any model. This type of error affects all predictive models, including ensemble methods. While bagging and boosting can significantly reduce variance and bias, respectively, they cannot affect the irreducible error. Understanding this limitation is crucial for realistic expectations of model performance and emphasizes the importance of high-quality, clean data for training predictive models.

8.3 Discuss the importance of "Feature Importance" in the context of Random Forest and how it is calculated.

8.3 Ans: Feature importance in Random Forest is crucial for understanding the influence of each feature on the prediction. It is calculated based on the decrease in node impurity (e.g., Gini impurity for classification tasks) that results from splits on each feature. Averaging this decrease across all trees in the forest gives a measure of the overall importance of each feature. This metric not only helps in model interpretation but also guides feature selection and engineering, allowing for more focused model optimization.

8.4 Describe the effect of "Learning Rate" in boosting algorithms like AdaBoost and XGBoost and how it influences model convergence.

8.4 Ans: The learning rate in boosting algorithms, such as AdaBoost and XGBoost, controls the impact of each subsequently added weak learner on the final model. A smaller learning rate means that each individual learner contributes more subtly to the final model, requiring more learners to converge but typically leading to a more robust and generalized model. Conversely, a larger learning rate allows the model to converge faster, with fewer learners, but increases the risk of overfitting. Proper tuning of the learning rate is crucial for balancing the trade-off between convergence speed and model accuracy.

8.5 Explain the role of "Loss Functions" in the formulation of boosting algorithms, specifically referring to Gradient Boosting and XGBoost.

8.5 Ans: In boosting algorithms like Gradient Boosting and XGBoost, the choice of loss function is critical for directing the training process. These algorithms minimize a specified loss function iteratively by adding weak learners that reduce the residual errors. For instance, Gradient Boosting can use loss functions like squared error for regression or logistic loss for classification, which directly influences the algorithm's approach to reducing prediction errors. XGBoost extends this by allowing custom loss functions, thus enabling more tailored model training that is optimized for specific types of data and prediction tasks.

8.6 Write down the mathematical expression for the general aggregation strategy used in ensemble learning as discussed in the chapter. Explain how this strategy works in the context of both Bagging and Boosting.

8.6 Ans: **Formulation and Explanation**: In ensemble methods, the general model output can be represented as

$$F(x) = \Phi\left(\sum_{m=1}^{M} \alpha_m f_m(x)\right),$$

where Φ is a function translating aggregated model outputs into the final prediction, α_m are the weights of individual models, and $f_m(x)$ are the predictions from individual models.

- **In Bagging**, each model has an equal weight, $\alpha_m = \frac{1}{M}$, reflecting the democratic voting or averaging principle among models, which effectively reduces variance.
- **In Boosting**, weights α_m are adjusted according to the model's performance, emphasizing models that correct the mistakes of predecessors, thus reducing both bias and variance in a sequential manner.

8.7 Provide a derivation of how bagging helps reduce variance without significantly affecting bias (assuming independence of individual models). Refer to the mathematical formulation of a bagged estimator and discuss its implications on a model's prediction error.

8.7 Ans: **Derivation and Explanation**: Consider a model's prediction as $f(x)$ and suppose the model's bias and variance are given by

$$\text{Bias}(f(x)) = \mathbb{E}[f(x)] - f_{\text{true}}(x),$$

$$\text{Variance}(f(x)) = \mathbb{E}[(f(x) - \mathbb{E}[f(x)])^2].$$

In bagging, the aggregated model prediction is

$$F(x) = \frac{1}{M} \sum_{m=1}^{M} f_m(x),$$

where each $f_m(x)$ is trained on a different bootstrap sample. Assuming the models are independent,

$$\text{Bias}(F(x)) = \text{Bias}(f(x)),$$

$$\text{Variance}(F(x)) = \frac{1}{M}\text{Variance}(f(x)).$$

The derivation shows that while the bias remains the same as that of any individual model, the variance is reduced by a factor of M, the number of models in the ensemble, demonstrating how bagging effectively decreases variance.

8.8 Derive the weight update rule used in the AdaBoost algorithm as mentioned in the chapter. Explain how this formula helps to focus learning on more difficult training examples.

8.8 Ans: **Derivation and Explanation**: AdaBoost aims to minimize an exponential loss function by updating the weights of training examples at each iteration. At iteration m, AdaBoost fits a weak classifier h_m to the weighted data, where each weight $w_{i,m}$ is associated with the i-th training example. The error rate ϵ_m of h_m is calculated as follows:

$$\epsilon_m = \frac{\sum_{i=1}^{N} w_{i,m}\mathbf{1}(y_i \neq h_m(x_i))}{\sum_{i=1}^{N} w_{i,m}},$$

where $\mathbf{1}(.)$ is the indicator function, activated when $y_i \neq h_m(x_i)$, and N is the total number of training examples.

Given ϵ_m, the weight update coefficient α_m for the classifier h_m is derived by aiming to balance the weights of correctly and incorrectly classified examples after the update:

$$w_{i,m+1} = w_{i,m}e^{-\alpha_m y_i h_m(x_i)},$$

where y_i is the true label, and $h_m(x_i)$ is the prediction. This formulation ensures that weights are increased for incorrectly classified instances ($y_i h_m(x_i) = -1$) and decreased for correctly classified instances ($y_i h_m(x_i) = 1$).

To find α_m, consider the goal of keeping the sum of weights constant before and after updating:

$$\sum_{i:y_i=h_m(x_i)} w_{i,m}e^{-\alpha_m} + \sum_{i:y_i\neq h_m(x_i)} w_{i,m}e^{\alpha_m} = \sum_{i=1}^{N} w_{i,m}.$$

Substituting and rearranging gives:

$$(1 - \epsilon_m)e^{-\alpha_m} + \epsilon_m e^{\alpha_m} = 1.$$

Solving for α_m yields:

$$\alpha_m = \frac{1}{2} \ln\left(\frac{1 - \epsilon_m}{\epsilon_m}\right).$$

This reweighting mechanism incentivizes the AdaBoost algorithm to concentrate subsequent training on the examples that are more challenging to classify, progressively refining the model's accuracy on these difficult cases.

8.9 Demonstrate mathematically how boosting can achieve lower bias with an increasing number of weak learners.

8.9 Ans: The ensemble model in boosting after M iterations is

$$F(x) = \sum_{m=1}^{M} \alpha_m h_m(x),$$

Assuming each weak learner corrects the residual errors from the previous model, the expected prediction of the ensemble becomes an increasingly accurate approximation of the true function (x), leading to

$$\text{Bias}(F(x)) = |\mathbb{E}[F(x)] - f(x)|.$$

As M increases, the collective corrections of weak learners reduce this bias, enhancing the ensemble's accuracy.

8.10 Implement a basic Random Forest classifier using Python's scikit-learn library on the Iris data set. Demonstrate the classifier's training process, prediction, and evaluate its accuracy on the test set.

8.10 Ans: **Solution**:

```python
from sklearn.datasets import load_iris
from sklearn.ensemble import RandomForestClassifier
from sklearn.model_selection import train_test_test_split
from sklearn.metrics import accuracy_score

# Load the Iris dataset
data = load_iris()
X, y = data.data, data.target

# Split the data into training and test sets
X_train, X_test, y_train, y_test = train_test_split(X, y,
                            test_size=0.3, random_state=42)

# Initialize and train the Random Forest classifier
```

```
clf = RandomForestClassifier(n_estimators=100)
clf.fit(X_train, y_train)

# Make predictions on the test set
predictions = clf.predict(X_test)

# Evaluate the accuracy of the classifier
accuracy = accuracy_score(y_test, predictions)
print("Accuracy:", accuracy)
```

Chapter 9

Performance Evaluation and Statistical Inference

Exercises

9.1 Suppose you are given a data set of 800 samples in total. Which of the following is/are among the best for the given options for evaluating an algorithm based on this data set?

(a) Use all samples as training and test sets.
(b) Based on single hold-out test, i.e., use M samples for testing and use $800 - M$ samples for training the algorithm.
(c) Based on multiple hold-out tests, i.e., perform hold-out test multiple times using randomly partitioned training and test sets.
(d) Based on N-fold cross-validation test.
(e) Based on leave-one-out test, i.e., use 1 sample for testing and use 799 samples for training the algorithm, and repeat this procedure until all samples have been tested. The final results are averaged.

9.1 Ans: (c), (d) and (e)

(a) This option does not consider untrained samples for evaluation.
(b) This option is subjected to biased partitioning.
(c) This option can reduce the partitioning bias.
(d) This option can also minimize partitioning bias and it uses all data for testing.
(e) This option is among the most thorough ones but it consumes heavy training effort for large data sets.

9.2 How can the hold-out test be performed using Python?

9.2 Ans: The hold-out test can be implemented based on the train_test_split function from the sklearn.model_selection module. This function takes in data as a NumPy array, and it returns two new arrays: one for the training data and one for the testing data.

9.3 What is the difference between regression error and classification error? Provide examples of their score measurements.

9.3 Ans:
- Regression error refers to the aggregation of distance difference between each sample output and the target output where the distance can be measured using squared error distance or absolute error distance. Suppose y_i denotes the sample target and $\hat{y}_i$ denotes the sample output (estimation) and m is the number of samples, then the mean squared error (MSE) can be written as $\frac{1}{m}\sum_{i=1}^{m}(\hat{y}_i - y_i)^2$, and the mean absolute error can be written as $\frac{1}{m}\sum_{i=1}^{m}|\hat{y}_i - y_i|$.
- Classification error refers to the aggregation of samples classified incorrectly which can be expressed as a classification error count or as an error ratio. The classification error ratio in general is the total number of misclassified samples over total number of samples, i.e., $(TP+TN)/(n^+ + n^-)$ (See Table 4.3).

9.4 How to do a hyperparameter tuning based on grid search in Python?

9.4 Ans: Grid search involves a systematic testing of different combinations of hyperparameter values in order to find the combination with best validation performance for the model. In Python, the GridSearchCV module from scikit-learn library can be utilized to perform the grid search under the cross-validation mode.

9.5 Under what situation cross-validation should be avoided?

9.5 Ans:
- When the data set is too small for splitting to get enough training size.
- When data is not independent and identically distributed.
- For time series data when data leakage occurs. In other words, data not intended for training could leak out and be used to train the model.

9.6 Is it possible to check if a classifier has underfit or overfit the data using only the training set? If yes, how? If no why?

9.6 Ans: Yes, it is possible to use only training set to check whether underfit or overfit occurs in the classifier model. The training set can be split into training and validation sets and use the validation set for checking. If the training accuracy is much higher than the validation accuracy, then it is likely that the model has overfit the training data. If the training accuracy is much lower than the validation accuracy, then it is likely that the model has underfit the training data.

9.7 A classifier that attains 100% accuracy on the training set and 70% accuracy on test set is better than another classifier that attains 80% accuracy on the training set and 75% accuracy on test set. True or False?

9.7 Ans: False (Reason: The 100% training accuracy with 70% test accuracy appears to have overfit the data.)

9.8 Suppose the hyperparameter η has been selected for a model using tenfold cross-validation. Among the following, the best way to decide a final model to use and estimate its error is to

(a) pick any of the 10 models from the 10 folds as the final model; use the average CV error for the 10 models as its error estimate.

    (b) pick any of the 10 models from the 10 folds as the final model; use its error estimate on the held-out data.

    (c) train a new model on the full data set using the hyperparameter η obtained; use the average CV error as its error estimate.

    (d) average all of the 10 models from the 10 folds; use the average CV error as its error estimate.

    (e) average all of the 10 models from the 10 folds; use the error of the combined model on the full training set.

9.8 Ans: (c)

1. It is best to utilize all the training data to find the final model.

2. As there is no mention of validation set or test set availability, the only best error estimate is the average CV error.

9.9 The_______ the p-value, the stronger the evidence against the null hypothesis provided by the data.

    (a) larger

    (b) smaller

9.9 Ans: (b)

According to https://en.wikipedia.org/wiki/P-value: "In null hypothesis significance testing, the p-value is the probability of obtaining test results at least as extreme as the result actually observed, under the assumption that the null hypothesis is correct. A very small p-value means that such an extreme observed outcome would be very unlikely under the null hypothesis".

9.10 One-sample t-test: To test the hypothesis that preparation study for more than 3 h for three days right before the test helps to score over 70 marks. Here are the results of a random sample of 12 students who had done the preparation study before the test: 86, 81, 71, 90, 69, 64, 76, 85, 77, 71, 80, 62. Test using the following hypotheses, report the test statistic with the p-value at 95% confidence level, then summarize your conclusion.

9.10 Ans: Hypothesis:

H_0: marks $\mu = 70$ (no effect: preparation study for 3 d does not help increase the mean mark)

H_1: marks $\mu \neq 70\,(> 70)$ (effect: preparation study for 3 d helps increase the mean mark)

Test statistic:

From the data, we obtain the mean $(\bar{x}) = 76$ and

$$s = \sqrt{\frac{1}{n-1}\sum_{i=1}^{n}(x_i - \bar{x})} = 8.8318. \text{ Then we get}$$

$$t = \frac{\bar{x} - \mu_0}{s/\sqrt{n}} = \frac{76 - 70}{8.8318/\sqrt{12}} = \frac{6}{2.5495} = 2.3534$$

<u>p − value:</u>

Because $n = 12$, we use the t distribution with $df = 11$ to find the probability. According to the t-Table (see a sample part of the table below), for $df = 11$, the p-value is 1.796 at statistical significance level of 0.05% (i.e., at 95% confidence level) for the one-sided (upper-tail) test.

<u>Conclusion:</u>

Since the p-value of 1.796 (at 95% confidence level) is below the threshold value $t = 2.35$, the null hypothesis is rejected in favor of the alternative hypothesis H_1.

(Note: we try not to over emphasize the statistics by concluding the preparation study of 3 days did make a significant increase in the mean marks.)

one-tailed α	0.10	0.05	0.025	0.01	0.005	0.0005
two-tailed α	0.20	0.10	0.05	0.02	0.01	0.001
df						
⋮	⋮	⋮	⋮	⋮	⋮	⋮
8	1.397	1.860	2.306	2.896	3.355	5.041
9	1.383	1.833	2.262	2.821	3.250	4.781
10	1.372	1.812	2.228	2.764	3.169	4.587
11	1.363	1.796	2.201	2.718	3.106	4.437
12	1.356	1.782	2.179	2.681	3.055	4.318
13	1.350	1.771	2.160	2.650	3.012	4.221
⋮	⋮	⋮	⋮	⋮	⋮	⋮

Bibliography

Åström, K. J., & Wittenmark, B. (1994). *Automatic control* (2nd ed.). Inc, Boston, MA, USA: Longman Publishing Co.

Ben-Israel, A., & Greville, T. N. E. (2003). *Generalized inverses: Theory and applications* (2nd ed.). New York: Springer.

Aizerman, M. A., Braverman, E. A., & Rozonoer, L. (1964). Theoretical foundations of the potential function method in pattern recognition learning. *Automation and Remote Control, 25*, 821–837.

Albert, A. (1972). *Regression and the Moore-Penrose pseudoinverse* (Vol. 94). New York: Academic Press Inc.

Amrhein, V., Greenland, S., McShane, B., & More than 800 signatories. (2019). Retire statistical significance. *Nature, 567*, 305–307.

Aronszajn, N. (1950). Theory of reproducing kernels. *Transactions of the American Mathematical Society, 68*(3), 337–404.

Baldi, P. (1988). Linear learning: Landscapes and algorithms. In *Advances in Neural Information Processing Systems (NIPS 1988)* (pp. 65–72).

Baldi, P., & Hornik, K. (1989). Neural networks and principal component analysis: Learning from examples without local minima. *Neural Networks, 2*, 53–58.

Baldi, P., & Lu, Z. (2012). Complex-valued autoencoders. *Neural Networks, 33*, 136–147.

Bamber, D. (1975). The area above the ordinal dominance graph and the area below the receiver operating characteristic graph. *Journal of Mathematical Psychology, 12*, 387–415.

Barnard, E. (1992). Optimization for training neural nets. *IEEE Transactions on Neural Networks and Learning Systems, 3*(2), 232–240.

Bartlett, P. L., Foster, D. J., & Telgarsky, M. (2017). Spectrally-normalized margin bounds for neural networks. In *Advances in Neural Information Processing Systems (NIPS 2017)* (pp. 6241–6250).

Barton, S. A. (1991). A matrix method for optimizing a neural network. *Neural Computation, 3*(3), 450–459.

Battiti, R. (1992). First and second order methods for learning: Between steepest descent and newton's method. *Neural Computation, 4*(2), 141–166.

Belkin, M., Ma, S., & Mandal, S. (2018). To understand deep learning we need to understand kernel learning. In *Proceedings of the 35th International Conference on Machine Learning (PMLR)* (pp. 541–549).

Benavoli, A., Corani, G., Demšar, J., & Zaffalon, M. (2017). Time for a change: a tutorial for comparing multiple classifiers through Bayesian analysis. *Journal of Machine Learning Research, 18*, 1–36.

K.-A. Toh et al., *Analytic Learning Methods for Pattern Recognition*,
https://doi.org/10.1007/978-981-96-2151-4

Bishop, C. M. (1995). *Neural networks for pattern recognition*. New York: Oxford University Press Inc.

Bishop, C. M. (2006). *Pattern recognition and machine learning*. Springer Science, Printed in Singapore.

Bobrowski, L., & Sklansky, J. (1995). Linear classifiers by window training. *IEEE Transactions on Systems, Man, and Cybernetics, 25*(1), 1–9.

Boser, B. E., Guyon, I. M., & Vapnik, V. N. (1992). A training algorithm for optimal margin classifier. In *Proceedings of the 5th ACM Workshop on Computational Learning Theory* (pp. 144–152). Pittsburgh, PA.

Boyd, S., & Vandenberghe, L. (2004). *Convex optimization*. Cambridge: Cambridge University Press.

Boyd, S., & Vandenberghe, L. (2018). *Introduction to applied linear algebra - Vectors, matrices, and least squares*. Cambridge: Cambridge University Press.

Breiman, L. (1996). Bagging predictors. *Machine Learning, 24*(2), 123–140.

Breiman, L. (2001). Random forests. *Machine Learning, 45*(1), 5–32.

Brent, R. P. (1991). Fast training algorithms for multilayer neural networks. *IEEE Transactions on Neural Networks, 2*(3), 346–354.

Bronshtein, I. N., Semendyayev, K. A., Musiol, G., & Muehlig, H. (2003). *Handbook of mathematics* (4th ed.). Berlin: Springer.

Broomhead, D. S., & Lowe, D. (1988). Multivariable functional interpolation and adaptive networks. *Complex Systems, 2*, 321–355.

Brunelli, R. (1994). Training neural nets through stochastic minimization. *Neural Networks, 7*(9), 1405–1412.

Burges, C. J. C. (1998). A tutorial on support vector machines for pattern recognition. *Data Mining and Knowledge Discovery, 2*(2), 121–167.

Campbell, S. L., & Meyer, C. D. (2009). *Generalized inverses of linear transformations (SIAM edition of the work published by Dover Publications Inc, 1991)*. Philadelphia, USA: Society for Industrial and Applied Mathematics, edition.

Chen, T., & Guestrin, C. (2016). Xgboost: A scalable tree boosting system. In *Proceedings of the 22nd ACM SIGKDD International Conference on Knowledge Discovery and Data Mining* (pp. 785–794). ACM.

Chervonenkis, A. Y. (2001). *Some properties of infinite VC-dimension systems*. Royal Holloway University of London. (Technical Report: CLRC TR 01-04).

Choromanska, A., Henaff, M., Mathieu, M., Arous, G. B., & LeCun, Y. (2015a). The loss surfaces of multilayer networks. In *Proceedings of the Eighteenth International Conference on Artificial Intelligence and Statistics* (pp. 192–204).

Choromanska, A., LeCun, Y., & Arous, G. B. (2015b). Open problem: The landscape of the loss surfaces of multilayer networks. In *Proceedings of The 28th Conference on Learning Theory* (pp. 1756–1760).

Cotter, N. E. (1990). The Stone-Weierstrass theorem and its application to neural networks. *IEEE Transactions on Neural Networks, 1*(4), 290–295.

Courant, R., & Hilbert, D. (1953). *Methods of mathematical physics*, Vol. I, First English edition, Translated and Revised from the German Original. New York: Interscience Publishers, Inc.

Cybenko, G. (1989). Approximations by superpositions of a sigmoidal function. *Mathematics of Control, Signals, and Systems, 2*, 303–314.

Dauphin, Y. N., Pascanu, R., Gulcehre, C., Kyunghyun Cho, S. G., & Bengio, Y. (2014). Identifying and attacking the saddle point problem in high-dimensional non-convex optimization. In *Advances in Neural Information Processing Systems (NIPS 2014)* (pp. 2933–2941).

Davis, P. J. (1970). *Circulant matrices* (p. 0471057711). New York: Wiley.

Demšar, J. (2006). Statistical comparisons of classifiers over multiple data sets. *Journal of Machine Learning Research, 7*, 1–30.

Derobert, L., & Thieriot, G. (2003). The Lorenz curve as an archetype: A historico-epistemological study. *The European Journal of the History of Economic Thought, 10*, 573–585.

Dinh, L., Pascanu, R., Bengio, S., and Bengio, Y. (2017). Sharp minima can generalize for deep nets. In *Proceedings of the 34th International Conference on Machine Learning (PMLR)* (Vol. 70, pp. 1019–1028).

Duda, R. O., Hart, P. E., & Stork, D. G. (2001). *Pattern classification* (2nd ed.). New York: Wiley Inc.

Dziugaite, G. K. & Roy, D. M. (2017). Computing nonvacuous generalization bounds for deep (stochastic) neural networks with many more parameters than training data. In *Proceedings of the Thirty-Third Conference on Uncertainty in Artificial Intelligence (UAI 2016)*.

Fan, J., & Li, R. (2001). Variable selection via nonconcave penalized likelihood and its oracle properties. *Journal of the American Statistical Association, 96*(456), 1348–1360.

Fisher, R. A. (1936). The use of multiple measurements in taxonomic problems. *Annals of Eugenics, 7*(Part II), 179–188.

Flach, P. (2010). ROC analysis. *Encyclopedia of machine learning* (pp. 869–875).

Flach, P. A. (2003). The geometry of ROC space: Understanding machine learning metrics through ROC isometrics. In *Proceedings of the Twentieth International Conference on Machine Learning (ICML 2003)* (pp. 194–201), Washington DC, USA.

Frank, I. E., & Friedman, J. H. (1993). A statistical view of some chemometrics regression tools. *Technometrics, 35*, 109–148.

Freund, Y., & Schapire, R. E. (1995). A decision-theoretic generalization of on-line learning and an application to boosting. *European Conference on Computational Learning Theory* (pp. 23–37). Berlin: Springer.

Freund, Y., & Schapire, R. E. (1996). Experiments with a new boosting algorithm. In *Machine Learning: Proceedings of the Thirteenth International Conference* (Vol. 96, pp. 148–156).

Friedman, J. H. (2001). Greedy function approximation: A gradient boosting machine. *Annals of Statistics*, 1189–1232.

Friedman, M. (1937). The use of ranks to avoid the assumption of normality implicit in the analysis of variance. *Journal of the American Statistical Association, 32*(200), 675–701.

Fu, W. J. (1998). Penalized regressions: The bridge versus the lasso. *Journal of Computational and Graphical Statistics, 7*(3), 397–416.

Fukunaga, K. (1990). *Introduction to statistical pattern recognition* (2nd ed.). New York: Academic Press, Morgan Kaufmann.

Funahashi, K.-I. (1989). On the approximate realization of continuous mappings by neural networks. *Neural Networks, 2*(3), 183–192.

Gauss, C. F., & Stewart, G. W. (1995). *Theory of the combination of observations least subject to errors: Part one.* Part two: Society for Industrial and Applied Mathematics, Supplement.

Gelman, A., Hill, J., & Yajima, M. (2012). Why we (usually) don't have to worry about multiple comparisons. *Journal of Research on Educational Effectiveness, 5*(2), 189–211.

Gini, C. (1914). Sulla misura della concentrazione e della variabilit'a dei caratteri. atti del reale istituto veneto di scienze (on the measurement of concentration and variability of characters). *Lettere ed Arti, 73*, 1203–1248. English translation (2005) in Metron, 63, 3–38.

Giorgi, G. M. (1990). Bibliographic portrait of the Gini concentration ratio. *METRON - International Journal of Statistics, XLVII, I*(1–4), 183–221.

Giorgi, G. M. (2005). Gini's scientific work: An evergreen. *METRON - International Journal of Statistics, LXII, I*(3), 299–315.

Girosi, F., & Poggio, T. (1989). Representation properties of networks: Kolmogorov's theorem is irrelevant. *Neural Computation, 1*(4), 465–469.

Gogus, A. (2012). Analytic learning. In N. M. Seel (Ed.), *Encyclopedia of the Sciences of Learning* (pp. 237–241). US, Boston, MA: Springer.

Golub, G. H. & Van Loan, C. F. (1996). *Matrix Computations.* Johns Hopkins. ISBN 978-0-8018-5414-9.

Goodfellow, I., Bengio, Y., & Courville, A. (2016). *Deep learning.* MIT Press. http://www.deeplearningbook.org

Gunasekar, S., Lee, J. D., Soudry, D., & Srebro, N. (2018). Implicit bias of gradient descent on linear convolutional networks. In *Proceedings of the 29th International Conference on Neural Information Processing Systems (NIPS 2016)* (pp. 1–22).

Gurwitz, C., & Overton, M. L. (1989). A globally convergent algorithm for minimizing over the rotation group of quadratic forms. *IEEE Transactions on Pattern Analysis and Machine Intelligence, 11*(11), 1228–1232.

Guyon, I. (2003). Design of experiments of the NIPS 2003 variable selection benchmark. (NIPS Feature Selection Challenge). http://clopinet.com/isabelle/Projects/NIPS2003/Slides/NIPS2003-Datasets.pdf.

Guyon, I., Hur, A. B., Gunn, S., & Dror, G. (2004). Result analysis of the NIPS 2003 feature selection challenge. In *Advances in Neural Information Processing Systems* (Vol. 17, pp. 545–552). MIT Press.

Hadamard, J. (1902). Sur les problèmes aux dérivées partielles et leur signification physique. *Princeton University Bulletin* (pp. 49–52).

Hanley, J. A., & McNeil, B. J. (1982). The meaning and use of the area under a receiver operating characteristic (ROC) curve. *Radiology, 143*, 29–36.

Hardy, G. H., Littlewood, J. E., & Pólya, G. (1934). *Inequalities*. London: Cambridge Univesity Press.

Hassoun, M. H. (1995). *Fundamentals of artificial neural networks*. Cambridge, MA: MIT Press.

Hastie, T., Montanari, A., Rosset, S., & Tibshirani, R. J. (2019). Surprises in high-dimensional Ridgeless least squares interpolation (pp. 1–43). https://arxiv.org/abs/1903.08560.

Hastie, T., Tibshirani, R., & Friedman, J. (2017). *The elements of statistical learning: Data mining, inference, and prediction*. Canada: Springer.

Hastie, T., Tibshirani, R., & Friedman, J. (2009). *The elements of statistical learning: Data mining, inference, and prediction*. Springer Series in Statistics: Springer.

Haykin, S. O. (2009). *Neural networks and learning machines*. New York: Prentice Hall.

Haykin, S. O. (2013). *Adaptive filter theory*. New York: Prentice Hall.

He, M., He, F., Shi, L., Huang, X., & Syykens, J. A. (2023). Learning with asymmetric kernels: Least squares and feature interpretation. *IEEE Transactions on Pattern Analysis and Machine Intelligence, 45*(8), 10044–10054.

Hecht-Nielsen, R. (1987). Kolmogorov's mapping neural network existence theorem. *In Proceedings of IEEE First International Conference on Neural Networks (ICNN), III*, 11–14.

Hecht-Nielsen, R. (1989). Theory of the backpropagation neural network. In *Proceedings of International Joint Conference on Neural Networks (IJCNN)* (pp. 593–605).

Hecht-Nielsen, R. (1990). *Neurocomputing*. Massachusetts: Addison-Wesley Publishing Company.

Herschtal, A. & Raskutti, B. (2004). Optimising area under the ROC curve using gradient descent. In *Proceedings of the Twenty-first International Conference on Machine Learning (ICML 2004)*. Banff, Alberta, Canada: ACM Press.

Hoerl, A. E., & Kennard, R. W. (1970). Ridge regression: Applications to nonorthogonal problems. *Technometrics, 12*, 69–82.

Hoerl, A. E., & Kennard, R. W. (1970). Ridge regression: Biased estimation for nonorthogonal problems. *Technometrics, 12*(1), 55–67.

Hornik, K., Stinchcombe, M., & White, H. (1989). Multi-layer feedforward networks are universal approximators. *Neural Networks, 2*(5), 359–366.

Hornik, K., Stinchcombe, M., & White, H. (1990). Universal approximation of an unknown mapping and its derivatives using multilayer feedforward networks. *Neural Networks, 3*(5), 551–560.

Hotelling, H. (1933). Analysis of a complex of statistical variables into principal components. *Journal of Educational Psychology, 24*, 417–441 and 498–520.

Hotelling, H. (1936). Relations between two sets of variates. *Biometrika, 28*, 321–377.

Huang, G.-B., Chen, Y.-Q., & Babri, H. A. (2000). Classification ability of single hidden layer feedforward neural networks. *IEEE Transactions on Neural Networks, 11*(3), 799–801.

Huang, G.-B., Zhu, Q.-Y., & Siew, C.-K. (2006). Extreme learning machine: Theory and applications. *Neurocomputing, 70*, 489–501.

Ioannidis, J. P. A. (2005). Why most published research findings are false. *PLoS Medicine, 2*(8, e124), 0696–0701.

Jaeger, H. (2002). Tutorial on training recurrent neural networks, covering BPPT, RTRL, EKF and the "Echo State Network" approach. *GMD Report 159, German National Research Center for Information Technology.*

Jaeger, H., & Haas, H. (2004). Harnessing nonlinearity: Predicting chaotic systems and saving energy in wireless communication. *Science, 304*, 78–80.

Jain, A. and Zongker, D. (1997). Feature selection: Evaluation, application, and small sample performance. *IEEE Trans. on Pattern Analysis and Machine Intelligence, 19*(2), 153–158.

Jain, A. K., Bolle, R., & Pankanti, S. (Eds.). (2002). *Biometrics: Personal identification in networked society.* Kluwer Academic Publishers.

Jain, A. K., Nandakumar, K., & Ross, A. (2005). Score normalization in multimodal biometric systems. *Pattern Recognition, 38*, 2270–2285.

Jain, A. K., Ross, A., & Prabhakar, S. (2004). An introduction to biometric recognition. *IEEE Transactions on Circuits and Systems for Video Technology, 14*(1), 4–20.

Jain, A. K., Ross, A. A., & Nandakumar, K. (2011). *Introduction to biometrics.* New York: Springer.

Jang, S.-I., Tan, G.-C., Toh, K.-A., & Teoh, A. B. J. (2020). Online heterogeneous face recognition based on total-error-rate minimization. *IEEE Transactions on Systems, Man and Cybernetics: Systems, 50*(4), 1286–1299.

Kabanikhin, S. I. (2008). Definitions and examples of inverse and ill-posed problems. *Journal of Inverse and Ill-Posed Problems, 16*(4), 317–357.

Kalman, R. E. (1960). A new approach to linear filtering and prediction problems. *Transactions of the ASME Journal of Basic Engineering, 82*, 35–45. (Series D).

Kam, A. C., & Kopec, G. E. (1996). Document image decoding by Heuistic search. *IEEE Transactions on Pattern Analysis and Machine Intelligence, 18*(9), 945–950.

Kaplansky, I. (1972). *Set theory and metric spaces.* Boston: Allyn and Bacon Inc.

Kawaguchi, K. (2016). Deep learning without poor local minima. In *Proceedings of the 29th International Conference on Neural Information Processing Systems (NIPS 2016)* (pp. 586–594).

Kawaguchi, K., & Bengio, Y. (2019). Depth with nonlinearity creates no bad local minima in ResNets. *Neural Networks, 118*, 167–174.

Kawaguchi, K., Kaelbling, L. P., & Bengio, Y. (2023). Generalization in deep learning. https://arxiv.org/pdf/1710.05468.pdf

Kaynak, C. (1995). Methods of combining multiple classifiers and their applications to handwritten digit recognition. Master's thesis, Institute of Graduate Studies in Science and Engineering, Bogazici University.

Ke, G., Meng, Q., Finley, T., Wang, T., Chen, W., Ma, W., Ye, Q., & Liu, T. -Y. (2017). Lightgbm: A highly efficient gradient boosting decision tree. In *Advances in Neural Information Processing Systems* (pp. 3149–3157).

Kelly, M., Longjohn, R., & Nottingham, K. (2024). The UCI machine learning repository. https://archive.ics.uci.edu

Kim, Y., Toh, K.-A., Teoh, A. B. J., Eng, H.-L., & Yau, W.-Y. (2012). An online AUC formulation for binary classification. *Pattern Recognition, 45*(6), 2266–2279.

Kimeldorf, G., & Wahba, G. (1971). Some results on Tchebycheffian spline functions. *Journal of Mathematical Analysis and Applications, 33*(1), 82–95.

Kleiber, C. (2008). The Lorenz curve in economics and econometrics. In *Advances on Income Inequality and Concentration Measures: Collected Papers in Memory of Corrado Gini and Max O. Lorenz* (pp. 225–242).

Kolmogorov, A. N., & Fomin, S. V. (1957). *Elements of the Theory of Functions and Functional Analysis* (Vol. 1). Rochester, N. Y.: Graylock Press (Translated from the first (1954) Russian edition by Leo F. Boron).

Kolmogorov, A. N., & Fomin, S. V. (1961). *Elements of the Theory of Functions and Functional Analysis* (Vol. 2). Baltimore, MD: Graylock Press (Translated from the first (1960) Russian edition by Hyman Kamel and Horace Komm).

Kůrková, V. (1991). Kolmogorov's theorem is relevant. *Neural Computation, 3*(4), 617–622.

Kůrková, V. (1992). Kolmogorov's theorem and multilayer neural networks. *Neural Networks, 5*(3), 501–506.

Langford, J., & Caruana, R. (2002). (Not) Bounding the true error. In T. G. Dietterich, S. Becker, & Z. Ghahramani (Eds.), *Advances in neural information processing systems 14* (pp. 809–816). MIT Press.

Lasko, T. A., Bhagwat, J. G., Zou, K. H., & Ohno-Machado, L. (2005). The use of receiver operating characteristic curves in biomedical informatics. *Journal of Biomedical Informatics, 38*(5), 404–415.

Lavrent'ev, M. M., Romanov, V. G., & Shishatskiĭ, S. P. (1986). *Ill-posed problems of mathematical physics and analysis* (Vol. 64). Providence, Rhode Island: American Mathematical Society. Translation of Mathematical Monographs.

LeCun, Y., Bengio, Y., & Hinton, G. (2015). Deep learning. *Nature, 521*, 436–444.

LeCun, Y., Bottou, L., Bengio, Y., & Haffner, P. (1998). Gradient-based learning applied to document recognition. *Proceedings of the IEEE, 16*(11), 2278–2324.

Leshno, M., Lin, V. Y., Pinkus, A., & Schocken, S. (1993). Multilayer feedforward networks with a nonpolynomial activation function can approximate any function. *Neural Networks, 6*, 861–867.

Li, H., Xu, Z., Taylor, G., & Goldstein, T. (2018). Visualizing the loss landscape of neural nets. In *Proceedings of the 6th International Conference on Learning Representations (ICLR 2018)* (pp. 1–17). Vancouver, BC, Canada.

Liao, Q., & Poggio, T. (2017). Theory of deep learning II: Landscape of the empirical risk in deep learning. *CBMM Memo No, 066*, 1–45.

Lin, J.-N., & Unbehauen, R. (1993). On the realization of a Kolmogorov network. *Neural Computation, 5*(1), 18–20.

Lorenz, M. O. (1905). Methods of measuring the concentration of wealth. *Quarterly Publications of the American Statistical Association, 9*, 209–219. (New Series, No. 70).

Lowe, D. (1989). Adaptive radial basis function nonlinearities, and the problem of generalisation. In *Proceedings of the first IEE International Conference on Artificial Neural Networks* (pp. 171–175).

Lowe, D., & Webb, A. (1990). Exploiting prior knowledge in network optimization: An illustration from medical prognosis. *Network: Computation in Neural Systems, 1*(3), 299–323.

Lu, J., Plataniotis, K. N., & Venetsanopoulos, A. N. (2003). Regularized discriminant analysis for the small sample size problem in face recognition. *Pattern Recognition Letters, 24*, 3079–3087.

MacAusland, R. (2014). The Moore-Penrose inverse and least squares. Lecture Notes in Advanced Topics in Linear Algebra (MATH 420).

Madych, W. R. (1991). Solutions of underdetermined systems of linear equations. In *Lecture Notes—Monograph Series, Spatial Statistics and Imaging* (Vol. 20, pp. 227–238). Institute of Mathematical Statistics.

Maor, E. (2010). *The pythagorean theorem: A 4,000-year history*. New Jersey: Princeton University Press.

Mason, S. J., & Graham, N. E. (2002). Areas beneath the relative operating characteristics (ROC) and relative operating levels (ROL) curves: Statistical significance and interpretation. *Quarterly Journal of the Royal Meteorological Society, 128*, 2145–2166.

McDermott, E., & Katagiri, S. (2004). A derivation of minimum classification error from the theoretical classification risk using Parzen estimation. *Computer Speech and Language, 18*, 107–122.

McNemar, Q. (1947). Note on the sampling error of the difference between correlated proportions or percentages. *Psychometrika, 12*(2), 153–157.

Mercer, J. (1909). Functions of positive and negative type, and their connection with the theory of integral equations. *Philosophical Transactions of the Royal Society of London Series A, Containing Papers of a Mathematical or Physical Character, 209*, 415–446.

Mitchell, T. (2023). CMU face images. https://archive.ics.uci.edu/dataset/124/cmu+face+images

Nemenyi, P. B. (1963). *Distribution-free multiple comparisons*. Princeton University.

Neter, J., Kutner, M. H., Nachtsheim, C. J., & Wasserman, W. (1996). *Applied linear regression models* (3rd ed.). Chicago: Irwin.

Neyshabur, B. (2017). *Implicit regularization in deep learning*. Ph.D. thesis, Toyota Technological Institute at Chicago.

Neyshabur, B., Bhojanapalli, S., McAllester, D., & Srebro, N. (2017). Exploring generalization in deep learning. In *Advances in Neural Information Processing Systems (NIPS 2017)* (pp. 5947–5956).

Neyshabur, B., Bhojanapalli, S., & Srebro, N. (2018). A PAC-Bayesian approach to spectrally-normalized margin bounds for neural networks. In *Proceedings of the 6th International Conference on Learning Representations (ICLR 2018)* (pp. 1–9). Vancouver, BC, Canada.

Nievergelt, Y. (2000). A tutorial history of least squares with applications to astronomy and geodesy. *Journal of Computational and Applied Mathematics, 121*, 37–72.

Nuzzo, R. (2014). Statistical errors: P values, the 'gold standard' of statistical validity, are not as reliable as many scientists assume. *Nature, 506*, 150–152.

Obuchowski, N. A. (2003). Receiver operating characteristic curves and their use in radiology. *Radiology, 229*(1), 3–8.

Osuna, E. E., Freund, R., & Girosi, F. (1997). *Support vector machines: Training and applications.* MIT Artificial Intelligence Laboratory and CBCL Department of Brain and Cognitive Sciences. (Technical Report: A.I. Memo No. 1602, C.B.C.L. Paper No. 144).

Patrick van der Smagt, P. (1994). Minimisation methods for training feedforward neural networks. *Neural Networks, 7*(1), 1–11.

Pan, V. Y. (1997). Solving a polynomial equation: Some history and recent progress. *SIAM Review, 39*(2), 187–220.

Pao, Y.-H., Park, G.-H., & Sobajic, D. J. (1994). Learning and generalization characteristics of the random vector functional-link net. *Neurocomputing, 6*(2), 163–180.

Pao, Y. H., & Takefuji, Y. (1992). Functional-link net computing: Theory, system architecture, and functionalities. *IEEE Computer, 25*(5), 76–79.

Park, C., & Yoon, Y. J. (2011). Bridge regression: Adaptivity and group selection. *Journal of Statistical Planning and Inference, 141*(11), 3506–3519.

Park, J., & Sandberg, I. W. (1991). Uniform approximation using radial-basis-function networks. *Neural Computation, 3*, 246–257.

Parzen, E. (1962). On the estimation of a probability density function and the mode. *The Annals of Mathematical Statistics, 33*(3), 1065–1076.

Paszke, A., Gross, S., Massa, F., Lerer, A., Bradbury, J., Chanan, G., et al. (2019). *Advances in Neural Information Processing Systems* (Vol. 32). Curran Associates, Inc.

Pearson, K. (1901). On lines and planes of closest fit to systems of points in space. *Philosophical Magazine, 2*, 559–572.

Plackett, R. L. (1950). Some theorems in least squares. *Biometrika, 37*(1/2), 149–157.

Poggio, T., & Girosi, F. (1990). Networks for approximation and learning. *Proceedings of the IEEE, 78*(9), 1481–1497.

Poggio, T., Liao, Q., Miranda, B., Banburski, A., Boix, X., & Hidary, J. (2018). Theory of deep learning IIIb: Generalization in deep networks. *CBMM Memo No, 090*, 1–37.

Posamentier, A. S. (2010). *The Pythagorean Theorem: The Story of Its Power and Beauty.* Amherst NY: Prometheus Books.

Rahimi, A., & Recht, B. (2007). Random features for large-scale kernel machines. In *Advances in Neural Information Processing Systems (NIPS 2007)* (pp. 1177–1184).

Rahimi, A., & Recht, B. (2008). Random features for large-scale kernel machines. In *Advances in Neural Information Processing Systems (NIPS 2008)* (pp. 1177–1184).

Rakotomamonjy, A. (2004). Optimizing area under ROC curve with SVMs. In J. Hernández-Orallo, C. Ferri, N. Lachiche, & P. A. Flach, (Eds.), *ROC Analysis in Artificial Intelligence, 1st International Workshop, ROCAI-2004, Valencia, Spain, Aug 22, 2004* (pp. 71–80).

Ramirez, C., Sanchez, R., Kreinovich, V., & Argaez, M. (2014). $\sqrt{x^2 + \mu}$ is the most computationally efficient smooth approximation to $|x|$: A proof. *Journal of Uncertain Systems, 8*(3), 205–210.

Rasmussen, C. E., & Williams, C. K. I. (2006). *Gaussian processes for machine learning*. The MIT Press. http://www.gaussianprocess.org/gpml/

Ripley, B. D. (1996). *Pattern recognition and neural networks*. Cambridge: Cambridge University Press.

Rosenblatt, F. (1958). The perceptron: A probabilistic model for information storage and organization in the brain. *Psychological Review, 65*(6), 386–408.

Rumelhart, D. E., Hinton, G. E., & Williams, R. J. (1986). Learning representations by backpropagating errors. *Nature, 323*(99), 533–536.

Santos, J. B., & Guerrero, J. J. B. (2010). Gini's concentration ratio (1908–1914). *Electronic Journal for History of Probability and Statistics, 6*(1).

Saunders, C., Gammerman, A., & Vovk, V. (1998). Ridge regression learning algorithm in dual variables. *International Conference on Machine Learning (ICML)* (pp. 515–521). Madison: WI.

Schmidt, W. F., Kraaijveld, M. A., & Duin, R. P. W. (1992). Feed forward neural networks with random weights. In *Proceedings of 11th IAPR International Conference on Pattern Recognition, Conference B: Pattern Recognition Methodology and Systems* (Vol. 2, pp. 1–4), The Hague.

Schölkopf, B., Herbrich, R., & Smola, A. J. (2001). A generalized representer theorem. In *Conference on Learning Theory (COLT), LNAI 2111* (pp. 416–426).

Schölkopf, B., & Smola, A. J. (2001). *Learning with kernels: Support vector machines, regularization, optimization, and beyond*. Cambridge: The MIT Press.

Schürmann, J. (1996). *Pattern classification: A unified view of statistical and neural approaches*. New York: Wiley.

Shawe-Taylor, J., & Cristianini, N. (2004). *Kernel methods for pattern analysis*. New York, NY: Cambridge University Press.

Sherman, J., & Morrison, W. J. (1950). Adjustment of an inverse matrix corresponding to a change in one element of a given matrix. *Annals of Mathematical Statistics, 21*(1), 124–127.

Sprecher, D. A. (1993). A universal mapping for Kolmogorov's superposition theorem. *Neural Networks, 6*(8), 1089–1094.

Strang, G. (2016). *Introduction to linear algebra* (5th ed.). Wellesley: Wellesley-Cambridge Press.

Student. (1908). The probable error of a mean. *Biometrika, 6*(1), 1–25.

Sun, L., Toh, K.-A., & Lin, Z. (2015). A center sliding Bayesian binary classifier adopting orthogonal polynomials. *Pattern Recognition, 48*(6), 2013–2028.

Swets, J. A. (1996). *Signal detection theory and ROC analysis in psychology and diagnostics*. Publishers, Mahwah, New Jersey, USA: Lawrence Erlbaum Associates.

Theodoridis, S., & Koutroumbas, K. (2009). *Pattern recognition* (4th ed.). Academic Press, Elsevier Inc., California, MA, USA.

Theodoridis, S., Pikrakis, A., Koutroumbas, K., & Cavouras, D. (2010). *Introduction to pattern recognition: A MATLAB approach* (4th ed.). Academic Press, Elsevier Inc., New York, USA.

Tibshirani, R. (1996). Regression shrinkage and selection via the lasso. *Journal of the Royal Statistical Society Series B, 58*, 267–288.

Tibshirani, R. (2011). Regression shrinkage and selection via the lasso: A retrospective. *Journal of the Royal Statistical Society Series B, 73*, 273–282. Part 3.

Tikhonov, A. N. (1963). Solution of incorrectly formulated problems and the regularization method. Doklady Akademii Nauk SSSR, 151:501–04. *Translated in Soviet Mathematics, 4*, 1035–1038.

Tikhonov, A. N., & Arsenin, V. Y. (1977). *Solutions of Ill-posed problems*. New York: Wiley.

Tipping, M. E. (2000). The relevance vector machine. In S. A. Solla, T. K. Leen, & K.-R. Müller (Eds.), *Advances in Neural Information Processing Systems* (Vol. 12, pp. 652–658).

Tipping, M. E. (2001). Sparse Bayesian learning and the relevance vector machine. *Journal of Machine Learning Research, 1*, 211–244.

Toh, K.-A. (2003). Deterministic global optimization for FNN training. *IEEE Transactions on Systems, Man and Cybernetics, Part B, 33*(6), 977–983.

Toh, K.-A. (2003b). Fingerprint and speaker verification decisions fusion. *International Conference on Image Analysis and Processing (ICIAP)* (pp. 626–631). Mantova: Italy.

Toh, K.-A. (2006a). Learning from target knowledge approximation. In *Proceedings of the First IEEE Conference on Industrial Electronics and Applications* (pp. 815–822), Singapore.

Toh, K.-A. (2006). Training a reciprocal-sigmoid classifier by feature scaling-space. *Machine Learning, 65*(1), 273–308.

Toh, K.-A. (2008). Deterministic neural classification. *Neural Computation, 20*(6), 1565–1595.

Toh, K.-A. (2014). Twisting key absolute space for stretchy polynomial regression. In *Proceedings of the 13th International Conference on Automation, Robotics and Computer Vision (ICARCV)* (pp. 953–957), Singapore.

Toh, K.-A. (2018a). Analytic network learning (pp. 1–28). arXiv:1811.08227

Toh, K.-A. (2018). Kernel and range approach to analytic network learning. *International Journal of Networked and Distributed Computing, 7*(1), 20–28.

Toh, K.-A. (2018c). Learning from the kernel and the range space. In *Proceedings of the 17th IEEE/ACIS International Conference on Computer and Information Science* (pp. 417–422). Singapore.

Toh, K.-A., & Eng, H.-L. (2008). Between classification-error approximation and weighted least-squares learning. *IEEE Transactions on Pattern Analysis and Machine Intelligence, 30*(4), 658–669.

Toh, K.-A., Kim, J., & Lee, S. (2008). Biometric scores fusion based on total error rate minimization. *Pattern Recognition, 41*(3), 1066–1082.

Toh, K.-A., Kim, J., & Lee, S. (2008). Maximizing area under ROC curve for biometric scores fusion. *Pattern Recognition, 41*(11), 3373–3392.

Toh, K.-A., Lin, Z., Li, Z., Oh, B., & Sun, L. (2018a). Gradient-free learning based on the kernel and the range space (pp. 1–27). arXiv:1810.11581

Toh, K.-A., Lin, Z., Sun, L., & Li, Z. (2018). Stretchy binary classification. *Neural Networks, 97*, 74–91.

Toh, K.-A., Molteni, G., & Lin, Z. (2023). Deterministic bridge regression for compressive classification. *Information Sciences, 648*, 1–22.

Toh, K.-A., & Tan, G.-C. (2014). Exploiting the relationships among several binary classifiers via data transformation. *Pattern Recognition, 47*(3), 1509–1522.

Toh, K.-A., Tran, Q.-L., & Srinivasan, D. (2004). Benchmarking a reduced multivariate polynomial pattern classifier. *IEEE Transactions on Pattern Analysis and Machine Intelligence, 26*(6), 740–755.

Toh, K.-A., & Yau, W.-Y. (2002). Multi-modal biometrics fusion: Beyond optimal weighting. In *Proceedings of the Seventh International Conference on Automation, Robotics and Computer Vision (ICARCV)* (pp. 788–792). Singapore. (invited paper).

Toh, K.-A., & Yau, W.-Y. (2004). Combination of hyperbolic functions for multimodal biometrics data fusion. *IEEE Transactions on Systems, Man and Cybernetics, Part-B, 34*(2), 1196–1209.

Toh, K.-A., Yau, W.-Y., & Jiang, X. (2004b). A reduced multivariate polynomial model for multimodal biometrics and classifiers fusion. *IEEE Transactions on Circuits and Systems for Video Technology (Special Issue on Image- and Video-Based Biometrics), 14*(2), 224–233.

Vapnik, V. N. (1998). *Statistical Learning Theory*. Wiley-Interscience Pub.

Vapnik, V. N. (1999). An overview of statistical learning theory. *IEEE Transactions on Neural Networks, 10*(5), 988–999.

Vapnik, V. N., & Chervonenkis, A. Y. (1991). The necessary and sufficient conditions for consistency in the empirical risk minimization method. *Pattern Recognition and Image Analysis, 1*(3), 283–305.

Vasudevan, A., Anderson, A., & Gregg, D. (2017). *2017 IEEE 28th International Conference on Application-specific Systems, Architectures and Processors (ASAP), Parallel Multi Channel convolution using General Matrix Multiplication*, pp. 19–24. https://doi.org/10.1109/ASAP.2017.7995254

Vempala, S. S. (2004). *The random projection method*. DIMACS Series in Discrete Mathematics and Theoretical Computer Science, American Mathematical Society, USA.

Vert, J.-P., Tsuda, K., & Schölkopf, B. (2004). A primer on kernel methods. In *Kernel methods in computational biology* (pp. 35–70).

Wang, Y., Yao, Q., Kwok, J. T., & Ni, L. M. (2021). Generalizing from a few examples: A survey on few-shot learning. *ACM Computing Surveys, 53*(3), 1–34.

Wasserstein, R. L., & Lazar, N. A. (2016). The ASA statement on p-values: Context, process, and purpose. *The American Statistician, 7*(2), 129–133. https://doi.org/10.1080/00031305.2016.1154108

Webb, A. R., & Copsey, K. D. (2011). *Statistical Pattern Recognition* (3rd ed.). Wiley

Welling, M. (2013). Kernel ridge-regression. In *Max Welling's classnotes in machine learning* (pp. 1–3). http://www.ics.uci.edu/~welling/classnotes/classnotes.html

Werbos, P. J. (1974). *Beyond regression: New tools for prediction and analysis in the behavioral sciences*. Ph.D. thesis, Beyond Regression: New Tools for Prediction and Analysis in the Behavioral Sciences.

Werbos, P. J. (1990). Backpropagation through time: What it does and how to do it. *Neural Networks, 78*(10), 1550–1560.

Widrow, B., Greenblatt, A., Kim, Y., & Park, D. (2013). The no-prop algorithm: A new learning algorithm for multilayer neural networks. *Neural Networks, 37*, 182–188.

Widrow, B., & Holf, M. E. (1960). Adaptive switching circuits. *IRE WESCON Convention Record, 4*, 96–104.

Wikipedia. (2015a). Johnson lindenstrauss lemma–Wikipedia, the free encyclopedia. https://en.wikipedia.org/wiki/Johnson%E2%80%93Lindenstrauss_lemma

Wikipedia. (2015b). Polynomial—Wikipedia, the free encyclopedia. http://en.wikipedia.org/wiki/Polynomial#History

Wikipedia. (2015c). Universal approximation theorem—Wikipedia, the free encyclopedia. http://en.wikipedia.org/wiki/Universal_approximation_theorem

Wikipedia. (2017). Archimedean spiral—Wikipedia, the free encyclopedia. https://en.wikipedia.org/wiki/Archimedean_spiral

Wikipedia. (2024). Metric space—Wikipedia, the free encyclopedia. https://en.wikipedia.org/wiki/Metricspace#Quasimetrics

Wilcoxon, F. (1945). Individual comparisons by ranking methods. *Biometrics Bulletin, 1*(6), 80–83.

Woodbury, M. A. (1950). *Inverting modified matrices* (Vol. 42). Memorandum Reptort: Statistical Research Group, Princeton University, Princeton, NJ.

Yan, L., Dodier, R., Mozer, M. C., & Wolniewicz, R. (2003). Optimizing classifier performance via an approximation to the Wilcoxon-Mann-Whitney statistic. In *Proceedings of the Twentieth International Conference on Machine Learning (ICML 2003)*, (pp. 848–855). USA: Washington DC.

Yan, R., Zhang, J., Yang, J., & Hauptmann, A. G. (2006). A discriminative learning framework with pairwise constraints for video object classification. *IEEE Transactions on Pattern Analysis and Machine Intelligence, 28*(4), 578–593.

Yang, L., Jin, R., Mummert, L., Sukthankar, R., Goode, A., Zheng, B., & Satyanarayanan, M. (2010). A boosting framework for visuality-preserving distance metric learning and its application to medical image retrieval. *IEEE Transactions on Pattern Analysis and Machine Intelligence, 32*(1), 30–44.

Yun, C., Sra, S., & Jadbabaie, A. (2018). Global optimality conditions for deep neural networks. In *Proceedings of the 6th International Conference on Learning Representations (ICLR 2018)* (pp. 1–14), Vancouver, BC, Canada.

Zhang, C., Bengio, S., Hardt, M., Recht, B., & Vinyals, O. (2017). Understanding deep learning requires rethinking generalization. In *Proceedings of the 5th International Conference on Learning Representations (ICLR 2017)* (pp. 1–15). Toulon, France.

Zhuang, H., Lin, Z., & Toh, K.-A. (2019). A low-memory learning formulation for a kernel-and-range network. In *Proceedings of International Joint Conference on Neural Networks (IJCNN)*.

Zhuang, H., Lin, Z., & Toh, K.-A. (2022a). Blockwise recursive moore-penrose inverse for network learning. *IEEE Transactions on Systems, Man, and Cybernetics: Systems (Part A), 52*(5), 3237–3250.

Zhuang, H., Weng, Z., Luo, F., Toh, K.-A., Li, H., & Lin, Z. (2021). Accumulated decoupled learning with gradient staleness mitigation for convolutional neural networks. In *Proceedings of 38th International Conference on Machine Learning (PMLR)* (pp. 12935–12944).

Zhuang, H., Weng, Z., Wei, H., Xie, R., Toh, K.-A., & Lin, Z. (2022b). ACIL: Analytic class-incremental learning with absolute memorization and privacy protection. In *Proceedings of 36th Conference on Neural Information Processing Systems (NeurIPS)*.

Zhuang, H., Lin, Z., Yang, Y., & Toh, K.-A. (2025). An analytic formulation of convolutional neural network learning for pattern recognition. *Information Sciences, 686*, 1–17.

Zou, H., & Hastie, T. (2005). Regularization and variable selection via the elastic net. *Journal of the Royal Statistical Society, Series B, 67*, 301–320. (Part 2).

Zou, H., & Li, R. (2008). One-step sparse estimates in nonconcave penalized likelihood models. *The Annals of Statistics, 36*(4), 1509–1533.

Zou, K. H., O'Malley, A. J., & Mauri, L. (2007). Receiver-operating characteristic analysis for evaluating diagnostic tests and predictive models. *Circulation, 115*(5), 654–657.

Zweig, M. H., & Campbell, G. (1993). Receiver-operating characteristic (ROC) plots: A fundamental evaluation tool in clinical medicine. *Clinical Chemistry, 39*(8), 561–577.

GPSR Compliance
The European Union's (EU) General Product Safety Regulation (GPSR) is a set
of rules that requires consumer products to be safe and our obligations to
ensure this.

If you have any concerns about our products, you can contact us on

ProductSafety@springernature.com

In case Publisher is established outside the EU, the EU authorized
representative is:

Springer Nature Customer Service Center GmbH
Europaplatz 3
69115 Heidelberg, Germany